Springer Series in Game Theory

Simon A. Levin
Editor

Games, Groups, and the Global Good

Springer

Editor
Professor Simon A. Levin
Department of Ecology and Evolutionary Biology
Princeton University
Eno Hall
Princeton, NJ 08544-1003
USA
slevin@eno.princeton.edu

ISSN 1868-517x
ISBN 978-3-540-85435-7 e-ISBN 978-3-540-85436-4
DOI 10.1007/978-3-540-85436-4
Springer Dordrecht Heidelberg London New York

Library of Congress Control Number: 2009926063

Cover design: WMXDesign GmbH, Heidelberg

Printed on acid-free paper

Springer is part of Springer Science+Business Media (www.springer.com)

Preface

No problem is more central to understanding biological organization than explaining cooperation. Indeed, the puzzles posed by extreme forms of cooperation were acknowledged by Darwin as challenges to his theories, and delayed his publication of the *Origin of Species* for 20 years. Today, we have learned a great deal about the evolution of cooperation, from quorum sensing in bacteria to coalitions among humans. Nevertheless, deep questions remain. How is cooperation maintained in large groups, where individuals help others they have never met, or whose identities are unknown to the helpers? Why will individuals apparently sacrifice their own welfare to sustain community norms, through charitable behaviors or punishment of norm offenders? How are institutions, from social norms to civil and religious laws, maintained? How do moral systems arise, and how are they maintained? These questions are central to understanding how societies maintain robustness, and they also are key to achieving a sustainable future for humanity.

Much of the formal theory of cooperation can be embedded within the theory of games, the origins of which are usually traced to John von Neumann's "Zur Theorie der Gesellschaftsspiele," (*On the Theory of Parlor Games*) published in 1928 in Mathematische Annalen, 100, pp. 295–300. Actually, Emile Borel published several papers that laid the foundations for game theory 7 years earlier, but it was von Neumann who really began to develop a comprehensive theory, culminating in his 1944 Princeton book with Oskar Morgenstern, *Theory of Games and Economic Behavior.* Von Neumann died in 1958; and on the 50th anniversary of his death, a symposium was organized at Princeton University with the sponsorship of the John Templeton Foundation to revisit progress in the theory of cooperation, and particularly to investigate what new advances in game theory might be stimulated by considering the broader question of the establishment of moral systems and the regulation of public goods. Most of the participants in that symposium then developed their papers into longer contributions for this book, and other authors were invited to complete the story.

Societies cannot exist without cooperation, or without norms, customs, laws, and other institutions that sustain cooperation. These provide collective benefits that maintain the groups, and provide them advantages in conflict with other groups. One of the great challenges facing humanity is in discovering whether those collective benefits can be extended to the global level, without the tribal conflicts that threaten our survival. First we must understand the dynamics of groups ... how they form, how they are maintained, and how they interact with their constituents and with other groups. What are the origins of the normative practices that sustain these groups, as well as the second-order systems of reward and punishment that sustain the norms themselves? What are the developmental and evolutionary trajectories that lead to the development of moral systems, from informal arrangements, to societies, religions, and other formal institutions with formal rules and practices?

This book examines how such approaches can be extended to consider the broader questions that cross scales of organization, from individuals to cooperatives to societies. By expanding traditional approaches, can we explain how heuristics, like concepts of fairness, arise, and how they become formalized into the ethical principles embraced by a society? How do those ethical systems themselves co-evolve with the societies and other institutions? What maintains the robustness of social contracts? This book brings together essays by a diverse group of scholars, providing a broad perspective on these questions, and hopefully suggesting promising new directions for game theory.

I am pleased to acknowledge the contributions of my co-organizers – Mary Ann Meyers, Martin Nowak and Steven Brams – in the organization of the conference, and in the development of many of the ideas in this Preface. I am also grateful to the authors of the chapters in this book, and to the other participants in the symposium – Robert Axelrod, Freeman Dyson and Ehud Kalai – for their intellectual contributions and discussions at the meeting. Terry Guthrie was tireless in organizing the conference, and Sandi Milburn was brilliant in helping this book come to fruition. I am especially grateful to Mary Ann Meyers and the Templeton Foundation for their inspiration, collaboration and support in creating the conference, and in the production of this book. Without Mary Ann, this book could never have appeared.

Princeton University
December 29, 2008

Simon Levin

Foreword

An iconic event in popular accounts of the history of ideas about evolution and the origin of species is the Huxley–Wilberforce debate in 1860. The Bishop of Oxford, "Soapy Sam" Wilberforce, was a gifted, though markedly self-satisfied, orator. Speaking first, he unwisely mocked Huxley by asking "Is it through grandmother or grandfather that you descend from a monkey?" This opened the door to Huxley's memorable riposte: "Would I rather have a miserable ape for a grandfather, or a man highly endowed by nature and possessed of great means and influence, and yet who employs these faculties and that influence for the mere purpose of introducing ridicule into a grave scientific discussion – I unhesitatingly affirm my preference for the ape". After that, the meeting dissolved into Pythonesque chaos. First, Fitzroy (who had captained the *Beagle*, gone on to become the Governor of New Zealand, but by 1860 was mentally unstable, and soon afterwards committed suicide) came raving down the aisle, brandishing a bible, and demanding that all return to "The Book". The meeting ended as Lady Brewster – in the words of one of Darwin's biographers, DeBeer – "employing an idiom now lost, expressed her sense of intellectual crisis by fainting".

What most accounts of these events overlook, however, is the fact that Wilberforce, had he possessed an all-encompassing knowledge of the science of his day, could have won the debate. The Darwin–Wallace theory of evolution, at that time, had three huge problems.

The first problem concerned the time available for evolutionary processes to operate. Fifty years were to elapse before the first glimmers of awareness of weak and strong nuclear forces were to appear. Of the four fundamental forces recognized by today's physics, only gravitational and electromagnetic ("chemical") forces were known in Darwin's day. But if the sun's energy source was gravitational, it could not have been burning for more than about 20 million years. And chemical fuels would give an even shorter life. A different calculation showed that it could not have taken more than roughly 20–40 million years for the earth to cool from molten rock to its present temperature. These two calculations meant that either the earth was at most a few tens of millions of years old, or that Victorian physics was fundamentally deficient. Faced with these arguments, Darwin removed all numerical references to geological time spans in the third and later editions of the *Origin of Species*, and you will look in vain for any explicit chronology in the later *Descent of Man*. Of course, the subsequent discovery of nuclear forces showed Victorian physics was indeed inadequate: the sun burns nuclear fuel; and the heat generated by the decay of radioactive elements inside the earth invalidates simplistic calculations about cooling rates. We now understand that evolutionary processes on earth have all the time they need.

The second problem stemmed from the conventional wisdom of the day, namely that inheritance worked by a blending of maternal and paternal characters. The essentials of this issue can be grasped by considering a trait (such as height or weight) that can be described by a single variable. Suppose the mother departs from the population average in this respect by an amount x and the father by an amount y. Then, under a scheme of blending inheritance, the progeny will depart from the mean by $\frac{1}{2}(x + y)$. Suppose further that, in the parental generation, the statistical scatter of the variable about its mean value is characterized by a variance σ^2; that is, the expectation values of x^2 andy^2 are both σ^2. It is then straightforward to show that, with blending inheritance, the variance of this trait in the next generation is halved. But persisting variability is the raw stuff upon which natural selection works to produce descent with modification; it was critical to Darwin's ideas. This fundamental difficulty was pointed out to him, most notably by the engineer Fleeming Jenkin. He acknowledged it as a problem, but – given the observed persistence of variability in natural populations – he simply put it aside. The resolution of this major difficulty lies, of course, in the fact that genes are inherited in particulate Mendelian fashion, not by "blending". And, as shown in 1908 independently by Hardy and by Weinberg, under Mendelian inheritance variability remains unchanged from generation to generation, unless perturbed by factors such as selection, mutation, statistical drift, or nonrandom mating.

In short, these first two of Darwin's three truly major difficulties have been entirely swept away by advances in our understanding of the natural world.

Darwin's third major unsolved problem, which he himself arguably saw as the most important, is not yet solved. This problem was, and still is, explaining how cooperative behavior among animals evolved. The present volume is devoted to recent advances toward a solution.

At first glance, the answer seems easy. You pay some small cost to gather a much larger cooperative benefit. For example, a prairie dog takes a personal risk in giving an alarm call, but all the colony benefits and, by taking turns as alarm giver, each individual's group benefit exceeds the occasional risk. But any such arrangement is immediately vulnerable to cheats who enjoy the benefits without paying the risk-taking dues. In evolutionary terms, such risk-avoiding cheats have a selective advantage. Today we would say their enhanced probability of survival, and consequent greater reproductive success, means their uncooperative behavior is more represented in the next generation (possibly via their genes, or alternatively by teaching their offspring – Dawkin's memes). It is thus unclear how such observed cooperative phenomena can arise, or if it does, how it can be maintained.

Following work on "kin selection" by Hamilton and others, a century after Darwin, we now understand how such cooperative associations can evolve and persist in relatively small groups of sufficiently closely related individuals. This would seem to solve the problem for many non-human groups of animals, which are indeed found in such small kin groups. In particular, haplo-diploid systems of genetic inheritance, where siblings share more genes than do parents with their offspring, further facilitate such kin selection, which can help explain apparent altruism among some social insects.

During the hundred-thousand years and more when humans existed as small bands of hunter-gatherers, such considerations of kin selection could well have promoted cooperative behavior. But for large aggregations of essentially unrelated individuals, as developed once agriculture appeared some ten millennia ago and cities began, the origin of cooperative associations – with group benefits which exceed the "cost of membership" – remains almost as puzzling today as it was for Darwin.

Nor is this some abstract, academic problem for evolutionary biologists. The past 150 years have seen the human population increase sevenfold, and the ecological footprint of the average individual also increase sevenfold, for an overall 50-fold rise in our impacts on the planet. And these impacts are still increasing. There are consequently huge and global problems – climate change, loss of biological diversity, pressure on water supplies, and much else – which demand globally cooperative solutions. These problems are further compounded by the fact that nations must cooperate, but – given past history – in equitable proportions.

These problems have recently received an increasing amount of attention in the scholarly literature, employing a variety of metaphors: the Tragedy of the Commons; the Free-Rider problem; the Prisoner's Dilemma; and others. These metaphors are allied to artificial games in which the subjects (usually undergraduates) trade small sums of money to test limits to altruism and tolerance of cheating. Given Research Councils' resources, the sums involved in these experiments are necessarily relatively small, and I think it likely that some of the results would be significantly different with larger stakes. Moreover essentially none of this work, either theoretical or experimental, deals with situations where the costs and benefits vary among the players, as it commonly does in the real world (actions to ameliorate future climate change provide one striking example).

This book on *Games, Groups, and the Global Good* thus brings together a diverse collection of evolutionary biologists and economists, reviewing recent advances and speculating on the way forward.

Early chapters set the stage, with Frank surveying the biological foundations of the problem and Nowak setting out a possible taxonomy of distinct mechanisms whereby cooperation may evolve in the absence of kin selection. Bowles and Gintis sketch a view based more on economic thinking about wider aspects of choice in situations involving other people. Maskin gives a beautifully clear account of evolutionary stable strategies in games where two players repeatedly interact, with symmetric pay-offs, but with the realistic additional feature that random mistakes can occur. Skyrms concludes this overview by suggesting it may be better to deal with ever-changing networks of interacting individuals.

The second broad section of the book considers cooperation and group formation. Here we largely move away from models and games, into what might be called a more humanistic idiom: Flack and Krakauer on the origins of moral systems; Levin on games, norms and societies; D. S. Wilson on prosociality and the evolution of institutions; Johnson on supernatural punishment and cooperation; Hare on moral motivation; and Appiah on explaining religion. I particularly like Levin's discussion of how social institutions and "norms" may arise, and how they can develop over time. He sees evolutionary strategies evolving as diffuse responses to collections of situations, which in turn points to "new territory for game theorists". I also enjoyed Johnson's answer to the often-asked question: if punishing cheats promotes cooperative behavior, who takes on the costly job of punishment? He suggests that inventing supernatural entities who are ultimately responsible can be a good answer, and he elaborates this with a correspondingly new framework for game theoretic models.

The third and final set of chapters address practical issues involving cooperation and problems of the commons: Ostrom on building trust to solve commons dilemmas; Brams and Kilgour on how democracy resolves conflict in difficult games; O'Neill on the duty to apologize as part of a normative regime; concluding with Sugden on team reasoning and market relationships. Specifically, Brams and Kilgour suggest that democratic processes can "stabilize cooperative outcomes" by giving voters a clear choice between a cooperative outcome and the inferior consequences of failing to cooperate. It is, however, far from clear to me that democratic processes work in this way in the real world, as distinct from pious idealizations of it. Apart from anything else, extensive work on "the theory of voting", dating back 200 years to Condorcet (and later Dodson, aka Lewis Carroll), show for instance that the result of democratic choices among three alternatives – A, B, C – can under certain circumstances result in any one of the three emerging as the winner, depending on how the voting system (or sequence of choices) is organized; Ken Arrow's Nobel Prize in economics was awarded essentially for his independent rediscovery of this fascinating phenomenon. O'Neill's discussion of the role of explicit apology is most intriguing, and several contemporary illustrations of his thesis come to mind (e.g. the recent and very positive outcome of Australia's new Prime Minister, Kevin Rudd, saying "sorry" to the Aboriginal population). Sugden's chapter was presumably written before financial markets collapsed, around the world. He notes that "economics cannot represent the idea that paradigm market relationships ... have moral content". It can be argued that recent events suggest that economic theory should explicitly represent such ideas, and do so in clearly enforceable ways!

My personal musings about how cooperative human societies evolved – which have some parallels with Johnson's thoughts about "supernatural punishment and cooperation" – are both less academic and analytic, and more gloomy. Once we move out of the mists of pre-history, we find stories of dreamtime, creation myths, ceremonies and initiation rites, spirits and gods, with a unifying theme that all seek simultaneously to help explain the external world and also to provide a "stabilization matrix" for a cohesive society. There are, moreover, some striking and unexplained similarities in belief systems and rituals from different times and places. Conscience, a simple word for a complex concept which helps foster behavior in accord with society's professed norms, has been memorably defined by H. L. Mencken as "the inner voice which warns us that somebody might be looking". And how helpful it is if that somebody is an all-seeing, all-knowing supernatural entity.

Common to these conjectured "stabilizing forces" in essentially all earlier societies are hierarchical structures, serving and interpreting the divine being or pantheon, along with unquestioning respect for authority. In such systems, faith trumps evidence.

But if indeed this is broadly the explanation for how cooperative behavior has evolved and been maintained in human societies, it could be very Bad News. Because although such authoritarian systems seem to be good at preserving social coherence and an orderly society, they are, by the same token, not good at adapting to change.

A fundamental principle emerging from the Neo-Darwinian Revolution of around a century ago is Fisher's Fundamental Theorem, which states that a population's potential rate of change of gene frequency (which measures its ability to adapt to changing circumstances) is proportional to the variance in gene frequency, which will usually be small if essentially all individuals are especially well-adapted to their current environment. That is, there is an inherent tension between adaptedness and adaptability. If there is any substance in my speculations about the answer to Darwin's problem in explaining cooperation in human societies, we again have a fundamental tension – at the level of the entire society – between on the one hand authoritarian "ties that bind" and permit stably cooperative aggregations, and on the other hand the ability to respond effectively to changing environmental circumstances. It could even be argued that the recent rise of fundamentalism, in both the East and the West, is an illustration of this meta-level version of Fisher's Fundamental Theorem, as complex faiths are reduced to intolerant ideologies to resist the challenge of societal change.

Be this as it may, the present collection of thoughts on *Games, Groups, and the Global Good* makes it clear that we are making progress in the journey toward understanding how complex human associations came into being, and what their strengths and vulnerabilities are. Given our many and pressing needs for globally cooperative actions against environmental and other problems, the book is hugely timely. And our journey to a full understanding, particularly of cooperation in equitable proportions, needs to accelerate.

Zoology Department
Oxford University
Oxford OX1 3PS, UK

Robert M. May

Contents

Contributors

Kwame Anthony Appiah Department of Philosophy, Princeton University, 219 1879 Hall, Princeton, NJ 08544, USA

Samuel Bowles Sante Fe Institute, 1399 Hyde Park Road, Pod C-7, Santa Fe, NM 87501, USA

Steven J. Brams Department of Politics, New York University, Room 309, 19 W. 4th Street, NY 10012, USA

Hannelore De Silva WU Wien – Institute for Banking and Finance, Heiligenstädter Strasse 46-48, 1190, Vienna, Austria

Jessica C. Flack Sante Fe Institute, 1399 Hyde Park Road, Pod B-6, Santa Fe, NM 87501, USA

Steven A. Frank Department of Ecology and Evolutionary Biology, University of California, Irvine, CA 92697-2525, USA

Herbert Gintis 15 Forbes Avenue, Northampton, MA 01060, USA

John E. Hare Yale University Divinity School, 409 Prospect Street, New Haven, CT 06511, USA

Dominic Johnson Politics and International Relations, School of Social and Political Science, University of Edinburgh, 4.27 Chrystal Macmillan Building, 15a George Square, Edinburgh EH8 9LD, UK

D. Marc Kilgour Department of Mathematics, Wilfrid Laurier University, Waterloo, ON N2L 3C5, Canada

David C. Krakauer Sante Fe Institute, 1399 Hyde Park Road, Main Annex-2, Santa Fe, NM 87501, USA

Simon Levin Department of Ecology and Evolutionary Behavior, Princeton University, Room 203, Eno Hall, Princeton, NJ 08544-1003, USA

Eric Maskin Institute for Advanced Study, Einstein Drive, Princeton, NJ 08540, USA

Martin A. Nowak Program for Evolutionary Dynamics, Harvard University, One Brattle Square, suite 6, Cambridge, MA 02138-3758, USA

Daniel Tumminelli O'Brien Department of Biology and Anthropology, Binghamton University, (State University of New York), P.O. Box 6000, Binghamton, NY 13902-6000, USA

Barry O'Neill Department of Political Science, University of California, 405 Hilgard Avenue, Los Angeles, CA 90095-1472, USA

Elinor Ostrom Workshop in Political Theory and Policy Analysis, Indiana University, 513 North Park Avenue, Bloomington, IN 47408-3895, USA

Karl Sigmund Faculty for Mathematics, University of Vienna, Nordbergstrasse 15, 1090 Vienna, Austria

Brian Skyrms School of Social Sciences, University of California, 3151 Social Science Plaza, Irvine, CA 92697-5100, USA

Robert Sugden School of Economics, University of East Anglia, Norwich NR4 7TJ, UK

Christine Taylor Program for Evolutionary Dynamics, Harvard University, One Brattle Square, Cambridge, MA 02138-3758, USA

David Sloan Wilson Department of Biology and Anthropology, Binghamton University, (State University of New York), P.O. Box 6000, Binghamton, NY 13902-6000, USA

Part I
The Evolution of Cooperation at the Level of Individuals

Evolutionary Foundations of Cooperation and Group Cohesion

Steven A. Frank

Abstract In biology, the evolution of increasingly cooperative groups has shaped the history of life. Genes collaborate in the control of cells; cells efficiently divide tasks to produce cohesive multicellular individuals; individual members of insect colonies cooperate in integrated societies. Biological cooperation provides a foundation on which to understand human behavior. Conceptually, the economics of efficient allocation and the game-like processes of strategy are well understood in biology; we find the same essential processes in many successful theories of human sociality. Historically, the trace of biological evolution informs in two ways. First, the evolutionary transformations in biological cooperation provide insight into how economic and strategic processes play out over time-a source of analogy that, when applied thoughtfully, aids analysis of human sociality. Second, humans arose from biological history – a factual account of the past that tells us much about the material basis of human behavior.

1 Introduction

People change their behavior in relation to what others do. The way in which individual behavior changes in relation to others calls upon understanding the evolutionary dynamics of populations. By "evolutionary," I simply mean the tendency for successful behaviors to increase in frequency.

Understanding the evolutionary dynamics of behavior developed through a long history of study in economics, in game theory, and in evolutionary biology. The common theme in all fields derives from analysis of self interested actors.

In economics and game theory, notions of self interest and utility can be problematic; the theory applies to the extent that one accepts certain assumptions about

S.A. Frank
Department of Ecology and Evolutionary Biology, University of California, Irvine
CA 92697–2525, USA
e-mail: safrank@uci.edu

S.A. Levin (ed.), *Games, Groups, and the Global Good*, Springer Series in Game Theory,
DOI: 10.1007/978-3-540-85436-4_1, © Springer Physica-Verlag Berlin Heidelberg 2009

these notions of the individual. By contrast, a simple measure of self interest arises inevitably in biology from the basic facts of heredity and reproduction: those traits associated with relatively less reproduction have been outcompeted and have disappeared. Heredity also provides a clear notion of continuity through time, an essential point in the study of behavioral dynamics.

The clear advantage of biology with regard to the application of evolutionary dynamics led the great statistician and evolutionary biologist R. A. Fisher to say in a 1928 letter to Darwin's son, Leonard:

> An engineer finds among mammals and birds really marvelous achievements in his craft, but the vascular system of the higher plants . . . has apparently made no considerable progress. Is it like a First Law, not a great engineering achievement, but better than anything else *for the price?* Are the plants not perhaps the real adherents of the doctrine of marginal utility, which seems to be too subtle for man to live up to? (Fisher 1983, p. 94)

In other words, evolutionary dynamics of individual interests works beautifully to explain biology, but for humans, the problem appears more complex. From which, many people conclude that the evolutionary dynamics of self interest teaches us little about humans. I draw different conclusions.

On the theoretical side, evolutionary dynamics achieves its greatest development and clarity in biology, because of the clear notions of self interest and continuity through time. I will therefore develop the evolutionary dynamics of conflict and cooperation within the biological frame, but in a general way that does not depend specifically on biology. The principles should therefore provide a solid foundation for the application to human behavior.

On the applied side, understanding the evolutionary dynamics of human behavior is not easy, but should not be abandoned. Self interest, for all the problems one may wish to raise, remains a powerful theoretical framework in which to analyze human behavior. Several chapters in this book discuss recent progress and application. A common view is that, to engineer a social environment that achieves a certain moral goal, such as reduction in hostility or design of fair laws, one must understand the social dynamics in play. In fact, engineering and dynamics always go together: to control the outcome of a system, one must understand the dynamics of that system. By this view, evolutionary dynamics and moral engineering are natural partners.

The first part of my paper, on the evolutionary dynamics of conflict and cooperation, provides basic tools that apply across the disciplines of biology, economics, and game theory. I then turn to a second aspect of evolutionary dynamics: the biological history of evolution. How has the tension between conflict and cooperation – between individual and group – shaped the history of life?

One may view this biological history in various ways. It may be a source of analogy about the dynamical processes that govern human sociality, but similarity arises only through a vague analogy of change in populations. Or biology may define our history in fundamental ways, because we derive from this history and have been shaped by it in nearly every aspect. Or, as many humanists prefer, we must maintain a sharp divide between biological history and our understanding of human morality.

I prefer to think of biological history as both a source of interesting analogy about human affairs and an essential part of our history. I sketch the connections and the limitations that arise from these lines of thought. I leave the reader with a series of questions about how much biological analogy and biological history explain modern human sociality.

2 Scope

Misunderstanding arises frequently with the words "evolutionary" and "moral." I delimit my scope before proceeding.

I described two distinct meanings of "evolutionary" in the introduction. The first meaning concerns the change in a population over time. Any economic or game theoretic study that aims to understand human behavior must, at least implicitly, be evolutionary. At any point in time, each individual has a certain probability distribution over possible behaviors. As time progresses, each individual's behavioral distribution may change in response to the factors under study. These simple points alone provide the necessary conditions for the population dynamics of behavior to form an evolutionary system. From a purely logical or formal perspective, such evolutionary dynamics of behavior do not differ from biological evolution, although the particular rules of continuity and change inevitably differ between particular economic and biological problems.

The second meaning of "evolutionary" concerns the specific facts of evolutionary history. How did humans evolve? How have our brains been shaped by our past history? What consequences does that past evolutionary history have for understanding the behavior of modern humans?

With regard to "moral," I consider two distinct positions that parallel the two types of evolutionary analysis. First, whatever one takes to be the scope of moral studies, most issues concern individual attitudes, beliefs, or behaviors that may change over time in response to those attributes in other members of the population – an evolutionary problem. Often the issues will turn on some aspect of life that comes down to conflict or cooperation, and what this book labels as the "global good."

Second, the particular facts of evolutionary history may help us to understand the dynamics of moral issues. Such insight may come purely by way of analogy. For example, the theory of justice introduced in Adam Smith's *Moral Sentiments* and developed by John Rawls forms a very close analogy with one of the fundamental processes that shaped cooperation in biological history (Alexander 1987; Leigh 1991; Skyrms 1996; Frank 2003). Alternatively, and more controversially, insight may follow from the particular ways in which humans have been designed by natural selection through our evolutionary history.

3 Evolutionary Dynamics

I start with the simplest force that favors cooperative evolution, the tendency for similar behaviors to interact. In the second section I discuss repression of competition, in which reduced opportunities for conflict make cooperation the only way in which to increase payoff. In the third section I consider correlated interests between individuals, in which an actor places some value on the consequences to those who receive the outcome of the behavior. In the fourth section I turn to synergism, the positive interaction or feedback between cooperative behaviors with respect to payoff.

3.1 Correlated Behaviors and Information About Social Partners

I start with a simple example to focus the problem. Suppose a group depends on a common resource – the commons. That resource may be land that supports farming or a forest that supports wild food products. The total success of the group depends on the long-term flow of goods from the common resource. Prudent exploitation maximizes long-term flow and group good; overexploitation reduces long-term flow and group success.

A self interested individual gains according to two distinct components of success. First, that individual gains a particular share of the local resource. Second, the value of the individual's share depends on the total value of the local resource.

The essential tension of sociality arises from the conflict between an individual's local share and the resource's total value. An individual always increases its share of the common resource by competing more strongly against neighbors. However, increased competition leads to over-exploitation of the resource, reducing long-term gain and lowering everyone's success. Self interested individuals tend to overexploit the common resource, leading to the tragedy of the commons (Hardin 1968).

I developed a simple evolutionary model of the tragedy of the commons (Frank 1994b, 1995a, 1998). This model highlights in a clear way the two key forces that can overcome the tragedy of the commons: correlated behaviors between social partners and repression of competition. In this section, I discuss correlated behaviors in terms of information about social partners. In the following section, I analyze repression of competition as the second key force that can promote group cohesion and prudent exploitation of shared resources.

3.1.1 A Simple Model

Suppose the world is divided into local groups. Each group has its own common resource, available only to members of that local group. We seek the behavior adopted by self interested individuals. We find that behavior by searching for a situation in which, if everyone adopted a particular behavior or nearly so, then no self interested individual could do better by deviating from the population norm (Maynard Smith 1982).

I measure individual behavior by the degree to which an individual exploits the common resource. Values range from 0 to 1. Higher values represent greater individual exploitation; lower values represent more prudent and cooperative individual behavior with regard to the long-term value of the shared resource.

I seek the population-wide value of behavior, z^*, such that anyone who deviates does worse. To find that value, I measure the behavior of individuals who deviate as $I - z^* = \delta$, where δ is a random variable that measures individual deviation. Each individual lives in a local group. If we focus on a particular individual within a group, and set that individual's deviation to $\delta = x$, then we can use the theory of least squares to write the optimal prediction for group deviation, G, given the individual deviation as (Frank 1998)

$$\mathrm{E}(G|\delta = x) - z^* = rx. \tag{1}$$

We read this as: given an individual's particular behavioral deviation, x, the expected deviation of that individual's group is rx, where r is the regression of the average group behavior on individual behavior. In this case, the regression is equivalent to the correlation between behaviors of members in a group. This correlation is just a description of pattern without implication about mechanism: we simply note that, given a particular individual deviation, the group deviates to the extent that individual and group behavior are correlated. Put another way, r measures an individual's information about social partners given the value of the individual's own behavior.

Individual success depends on the product of two components. First, measure an individual's share of the local resource as $f(I,G)$, where I is the individual's competitive grab for local share, and G is the local average competitiveness. The function f rises with I and declines with G, because an individual's competitiveness raises its local share, and group competitiveness shrinks its local share. Second, measure the long-term value of the local resource as $h(G)$, in which long-term value declines as group competitiveness increases. Thus, we can write individual success as

$$W = f(I, G)h(G), \tag{2}$$

where it would be better to write this expression as the expected payoff given an individual's behavioral deviation, $\mathrm{E}(W|\delta = x)$, but for simplicity I just write W.

We analyze how payoff changes with individual behavior by

$$\frac{\mathrm{d}W}{\mathrm{d}x} = f_x h + r(fh_y + f_y h) \tag{3}$$

$$= -C_m + rB_m, \tag{4}$$

where subscripts denote partial differentiation with respect to that variable, and $y = rx$ is the group deviation (Taylor and Frank 1996; Frank 1998). In (3), the first term, $f_x h$, is the marginal change in an individual's share of the local resource as the individual behavioral deviation, x, changes. This measures the direct effect of an individual's behavior on success, holding constant how the individual's behavior correlates with the average competitiveness of neighbors and the value of the group resource. By convention, we call this direct effect of an individual's behavior on success the marginal cost of cooperation, C_m. In this case, $C_m = -f_x h$, where the minus sign arises because cooperation means a decrease in competitiveness, that is, a decrease in x.

The second term on the right side of (3) has two parts. First, fh_y measures the consequences of the marginal increase in the group resource as competition within the group decreases, that is, as the group deviation, $y = rx$, decreases. Second, $f_y h$ measures the consequences of the marginal decrease in the competitive pressure imposed by neighbors as the group deviation decreases. By convention, B_m measures the way in which a marginal increase in cooperative behavior among neighbors affects marginal change in individual success. This marginal benefit term, B_m, is weighted by r, because group behavior changes at a rate r relative to a change in individual behavior. Thus, r functions as an exchange rate between the marginal costs of individual cooperative behavior and the marginal gains of group cooperative behavior, rendering the costs and benefits on the common scale of individual payoff.

The condition for evolutionary dynamics to favor an increase in an individual behavior requires that the change in payoff with an increase in behavioral deviation be greater than zero, that is, $\mathrm{d}W/\mathrm{d}x > 0$, which also means that

$$rB_m - C_m > 0, \tag{5}$$

an inequality known as Hamilton's rule in biology (Frank 2006, see (15) below). In a moment, I discuss the importance of r, the group correlation. But first, I look at the outcome of the tragedy of the commons in a very simple case.

In (2), let $f = I/G$, which means that an individual's share of the local resource is proportional to its competitiveness, I, divided by the average competitiveness of members of the local group, G. Let $h = 1 - G$, which means that the value of the group resource decreases linearly with the average competitiveness of group members, yielding

$$W = \frac{I}{G}(1 - G), \tag{6}$$

or

$$\mathrm{E}(W|\delta = x) = \frac{z^* + x}{z^* + rx}(1 - z^* - rx). \tag{7}$$

With these assumptions, the behavior that, once adopted by nearly everyone, cannot be improved with regard to individual payoff is

$$z^* = 1 - r, \tag{8}$$

obtained by solving $\mathrm{dE}(W|\delta = x)/\mathrm{d}x = 0$ evaluated at $x = 0$, as described in Frank (1994b, 1995a, 1998). Clearly, as the correlation between social partners, r, increases, individual competitiveness, z, declines, or, equivalently, individual cooperative behavior increases and enhances the long-term prudent harvesting of the common resource.

3.1.2 Interpretation of Group Correlation

In this model, the severity of the tragedy depends on the behavioral correlation between group members. If, for example, group members are perfectly correlated, then they all have the same behavioral level of competitiveness, and no one can outcompete a neighbor. If no gains can be had at a neighbor's expense, then the only way to increase individual gain is by increasing the value of the common good. As the correlation, r, between neighbors declines, opportunity to outcompete neighbors rises, and each individual is favored to raise its competitive efforts even though the outcome is worse for all.

This model does not assume or depend on any particular mechanism that imposes correlation in the behavior between group members. In biology, the classical interpretation is that correlated behavior arises from correlated genes, usually between genetic relatives derived from recently shared ancestors. In a simple case, siblings would be genetically correlated by one-half. To calculate r in this case, note that, in a group of size N, the individual itself composes a fraction $1/N$ of the group, and an individual is correlated to itself by one. So, for siblings, $r = (1/N)1 + [(N-1)/N](1/2)$.

Correlation does not require genetics and shared genealogy. Individuals may choose correlated social partners. Individual choice of where to live may be correlated with behavior, so that those living in a particular place tend to behave in a correlated manner. Or, there may be some extrinsic force that imposes behavioral correlation.

The notion of information about social partners is very general (Aumann 1974, 1987). With genetic relatedness in biology, individuals do not necessarily "know" or have direct information about the behavior of their partners. Rather, if an individual happens to live near correlated individuals, then natural selection will favor those behaviors that exploit the correlations. By the evolutionary process, the existing correlation becomes exploited as information, and the resulting behavior is shaped in accord with that information (Binmore 1994; Pollack 1996; Skyrms 1996; Frank 1998, 2006). In humans, the evolutionary dynamics of behavioral adjustment may be complex. But, as long as individuals seek self interest by some process of trial and error, they may often come to settle on behaviors that exploit existing correlations: the invisible hand may come to discover and use information about social partners without conscious knowledge of those associations. Alternatively, direct and conscious information may come into play in some cases.

The point here is that behavioral correlations often shape conflict and cooperation more powerfully than any other process. The next section turns to mechanisms that may escape the tragedy of the commons when the intrinsic behavioral correlations are low. In that case, some secondary force must impose correlation to bring individual interests in line.

3.2 Repression of Competition

The simple tragedy of the commons model in (6) illustrates well the great importance of behavioral correlation. Self interested individuals compete at a level $z^* = 1 - r$, where r is the behavioral correlation among members of a group. As the correlation declines, competition becomes more severe, and shared resources become over-exploited to the detriment of all.

If some mechanism creates strong correlations within groups, then self interested individuals naturally adjust their behavior to cooperative ends. In the absence of intrinsic correlation, behavior tends to be competitive and mutually destructive. So, in the absence of an intrinsic correlation, what additional force can bring the interests of the competitive group members into line and thereby improve everyone's lot?

One possibility is that the self interested members of the group would gain by investing some of their own resources in mechanisms that repress competition in their group. Such policing of competition, by reducing the opportunities for individual gain against neighbors, would have the effect of imposing greater correlation among group members in the payoffs they receive. As mechanisms that level opportunities for individual gain intensify, individual payoffs become increasingly correlated with other group members independently of the resources that each individual invests in selfish and competitive behaviors (Alexander 1979, 1987; Leigh 1991; Skyrms 1996; Frank 2003, see these references for connections to notions of fairness and justice discussed by Adam Smith and John Rawls).

I focus on a simple extension of the tragedy of the commons model from the previous section, in which I add a second behavioral character that determines the extent to which individuals contribute their own resources to repressing selfish, competitive behaviors within their group (Frank 1995a, 1996c).

In the previous section, I used a simple payoff function to describe the tragedy of the commons

$$W = \frac{I}{G}(1 - G), \tag{9}$$

where I is the intensity at which an individual competes against neighbors for a share of the local resources, and G is the average intensity of competition within a group. I extend that model by adding a second behavior expressed by each individual, A, the amount an individual invests in mechanisms that police and repress local competition in the group, with $0 \leq A \leq 1$. The average investment in policing per group member is P. With this second character that represses competition, we can now express individual payoff as

$$W = (1 - cA)[P - (1 - P)(I/G)][1 - (1 - P)G]. \tag{10}$$

The first term applies a discount to individual success for the cost of investment in the public good through the policing mechanism, where c is the cost per unit investment in policing, A.

The second term is the individual's competitive success against neighbors for obtaining a share of local resources: a fraction P of local resources are distributed evenly to all group members, where $0 \leq P \leq 1$ is the average level of investment in the mechanisms that repress competition; a fraction $1-P$ of local resources remains available for splitting by competitive interactions, of which the focal individual acquires its share in proportion to I/G, given by the relative competitiveness of an individual, I, compared with the average level of competitiveness in the group, G.

The third term quantifies the long-term value of the shared resource. As before, the resource value declines with local competition. In this case, G is the average latent competitiveness of individuals, but only a fraction $1 - P$ of that competitiveness can be expressed, because local policing represses a fraction P of competitive behavior.

We need to find the values of the two behaviors, competitiveness and policing, such that when the population adopts the values (z^*, a^*), no individual that deviates can do better. I discussed the details in Frank (1995a, 1996c); here I give a brief summary. As before, the correlations in behavior among group members play a key role. Here, r_z is the correlation in competitive values between a randomly chosen individual and the group average, and r_a is the correlation in the amount invested in policing between a randomly chosen group member and the group average.

With regard to competitive behavior, self interested individuals are favored to express a level

$$z^* = \frac{1-r_z}{1-a^*(1-r_z)}. \tag{11}$$

The numerator is the solution for the simple tragedy of the commons model as given in (8). The denominator term, $a^*(1 - r_z)$, accounts for the amount of competition that is repressed, expressed as uncorrelated behavior $1-r_z$ that is repressed at a level a^*; this amount of reduced competition does not lower the long-term value of the shared resource. In this simple model, competition has a cost only through its affect on the value of the shared resource. So as mechanisms that repress expression of competition rise, the competitive tendency of individuals also rises. As I mentioned in Frank (1996c):

> The high competitiveness in a policing situation is no different from high internal pressure in a fish that lives at great depth. The fish brought to the surface explodes; intense competition and avoidance of repressive policing cause chaos when the same amount of energy is devoted to competition in the absence of repressive policing.

We might add an additional direct cost of competitiveness, as in Frank (1996c), but I do not include that here.

With regard to repression of competition, self interested individuals are favored to invest in policing the group at a level

$$a^* = \frac{r_a(1-r_z) - cr_z}{cr_a(1-r_z)} = \frac{1}{c} - \frac{r_z}{r_a(1-r_z)}, \tag{12}$$

with the constraint that $0 \leq a^* \leq 1$. The investment in policing to enforce group cohesion: declines with cost of effective repression, c; declines with the intrinsic correlation in competitive behavior, r_z, because increased competitive correlation reduces the ability of one individual to outcompete another and thus favors individuals to reduce their competitive tendencies without repression; and rises with the intrinsic correlation in investment in policing, r_a, because greater correlation reduces the loss an individual pays relative to neighbors for contribution to policing.

This simple model captures well how two opposing behaviors together shape the nature of group cohesion. On the one hand, individual competition within groups inevitably leads to the tragedy of the commons unless checked by some opposing force. Intrinsic correlation in the competitive tendency between group members, r_z, can alleviate the tragedy, because correlated group members cannot outcompete their neighbors and so gain by lowering their competitive tendencies. On the other hand, if the intrinsic correlation in competitive behavior is low, then selfish individuals are often favored to contribute to their own good by preventing the local devastation of their shared resource. They may accomplish this by investing in mechanisms that repress local competition, such as aspects of policing behavior.

The two distinct behaviors – individual competitiveness and contribution to group mechanisms that suppress competition – lead to an interesting duality in individual behavior. The most competitive groups, with low intrinsic correlation, r_z, between group members, most strongly favor competitive individuals to contribute resources to the group good through investment in the policing mechanisms. Increased policing favors individuals to become even more competitive, because competition is often suppressed as fewer shared resources become available for open competition. So behavioral dynamics tend to favor both greater contribution to policing mechanisms that promote the global good by preserving the shared resource and greater competitiveness of individuals. Ultimately, the outcome depends on c, how costly it is to develop an effective mechanism to repress competition, and on the intrinsic correlations in behavior that tie the success of individuals to other members of the group.

We may think of repression of competition as a mechanism that enhances local correlation. In a group that invests a^* to suppress competition, the effective correlation in competitive behavior between group members becomes $a^* + (1 - a^*)r_z$. In words, a fraction a^* of local resources is distributed fairly and without disruptive competition that degrades the common resource, and a fraction $1 - a^*$ of local resources remains available for local and destructive competition. Among that open fraction, $1 - a^*$, the intrinsic correlation r_z comes into play, leading to overall correlation in competitive success against neighbors as express by $a^* + (1 - a^*)r_z$.

In the models here, I have assumed that each group member begins with the same amount of resources. Interestingly, if individuals vary in their available resources, even by small amounts, behavior tends to diverge between individuals (Frank 1996c). The relatively stronger individuals allocate much of their excess resources in policing mechanisms that promote the global good, whereas relatively weak individuals allocate nothing to policing mechanisms that preserve shared resources.

Put another way, small variations in individual resources cause the well endowed to take over social control.

3.3 Correlated Interests

The previous sections focused on correlated behaviors. Such correlation plays a particularly important role when the individuals under study take actions and also receive the consequences of similar actions by others. Typical games, the tragedy of the commons, and mutual coercion fall into this class in which individuals are both actors and recipients.

In many behavioral situations, an individual acts to affect a recipient, but the recipient does not take any action. For example, an individual may provide aid to a brother or offspring without reciprocation. Theories based on correlated behaviors do not apply in these sorts of one-sided interactions. So, how may we account for altruistic behaviors in these cases?

Presumably, an actor who makes a costly behavior in favor of a recipient must value the recipient's interests. In biology, we have an extensive theory by which we can calculate how much an actor values different recipients (Frank 1998). In this case, individuals do not consciously put different values on different recipients. Instead, natural selection shapes the behaviors of actors in relation to different recipients. Outside of a biological framework, no theory provides an absolute basis for assigning relative values. Because I focus in this section on general forms of the theory that transcend biology, I limit my discussion here to how we may describe relative valuation between actors and recipients, without regard to what causes such valuations.

Suppose that, in valuing the total payoff to an individual in return for some behavior, z, we consider the individual's relative regard for others and self as

$$W(z) = vW_o(z) + W_s(z), \tag{13}$$

where W_o is the valuation of others affected by an actor's behavior, W_s is the valuation to self as a consequence of an actor's behavior, and v is an exchange rate between self valuation and valuation of others in a particular behavioral situation. The change in an actor's total valuation in return for a small change in behavior can be written as $\mathrm{d}W/\mathrm{d}z$, and, using primes to denote differentiation with respect to z, we may write the condition for an increase in the particular behavior to be favored as

$$W' = vW_o' + W_s' > 0. \tag{14}$$

It is often useful to write this condition equivalently as

$$vB_m - C_m > 0, \tag{15}$$

where B_m is the marginal benefit to the recipient, and C_m is the marginal cost to the actor. The behavior is favored when the value-weighted marginal benefits to others are greater than the marginal costs to self. We could apply this method of valuation to any sort of game or economic analysis of self interest.

In biology, the condition in (15) for a behavior to be favored by natural selection is known as Hamilton's rule (Hamilton 1970). The identical form of (5) misleads (Frank 1998, 2006). In (5), r measures the effect of behavioral correlation between neighbors on the direct success of the actor. The correlation may arise by genetic similarity, but other processes that impose correlation work in the same way. By contrast, v in (15) measures an actor's regard for the success of a recipient of the behavior – often, the recipient does not express any behavior in return and has no direct affect on the actor. Biology values v by the genetic similarity of the actor to the recipient.

3.4 Synergism and the Origin of Mutually Beneficial Behaviors

Different groups with complementary skills or resources can achieve synergistic gains by cooperating. However, if few tend to join cooperative ventures, then an individual who puts forward its potentially complementary resource may end up losing that resource. By contrast, if everyone tends to join synergistic activities, then no one gains by withholding their complementary skill, and cooperation is easily maintained. In this case, the difficulty concerns how to start synergistic partnerships which, once they become common, are easily maintained by advantages to self interested individuals (Axelrod and Hamilton 1981).

Much of the cooperative structure of life in biology and in human behavior arises from such synergistic interactions. The positive feedbacks and consistency of cooperation often become so deeply embedded that their very existence can be difficult to discern. The more cooperative and nonvarying the interaction, the less one tends to notice it.

For example, many animals depend on the numerous bacteria that they carry in their bodies: the bacteria provide essential dietary products to the host. The bacteria, in turn, sometimes cannot live without their hosts. This modern synergism is easy to understand: mutual dependence and mutual gain, although the potential for conflict remains within the alliance. In humans, specialized production and trade engenders mutual dependence and enhanced alliance; subsequent conflict runs within the constraints of synergistic benefits.

How do transitions occur between an initially uncooperative situation and a final situation in which cooperation reigns? Often, in the initially uncooperative state, no one gains by offering their special skills or resources if the behavioral or structural situation does not return synergistically matching skills or resources. So, the difficulty is how to get things started.

The transition from an initially uncooperative state to a cooperative one often turns on the behavioral correlation between potential partners (Axelrod and

		II	
		C	D
I	C	a	0
	D	1	1

Fig. 1 Matrix for a two-player game. The cells show the payoff to player I given strategies by two players in an encounter. The *C* and *D* strategies correspond to cooperation and defection. The payoff to player II in this symmetric game can be obtained by transposing the matrix. I assume $a > 1$

Hamilton 1981). For example, if cooperation is rare, but the behavioral correlation is high, then those rare individuals who tend to cooperate will often meet cooperative partners, and so mutually beneficial synergism can get started. As before, the cause of such correlation does not matter: cooperative individuals may be able to recognize each other and seek each other, or the few cooperative individuals may for accidental reasons tend to live near each other.

A simple game captures the way in which behavioral correlations influence transitions from uncooperative beginnings to synergistically cooperative and mutually beneficial social structures (Frank 1998). Consider, for example, a particular interaction between an individual from group I and an individual from group II.

The game matrix in Fig. 1 shows the payoffs for three different outcomes. First, an individual keeps the initial resource if it does not enter a joint venture with a potential partner. In the figure, withholding cooperative behavior is the D or defect strategy. A defector keeps the initial resource, in this case equal to a payoff of one unit, no matter what the partner does. Second, if an individual puts forward its resource in the C or cooperative strategy, and the partner does not reciprocate, then the individual loses its resource and receives a payoff of zero. Third, if both cooperate, then both gain the synergistic benefits with a payoff of $a > 1$.

In a particular encounter, player I cooperates with probability p, and player II cooperates with probability q (mixed strategies allowed). The payoff to player I is

$$w(p, q) = 1 + p(aq - 1), \tag{16}$$

and the payoff to player II is $w(q, p)$ by the symmetry of the payoff structure. Here, the players are drawn from separate populations, with average strategies $\overline{p}$ and $\overline{q}$, respectively.

From (16), we can see that player I is favored to increase its level of cooperation, p, when, on average,

$$q > 1/a. \tag{17}$$

To understand this condition, we must consider what information player I has about the expected behavior, q, of its partner, player II.

Player I has information about player II's strategy to the extent that interacting pairs have correlated behaviors. Suppose p deviates from its population average by

$\delta = p - \overline{p}$. Then we can describe the information player I has about the expected behavioral deviation of its partner by a regression equation

$$\mathrm{E}(q|p) - \overline{q} = r\delta, \tag{18}$$

where r is a regression coefficient, because the players are drawn from different populations (Frank 1994a). Given this regression equation, we can express the expected value of player II's behavior given information about player I's behavior as $\overline{q} + r\delta$, and so the condition in (17) for player I to be favored to increase its cooperative behavior becomes

$$\overline{q} + r\delta > 1/a. \tag{19}$$

If we start with the absence of cooperation, $\overline{p} = \overline{q} = 0$, then full cooperation can spread only when

$$r > 1/a. \tag{20}$$

Thus, a significant behavioral correlation is needed to make the transition to a cooperative state. Once the populations have moved to full cooperation, they gain from synergistic benefits because $a > 1$. At that point, no correlation is needed to maintain cooperation. Thus, correlation drives the initial transition, but does not play a role in subsequent maintenance.

With full cooperation, each population may become dependent on the skills and resources of its partner population. At that point, mutual dependence causes cooperation to become essentially irreversible (Frank 1995b). Much of cooperation and the evolution of social structure may follow such a path, through which brief periods of information about social partners allow mutually beneficial traits to flourish. As those traits flourish, they become embedded in the structure of opportunities available and payoffs gained. Such traits of mutual dependence may come to seem more as fixed aspects of the social environment than as interesting characteristics reflecting the social tension between cooperation and conflict.

4 Biological History

Some people may believe that biological history can teach us little about our own species' conflicts, cooperative associations, and moral dilemmas. But, upon study, one has to be surprised by how often the basic forces of social tension in human life have deeply and inexorably shaped biological history. The lessons drawn from such similarity are, certainly, points of debate. But before we can consider the debate, we need some facts to set a common ground for discussion.

I organize biological aspects of cooperation along the lines of the four major forces that shape the nature of conflict and cooperation among self interested individuals. Those forces are correlation between social partners, repression of competition, correlated interests between actor and recipient, and synergism. In this section, I consider the role played by each of those forces in the history of life.

The broad topic is, of course, too great to cover fully. So I use a few examples to illustrate the key points.

4.1 *Correlated Behavior and the Tragedy of the Commons*

The tragedy of the commons arises because self interested individuals gain by competing against neighbors. Rapacious individuals outcompete their neighbors and gain a larger share of the local resource. But rapacious behavior depletes the local resource in a way that reduces the long-term yield, causing harm to all members of the group. A mechanism that causes correlation in behavior between group members favors prudent behavior, greater sustainable productivity, and benefit to all group members.

I applied the tragedy of the commons to problems in biology (Frank 1995a, 1996b), extending a long history of work on group selection (Hamilton 1967, 1972, 1975; Lewontin 1970; Leigh 1977, 1991; Wilson 1980; Colwell and Wilson 1981; Szathmáry and Demeter 1987). My work arose from study of biological problems such as the sex ratio of progeny produced by mothers within small isolated groups and the amount of harm a parasite causes to its host. I discuss how these applications and related problems have grown into a major field of study in evolutionary biology (Rankin et al. 2007).

4.1.1 Sex Ratio: The Production of Males as a Competitive Trait

In almost all animals, females produce babies and males do little except compete, mate, and provide sperm. With regard to reproduction, females are productive and males are competitive. In some animals, males contribute more that just matings, and the situation is more complex. But the vast majority animals follow the simple dichotomy. For example, one can think of most male insects as the mother's competitive winged sperm.

Consider the male-female distinction from a mother's point of view. She can make a daughter, who produces babies. Or she can make a son, who competes with other males for matings but produces nothing directly.

A mother's investment in sons is an investment in a trait to compete against other mothers in the local mating group (Hamilton 1967). The more a mother invests in sons, the more grandchildren she will have through her sons. Those extra grandchildren come at the expense of reduced numbers of grandchildren through sons by other mothers in the group, because the total number of grandchildren is fixed by the number of productive daughters that are made.

This problem of allocation of resources to competitive sons is formally equivalent to the tragedy of the commons problem that I introduced earlier in (6) (Frank 2006). In particular, suppose a mother allocates a fraction I of her reproductive resources to sons and a fraction $1 - I$ to daughters. Assume that, in a local group, daughters and sons grow up and mate with each other, and then the mated daughters disperse to find new reproductive opportunities – a pattern of mating and dispersal that occurs in many insects (Hamilton 1967). In a local group, suppose that, on average, mothers allocate G of their resources to sons and $1 - G$ to daughters.

A mother's payoff is the combination of her success through her sons, W_s, and her success through her daughters, W_d, yielding total success as $W = W_s + W_d$. For sons, the payoff follows exactly the tragedy of the commons expression for payoff in (6) as

$$W_s = \frac{I}{G}(1 - G). \tag{21}$$

The first term, I/G, accounts for the relative success of a mother through the matings obtained by her sons. For example, a mother may make K babies of which a fraction I are sons, giving her KI sons. In the group, there are N mothers who, on average, each make KG sons. So the fraction of all males in the group by the focal mother is $I/(NG)$. Those males compete for matings among the local resource, the $KN(1 - G)$ daughters produced by all mothers in the local group. Combining the terms and dropping K as an arbitrary proportionality constant gives W_s. Success through daughters is the number of daughters produced by a mother, $K(1 - I)$, and again we drop K as a proportionality constant. Combining success through sons and daughters yields total payoff as (Hamilton 1967)

$$W = \frac{I}{G}(1 - G) + 1 - I. \tag{22}$$

We apply the same methods used to obtain (8). The solution is $z^* = (1 - r)/2$, where z^* is the fraction of resources a mother allocates to sons such that, if nearly everyone adopts this behavior, no behavior that deviates from it can obtain a higher payoff, and r is the correlation in the sex ratio produced by mothers within a local group (Frank 1985, 1998).

The solution parses more easily when we write the best allocation to sons and daughters as a ratio $1 - r : 1 + r$. Here, the term $1 - r$ for male allocation arises from the equivalence between male allocation and the tragedy of the commons (Frank 2006). Sons are the direct expression of a mother's competition against neighboring mothers. Sons are made at the expense of daughters. Daughters contribute to the success of all mothers in the group by providing mates for those mothers' sons.

The term $1 + r$ for the relative value of female allocation arises as follows. Each extra daughter made by a mother contributes to the mother's success directly through the grandchildren produced by the daughter – the valuation of one. In addition, an extra daughter provides an extra mate to males in the local group. That extra mate for sons accrues to the strategy pursued by the focal mother in proportion to the correlation between the mother's strategy (her sex ratio) and the average strategy in the local group (Frank 1998).

This sex ratio model makes a simple qualitative prediction. As the correlation between mothers declines, mothers compete more intensely with each other by raising their relative allocation to sons. If we assume that behavioral correlation in a small group arises mainly from a mother's correlation to herself, then in the simplest case $r = 1/N$, where N is the number of mothers in the local group. This gives us Hamilton's (1967) famous model of the sex ratio under local competition for mates. With this expression, r declines as the number of mothers in the group, N, rises. So the prediction becomes: as N rises, the fraction of males produced by mothers should rise.

Many studies show that as more mothers contribute to a local group, the competitive allocation to sons rises (Godfray and Werren 1996; Hardy 2002). My own study on fig wasps provides a simple and direct demonstration (Frank 1985, see also Herre 1985). In the species I studied, a mated fig wasp female gets inside a fig, lays her eggs, and dies. More than one mated female may lay her eggs within a fig during a short window of a few days. About 4 weeks later, the male offspring emerge first within the dark cavity in the center of the fig. The males mate with the quiescent females. After a few days, one of the males chews a tunnel through the wall of the fig, stimulating the mated females to emerge, exit, and find another fig to start the cycle anew.

Fig biology imposes exactly the life course assumed by the sex ratio model of local mating and competition among males (Hamilton 1979). To test the theory, I manipulated the number of mothers that enter each fig. Do these tiny mothers, each less than 2 mm, detect the number of other mothers in the dark fig cavity and adjust their allocation to competitive sons? Figure 2 shows that they do.

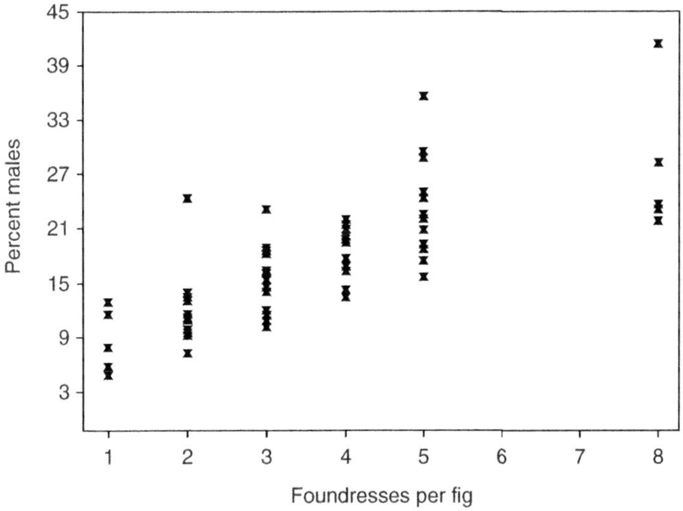

Fig. 2 Sex ratio of fig wasps. Foundresses per fig represents the number of mothers laying eggs in each fig. The numbers of foundresses were controlled experimentally. Each "X" marks the sex ratio of all foundresses in a single fig. The percent males rises with the number of foundresses, matching the tragedy of the commons prediction that mothers increase their competitive allocation to sons in response to an increase in the number of other mothers that compete in the local group. Redrawn from Frank (1985)

4.1.2 Parasite Virulence: A General Model for Prudent Versus Rapacious Exploitation of Resources

Some parasites exploit their hosts in a prudent way, taking the resources that they need without causing noticeable damage. Prudent exploitation yields sustainable benefits to the parasite as long as the host remains healthy. Other parasites attack their host more quickly and vigorously. Rapid exploitation may allow the parasites to achieve higher reproductive rates, but damage to the host reduces the parasites' opportunity for sustainable yield (Frank 1996b).

Following this economic line of thought, each parasite faces a tradeoff when increasing the rate at which host resources are used. Greater exploitation has the benefit of more rapid reproduction and transmission to new hosts, but carries the cost of reducing the host's ability to procure more resources in the future. For each host-parasite interaction, there may be a particular optimum schedule of host utilization that maximizes the parasites' balance between rapid transmission and the time before the host dies (Fenner et al. 1956; Levin and Pimental 1981; Anderson and May 1982; Levin 1983).

One process missing from the tradeoff between transmission and virulence concerns the "social" aspect of parasite interactions. Suppose that prudent exploitation of a host maximizes a parasite's reproduction. Natural selection then favors each parasite, when alone in a host, to follow the prudent strategy. There is, however, a problem when two or more parasites with different strategies occupy the same host. If one strategy extracts host resources rapidly and reproduces quickly, then the host may die in a short time. A prudent strategy would have relatively low reproduction when paired in a host with a rapacious strategy because, for both strategies, the host is short-lived, and the rapacious strategy reproduces more rapidly than the prudent one. This is the tragedy of the commons.

Correlation in behavior between members of a group mitigates the tragedy of the commons. In biology, the correlation typically arises by genetic relatedness within a group. In the case of parasite virulence, the prediction is that more related parasites within a host will behave more prudently, competing less intensely among themselves and causing less harm to the host (Hamilton 1972; Bremermann and Pickering 1983; Frank 1992, 1996b). Figure 3 supports the predicted trend: reduced correlation among parasites increases the damage caused to the host.

The pattern in Fig. 3 leaves open the issue of whether competition between unrelated parasite lineages plays a direct role in causing harm to the host. de Roode et al. (2005) showed that, in the parasite that causes malaria, the more competitive parasite lineages did outcompete other parasites within the host and did cause greater harm to the host. This study of the malaria parasite ties the direct competition over local resources to the harm caused to the public good – the health of the host that provides resources to the parasites.

The problem of parasite virulence captures well the essence of many biological examples of the tragedy of the commons (Frank 1996b). For example, the most primitive cells probably contained several molecules that could make copies of themselves – the primitive genes. Each trait of those replicating molecules was

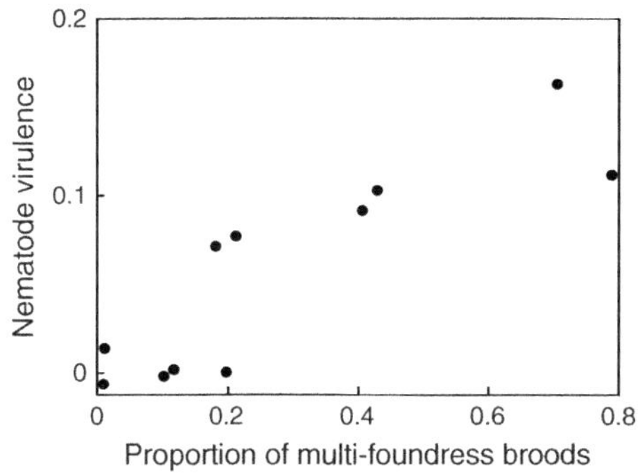

Fig. 3 Virulence of nematodes infecting fig wasps. The fig wasps I described in the sex ratio section often carry parasitic nematodes. Herre (1993) studied how much harm these parasitic nematodes cause to their hosts. He predicted that greater mixing between nematode lineages would reduce the correlation of behavior (relatedness) within hosts and lead to greater virulence. The data support the prediction. Here, multi-foundress broods measure the fraction of figs in which more than one wasp entered. The more often multiple foundresses enter a fig, the more often the nematode lineages will likely mix, reducing within-host correlation. Herre measured virulence by $1 - f_i/f_u$, where f_i and f_u are the number of babies produced by infected and uninfected wasps, respectively. Lower productivity of infected wasps corresponds to higher virulence. Redrawn from Herre (1993)

selected according to the balance between individual benefit from rapid exploitation of local resources and group benefit from prudent exploitation of local resources. In other words, the problem of cooperation versus conflict in groups arose in the earliest stages of biological history.

4.1.3 Scale of Competition and the Role of Group Productivity

I presented a simple tragedy of the commons model in (6). In that model, individuals can potentially gain by restraining competition in order to enhance the productivity of their group. Individuals in more productive groups benefit by getting a piece of a greater local resource, even if their piece may be smaller than their neighbors'.

This simple model for the tragedy of the commons makes implicit assumptions about how individuals compete and how we measure success. In evolutionary models, we generally measure the success of an individual relative to some base population, because we are interested in whether a behavior is gaining or losing in frequency in response to its success relative to other behaviors in the comparison population.

Suppose we regard the local group as part of a broader population, and we measure the success of each individual relative to the broader population. Then a cooperative individual can gain in the population by trading a smaller share of the local resource in return for a greater total value of the local resource. In other words, a prudent group gains greater total productivity, benefiting all members of the group.

The interpretation of differential group productivity benefiting members of prudent groups matches the tragedy of the commons model that I gave earlier. In this case, an individual's success is measured relative to the broad population composed of many local groups.

But what if an individual's success is measured relative only to other members of its local group? For example, if there is no competition between members of different groups, then each individual's relative success arises only from its advantage or disadvantage compared to its neighbors in its local group (Wilson et al. 1992; Taylor 1992a, b; Queller 1994).

We can, in general, define the problem by the scale of competition (Frank 1998). Let the scale of competition, s, be the probability that an individual ultimately competes against and measures its relative success against only local group members, and $1-s$ be the probability that an individual ultimately competes against and measures its success against members of the broader population. The base population for measuring success determines how a particular level of success translates into change in the frequency of a behavioral strategy.

With this definition for the scale of competition, we can extend the tragedy of the commons model in (6) to

$$W = \frac{I}{G}\left(\frac{1-G}{s(1-G)+(1-s)(1-z^*)}\right), \tag{23}$$

where z^* is the average level of competitive behavior in the population, $I = z^* + \delta$ is the deviation from the average by a randomly chosen focal individual, and $G = z^* + r\delta$ is the deviation of the group average by the focal individual's local group. By following the approach given earlier, we can find the value of z^* that, once adopted by the population, cannot be beat. Because z^*, the level of competitiveness of individuals, varies in this model between zero and one, we can write $1 - z^*$ for the level of individual cooperation, yielding

$$1 - z^* = r\left(\frac{1-s}{1-rs}\right), \tag{24}$$

where r is the correlation in behavior within local groups. Increased behavioral correlation, r, favors cooperation. By contrast, increased local competition, s, reduces cooperation. An individual cannot gain by providing benefit to a neighbor if the individual's ultimate success is measured only against neighbors. Cooperation can increase only to the extent that an individual ultimately competes against and measures success against members of other groups.

Griffin et al. (2004) studied how behavioral correlation and the scale of competition jointly determine cooperative behavior in bacteria. Pathogenic bacteria often face iron limitation when living within a host; hosts often withhold iron as a defense against bacteria. Some bacteria can secrete a molecule – a siderophore – that scavenges iron from the host. The bacteria then take up siderophore-iron complexes to overcome their deficiency.

Siderophore production is a public good: costly for individuals to produce and equally beneficial for all members of the local group. In particular, any member of the local group can take up a siderophore-iron complex independently of who originally secreted the siderophore.

Griffin et al. experimentally varied behavioral correlation by changing the amount of mixing between different bacterial clones. Relatively pure clones cause high behavioral correlation through genetic similarity. Mixed clones have lower behavioral correlation because of greater genetic diversity. They varied the scale of competition by altering the constraint placed on the contribution of local groups to the following generation. If all groups contribute equally by constraint, then individuals compete only locally within their group, and the scale of competition is entirely local. If groups contribute in proportion to their productivity, then individuals compete fully with members of other groups, and the scale of competition is global.

The experiment set the conditions that determine behavioral correlation and the scale of competition. Then, over time, the evolutionary change in populations was followed with regard to siderophore production, which measures production of a public good and the degree of local cooperation. Figure 4 shows that the experiment supports the predictions of (24): greater behavioral correlation and global competition increase cooperation.

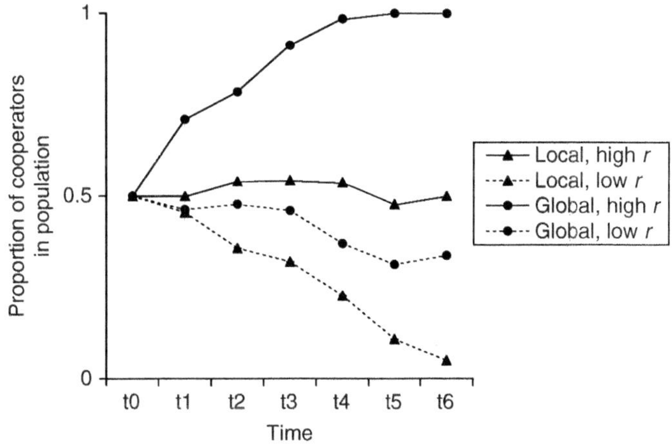

Fig. 4 Evolutionary change in siderophore production. Time moves from left to right, each unit representing 1 day and one round of mixing of the bacteria to impose either local or global competition. The behavioral correlation, r, was controlled by the degree of mixing between different bacterial clones. From Fig. 3 of Griffin et al. (2004)

4.2 *Repression of Competition*

Correlated behaviors align interests and favor reduced competition. But highly cooperative behavior often occurs in nature with little correlation in the intrinsic tendency of individuals. To achieve that high level of cooperation in the absence of intrinsic correlation, there must be some force extrinsic to each individual that tends to align interests and behaviors (Frank 2003).

Reduced opportunity for competition can align interests. If an individual cannot compete against neighbors, then that individual can increase its own success only by enhancing the efficiency and productivity of its group (Leigh 1977; Alexander 1979). In the first example, a randomization process assigns success to individuals independently of their behavior within the local group, preventing any individual from gaining by competing against neighbors. In the second example, powerful individuals within the local group repress competition between lower ranking individuals; such policing of competition appears to play a key role in social integration within the group. In the final example, one species domesticates and essentially enslaves another species. The master species gains by preventing competition between enslaved members of the group, in order to prevent wasted energy devoted to internal competition.

4.2.1 Randomization and Fairness

Sexual reproduction mixes the genes from two parents to make an offspring. Each parent contributes one half of the genes. Meiosis is the process by which each parent selects one half of its own genes for transmission to the child. Biologists often refer to the process as "fair meiosis," to emphasize that each copy of a gene has an equal chance of being chosen. This randomization process means that no gene copy can gain an advantage over other gene copies in being transmitted to the offspring.

With no opportunity for local competition, all gene copies gain only with the enhanced success of the whole group (Leigh 1971). In this case, we call the group of the genes the "genome." The unity of the genome, and thus the unity of the individual, is so nearly complete that one often thinks of the genome in a unitary way rather than as a collection of cooperating genes (Maynard Smith and Szathmáry 1995). However, competition between gene copies does occur in nature, in which one gene copy increases its chance of getting transmitted to the offspring at the expense of other gene copies (Crow 1979).

Competition between gene copies reminds one that the reproductive fairness and the near unity of the genome evolved in the face of competitive pressure between neighbors. The puzzle concerns how the process of natural selection, acting on the interests of the individual gene copies, led to particular biochemical mechanisms that typically repress internal competition (Frank 1995a), and how those mechanisms can sometimes be subverted by certain competitive types.

This example shows that the very foundations of sex, reproduction, and the genetic transmission of information arose from the group cohesion of a collection of genes. That cohesion was created by processes that repress internal competition and bind the interests of the separate genes to the group (Maynard Smith and Szathmáry 1995).

As I mentioned, the cohesion of the genome often appears so complete that we use the word "individual" to refer to a single genome – a single collection of genes in an organism. But that "individual" is a constructed group, cohesive only because of the high reliability of the internal mechanisms that ensure reproductive fairness among gene copies.

The occasional gene copies that subvert these fairness mechanisms and outcompete neighbors emphasize that unity in biology must always be constructed and maintained. Such unity arises solely from self interested actors. The system of cohesion built by those self interested actors must enforce against the competitive tendencies of those same actors. And so it always goes: conflict and cooperation in constant tension and never separable.

4.2.2 Policing and Repression of Competition

Individuals in a group may prevent others from competing. Such repression of competition by third-party policing reduces opportunity for individuals to gain at the expense of their neighbors. Once again, in the absence of opportunity to outcompete a neighbor, an individual can increase its own success only by enhancing group efficiency and productivity.

Policing of competition can be a very effective mechanism to promote group cohesion. However, policing competition between others can be dangerous or costly. Why should a self interested actor take on the costs of the policing role? In Sect. 3.2, I showed the conditions under which an individual may gain more by its benefit from living in a more productive group than it loses by the costs of policing. I also mentioned how the theory predicts an interesting asymmetry with regard to policing: those individuals with relatively greater vigor or resources are favored to take on the policing role, whereas those with relatively lower vigor or resources do not gain from policing (Frank 1996c). Thus, the theory predicts that the relatively powerful individuals impose social control on the group when effective mechanisms exist for dominant individuals to repress competition and promote group cohesion.

Flack et al. (2005a, b) studied policing and group cohesion in pigtailed macaques. Dominant males intervene to control disputes between pairs of lower ranking individuals. Those policing acts usually do not favor one competitor over the other, but rather the intervention puts an end to the conflict.

Flack et al. (2005a, b) analyzed the consequences of policing interventions for various aspects of group cohesion. They compared two situations in a semi-natural captive colony. In the baseline case, the dominant males were present and acted in their normal way to settle disputes. In the "knockout" experiment, Flack et al. removed the dominant males and placed them just outside a wall that bounded the colony. The dominant males were visible to the colony members but could not intervene.

When policing males were removed, the amount of aggressive behavior in the colony increased. Measures of aggression included initiation of conflicts, intensity of conflicts, biting, and joining a conflict. With the rise in conflict, there was also a decline in affiliative behaviors: reconciliation, play, grooming, and physical proximity. From these observations, Flack et al. (2005a) concluded:

> The extent to which policing is important to organizational robustness is surprising considering that actual policing behavior occurs relatively rarely. This suggests that the simple presence of individuals responsible for conflict management can change the way group members are willing to interact with one another.

In my own work (Frank 1996c), I developed the theoretical prediction with regard to policing when individuals vary in vigor or resources:

> Small variations in individual vigour or resources can lead to large variations in individual contributions to policing the group. Stronger individuals often invest all of their excess resources into policing, but weaker individuals do not contribute to group cohesion.

Flack et al. (2005b) directly addressed this prediction:

> The primary finding of this study is that heterogeneities in power, by producing heterogeneities in the cost of conflict management for individuals, lead to heterogeneities in the tendency to police.

In pigtailed macaques, the well endowed make essentially all the investment in social control.

In a subsequent paper, Flack et al. (2006) concluded that policing by dominant individuals plays a key role in group cohesion:

> We observe that when policing is operational, group members build larger social networks characterized by greater partner diversity and increased potential for socially positive contagion and cooperation. Without policing, high conflict frequency and severity leads to more conservative social interactions and a less integrated society.

4.2.3 Domestication and Repression of Competition

Humans have domesticated various animal and plant species for food production. Ants began farming much earlier – about 50 million years ago (Mueller 2002).

Fungus growing ants collect plant material to feed their crops. The ants weed their gardens to protect against fungal parasites that specialize in attacking fungus gardens. The ants also grow specialized cultures of bacteria on their bodies in order to use the antibiotic secretions produced by their partner bacteria (Zhang et al. 2007). The antibiotics protect the fungal gardens from bacterial diseases.

Domesticated fungal species were once wild, free living species. New domesticates carry with them their own evolved tendencies for competition in local groups – their tragedy of the commons – in which such competition reduces the efficiency and productivity of the domesticates.

I developed the general prediction that nonhuman masters gain by repressing competition among domesticates in order to elicit the most efficient domestic productivity (Frank 1996a). In natural, unregulated situations, mixture between genetically unrelated strains often leads to greater competition between individuals and a decline in productivity. Such competition develops between mixed lineages because mixture reduces the behavioral correlation between individuals. Following this logic, the easiest way for ant farmers to reduce conflict between fungal domesticates would be to prevent mixing of fungal lineages in their gardens. Homogeneous domesticates have high correlation in their cooperative tendencies, leading to an intrinsic tendency to reduce competition and enhance group productivity.

Ants do in fact prevent mixing of domesticate fungal lineages (Bot et al. 2001; Mueller et al. 2004; Zhang et al. 2007). When a newborn queen leaves her birthplace to found a new colony, she brings with her the fungal lineage from her natal colony. The fungi produce chemicals that inhibit growth by competing lineages; the ants spread those anti-competitor chemicals in their feces, which fertilize the growing fungal garden.

4.3 Correlated Interests

An individual may value payoff to another individual. Such other regarding behavior arises commonly in biology from genetic relatedness. For example, life depends on the regard a parent has for its offspring.

In biology, we can tally the payoff to a parent for its various behaviors directed at offspring. We first count the benefits of those behaviors for the offspring discounted by the genetic correlation between parent and offspring. We then subtract off the direct cost of the behaviors for the parent. This calculation yields Hamilton's rule, by which a behavior is favored when $rb - c > 0$, where r is the genetic correlation (or regression) between actor and recipient, b is the benefit to the recipient, and c is the cost to the actor – see (15).

The actor's valuation of the recipient in proportion to genetic correlation arises from the fact that transmission of strategies through time ultimately determines the evolutionary dynamics of behavior. A recipient of a behavior carries the actor's deviation from the average strategy in proportion to the genetic correlation between actor and recipient.

Other forces may sometimes affect the valuation of a recipient by the actor. For example, the recipient may return a beneficial behavior at a later time (Trivers 1971). In this case, valuation of another arises as an investment in an expected future payback. Most often, however, other regarding behavior in biology arises from genetic relatedness.

Parental altruism toward offspring occurs so widely in nature that I will not elaborate further on that case. Instead, I focus on two interesting examples in which individuals weigh their comparative regard for different classes of genetic relatives. The comparative aspect highlights the quantitative nature by which an actor regards the payoff to others.

4.3.1 Competition for Being the Queen

In many social insect colonies, a newborn female may develop into a reproductive queen or a partially sterile worker (Wilson 1971). A queen directly produces offspring; a worker helps to raise sisters and brothers. A newborn female often would gain more by developing into a queen rather than a worker. Here, I measure success by the biological standard of genetic contribution to future generations. I tabulate total genetic success by the effect of a behavior on direct reproduction and on the reproduction of genetic relatives weighted by the genetic relatedness of actor to recipient.

The queen or the older workers usually control the fate of newborn females – development into either the queen or worker caste. The elders control the caste of newborns by manipulating offspring size, by varying chemical stimulus, and by altering the provisioning of food. The elders' control over caste represses competition between newborns over development into queens.

In a particular type of social bee, elders do not coerce the caste of newborns (Wenseleers and Ratnieks 2004). Each newborn female may develop unconstrained into either a queen or a worker. However, the number of available opportunities for newborn queens to lead a colony is limited. Those who develop into queens compete with each other for those limited slots. Only a small number of queens succeed in this competition; the losers die and do not contribute to the colony productivity.

The competition between newborn queens causes a tragedy of the commons (Wenseleers and Ratnieks 2004). To increase group efficiency and productivity, the colony should avoid costly overproduction of queens, most of whom die in competition with their neighbors. Those queens who die in competition could have developed into workers who contribute to common productivity. However, individuals may gain by competing for the extra individual payoff of being a queen.

The calculation of payoffs for being a worker or queen is more complex in this case than for the simple tragedy of the commons models I discussed earlier. In this case, we must account for other regarding valuations ascribed to different kinds of genetic relatives. In particular, we want to know how the behavioral choice of being a queen or a worker affects others in the colony, and how the actor, faced with the choice of alternative development, regards those who are affected by the choice.

A special aspect of bees, ants, and wasps arises from their asymmetric inheritance system. A mother's unfertilized egg develops into a son; her fertilized egg develops into a daughter. Queens mate with males and produce both unfertilized sons and fertilized daughters. Workers do not mate, but can lay unfertilized sons.

Workers will typically be offspring of the queen. The workers help to rear the new eggs laid in the colony. Most of those new eggs will be laid by the queen and will be the workers' sisters and brothers. However, some of the workers will directly lay their own sons. Thus, in rearing new eggs, a worker will also be helping to rear some of her sisters' sons – that is, her nephews.

Here is a central point: a female is more closely related to her nephew than to her brother because of the peculiar asymmetry in inheritance (Hamilton 1972). So the ratio of nephews to brothers determines the valuation a worker gains from her effort to rear the eggs produced by the colony. The more nephews produced directly by her sister-workers, the more a female values the eggs she helps to rear as a worker.

Now we return to the key behavioral choice. A newborn female can become a queen and rear sons and daughters. Or she can become a worker and rear sisters, brothers, and nephews. The greater the ratio of nephews to brothers, the greater the valuation of being a worker via accounting for genetic regard for others.

Thus, we come to a simple prediction (Wenseleers and Ratnieks 2004). The fraction of females who develop into queens should decline as the amount of egg laying by workers rises. To state the reasoning again: more egg laying by workers means that a worker will raise an increased ratio of nephews to brothers. A worker values a nephew more highly than a brother. As the fraction of nephews increases, the relative value of being a worker rises, and the relative gain of competing for a queenship drops compared to the expected gain of being a worker. Figure 5 shows data that support the prediction. More egg laying by workers is associated with a lower fraction of newborn females developing into queens.

4.3.2 Worker Valuation of Egg Production by Other Workers

I stated that, in the bees, ants, and wasps, a sister is more closely related to her nephew than to her brother. This asymmetry occurs when a colony has a single

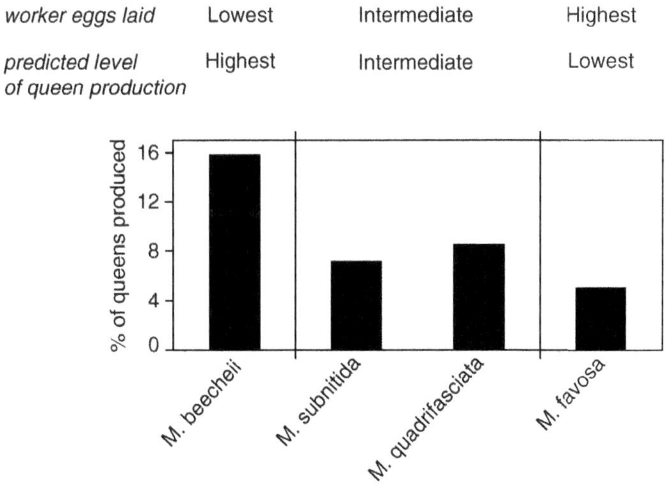

Fig. 5 Fraction of newborn females who develop into queens. More worker eggs laid means that a worker rears a higher fraction of more valuable nephews compared with brothers. Theory predicts that as the fraction of more valuable nephews increases, more newborn females will develop into workers rather than queens, causing the percentage of queens produced to decline. Data from four bee species of the genus *Melipona* support the prediction. The percentage of queens produced in *subnitida* and *quadrifasciata* do not differ significantly, so those two species are lumped into a single intermediate category. Redrawn from Wenseleers and Ratnieks (2004)

queen who mates with only one male, as in the particular bees discussed in the previous example. However, in some other species, a queen may mate several times, or there may be multiple queens. The number of queens, the number of times a queen mates, and the level of inbreeding affect the asymmetry in relatedness of a worker to nephews and brothers (Hamilton 1972).

We can avoid complexity by considering a simple prediction. When a worker is more related to the sons of other workers than to sons produced by the queen, she will allow other workers to produce sons without interference. By contrast, when a worker is more related to the queen's sons than to sons produced by other workers, she will interfere with reproduction by other workers (Wenseleers and Ratnieks 2006).

Figure 6 supports the prediction that relatedness asymmetry determines worker behavior. When workers are more related to other workers' sons than to the queen's sons, then the percentage of males produced by workers rises significantly above zero. By contrast, when workers are more related to the queen's sons than to other workers' sons, production of sons by workers is almost always very close to zero. In this case, the workers prevent other workers from producing sons by eating the eggs laid by workers.

These two examples from social insects show that the biological theory of other regarding valuation based on genetic kinship provides precise quantitative understanding of behavior.

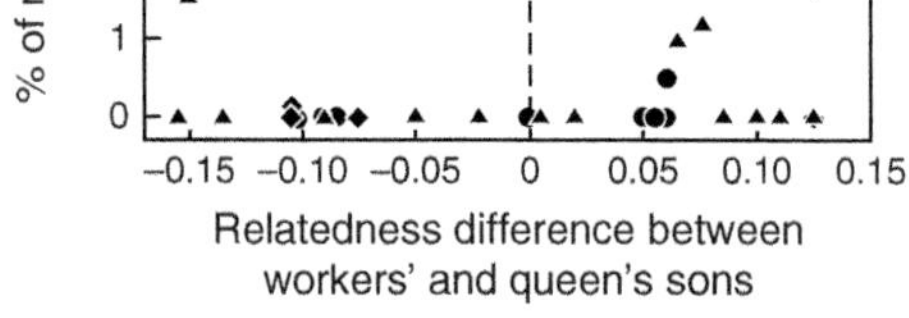

Fig. 6 Relatedness asymmetry determines whether a worker allows other workers to reproduce. Negative values of the relatedness difference mean that a worker is more closely related to the queen's sons than to the workers' sons. Positive values mean the opposite. The height of each point shows the percentage of all males produced by a colony that derive from workers (scaled logarithmically). The plot shows 90 different species of ants (*circles*), bees (*squares*), and wasps (*triangles*). Redrawn from Wenseleers and Ratnieks (2006)

4.4 *Synergism and the Origin of Mutually Beneficial Behaviors*

The prior examples concerned behaviors that can be directly observed: sex ratio, parasite virulence, or tolerance of egg laying by sister-workers. By contrast, synergism concerns the origin of mutually beneficial behaviors at some time in the past.

Mutually beneficial behaviors often require complementary specialization. Such specialization frequently does not exist in advance of the cooperative venture (Frank 1995b). So the first problem concerns how the mutually beneficial behavior got started and became sufficiently integrated to provide fully synergistic benefits.

As parties come to depend on each other's complementary specializations, they may over time become mutually dependent. If so, then a second problem concerns the irreversibility of synergistic behavior: neither party can succeed without the partnership. Put another way, as each player becomes adjusted to the presence of the other, the other takes on the role of an essential part of the environment without which the individual cannot succeed. Indeed, the integration can become so complete that it is hard to see the past history – in the present, the players have become so completely interdependent that we tend to view them as a unit.

Much of the deep structure of life may have followed a path in which separate entities interacted synergistically and became mutually dependent. Synergism may be the most important of topics in the study of group integration, but it is also the most difficult of topics to analyze. Past separation becomes hidden in present integration.

4.4.1 The Origin of Integrated Individuals

It is difficult to identify past synergism in current behaviors. So I will start with a theoretical example. The example concerns how the earliest kind of life may have become integrated into more complex cooperative groups.

Life depends on molecules that copy themselves. Those molecules that replicate at the highest rate increase in abundance. If the error rate in replication is sufficiently low, then a progeny molecule is mostly like its parent, and carries the same information that provided a replicative edge to its parent (Eigen 1971).

All of life became structured into cells early in history. A cell contains the informational molecules that copy themselves. Those informational molecules direct the biochemical physiology of the cell. The physiology runs the program by which the cell acquires resources, protects itself against perturbation, and copies its informational molecules.

A cell is a complex cooperative consortium of multiple informational molecules, each informational component directing a part of the physiology needed to run the collaborative enterprise. Because cells require complex integration of components, simple informational molecules that copy themselves must have preceded the earliest cells (Eigen and Schuster 1979; Maynard Smith 1979).

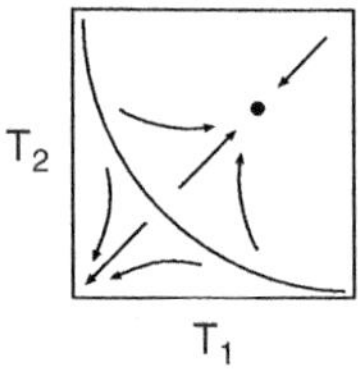

Fig. 7 Thresold model for the origin and evolution of synergistic cooperation. Individuals of population 1 have a trait, T_1, that enhances the reproductive rate of members of population 2, but the trait that benefits the partner also reduces the actor's own fitness. Likewise, members of population 2 have a trait, T_2, that enhances the reproduction of individuals of population 1 at a cost to the actor. Larger values of T provide more benefit to the partner at a higher cost to the donor. When both populations have low trait values, as would be expected when the partners first meet, natural selection continually pushes the traits to lower values. If, however, the pair of traits is above a threshold upon first meeting, then cooperation can increase because of synergistic feedback. Correlation in the traits between populations increases the probability that a particular group will have a pair of individuals above the threshold. Such correlation may arise from spatial associations. From Frank (1995b)

How can different kinds of self interested molecules come to be associated in a mutually beneficial synergism?

It is easy enough to imagine that if two different types express complementary information, then their interaction produces synergistic benefits. But how did the two types come to express complementary information, if each initially evolved in isolation? The problem turns on a threshold, as in the earlier theoretical section on synergism (Frank 1995b). If both populations of alternative molecules express, on average, a level of complementary information above some threshold, synergism follows easily. But the initial state is inevitably below the threshold.

Figure 7 shows this threshold model for synergism. Once both partners are above the threshold, mutually beneficial interactions strongly enhance the cooperative traits. After enhancement by positive feedback, the partners may be investing heavily in traits that benefit each other in order to receive enhanced return benefits. At that point, the partners may become fully dependent on each other for survival. Once above the threshold, the likely path is: complementation and positive feedback $\rightarrow$ specialization $\rightarrow$ irreversible transition to a highly integrated and cooperative state (Eigen and Schuster 1979).

How can partners pass the threshold? Suppose, initially, that the average traits of the two populations are below the threshold. However, there is variability in each population. So, by chance, some pairs of individuals will be above the threshold. Those chance pairs will do better than average, because they gain the synergistic benefit from their positive feedback on each other. Each will have more offspring than average, spreading the cooperative trait values in their populations. However, if their progeny associate randomly with partners, then on average they will be investing highly in cooperation but will be matched with partners who do not reciprocate. So those cooperative progeny will do less well than average, and no net progress in cooperative evolution ensues.

Spatial associations may help (Axelrod and Hamilton 1981; Nowak et al. 1994). Suppose, by chance, that a pair of cooperators comes together. They do well and leave more progeny than average. If their progeny tend to associate rather than mix randomly, then the synergism continues, and the paired lineages of cooperators expand. The spatial association, extended over time, allows the cooperative pairing to continue long enough to increase significantly, possibly pushing the average trait values over the threshold (Frank 1994a, 1995b). Once the average values pass the threshold, both populations rapidly enhance their synergistic traits, and the spatial associations are no longer required.

Cells are membrane bound structures that naturally impose spatial associations (Maynard Smith and Szathmáry 1995). Perhaps the spatial associations imposed by the early cells helped to push various biochemical synergisms over their cooperative thresholds. As those thresholds were passed, the mutual dependence between molecules became fixed. At that point, the biochemical integration became so deep that we would have a hard time recognizing the self interested histories behind the cohesive group.

4.4.2 Irreversible Thresholds in the Social Evolution of Insect Colonies

Positive feedback, historical change, and irreversibility may often play important roles in the evolution of complex and highly integrated groups. In this section, I consider Bourke's (1999) interesting analysis of social insect colonies (Frank 2003). Although Bourke's study transcends my simple models of synergism, his analysis does emphasize strongly the role of positive feedback and irreversible thresholds in cooperative evolution.

Bourke (1999) began by noting that, across different species of social insect, small colonies tend to have relatively little morphological differentiation between queens and workers. In addition, the workers have a relatively high degree of reproductive potential. By contrast, large colonies tend to have strong morphological differentiation between queens and workers and reduced reproductive potential of the workers.

Alexander et al. (1991) argued that in small colonies each worker has a significant probability of replacing the queen because there are relatively few competitors. By contrast, workers in large colonies have relatively little chance of succeeding to become queen. Thus, workers in large colonies are favored to reduce investment in reproductive potential and become more specialized for their worker roles. This reduction leads to strong morphological differentiation between workers and queens and low reproductive potential of workers. Absence of potential reproduction by workers reduces conflict between workers and other colony members, because the workers can enhance their fitness mostly by increasing the success of the colony.

Ratnieks and Reeve (1992) suggested that worker control of reproduction by other workers (policing) may be ineffective in small colonies. If there are few other workers, then a single worker may be able to dominate her neighbors and succeed in producing sons. As the number of workers rises, policing becomes more effective because a single worker cannot dominate the collective.

Bourke (1999) combined these ideas to argue that positive feedbacks occur between colony size, policing, reproductive potential of workers, and morphological differentiation between workers and queens. As colony size rises, policing becomes more effective, which favors reduced allocation to reproduction by workers. As workers concentrate more on their colony-productive roles, conflict subsides and the colony becomes more efficient. Greater efficiency may drive colonies to larger size, further specializing workers for nonreproductive tasks and aligning the interests of the workers with the interests of the colony.

5 Historical Analogy

The tension between conflict and cooperation of self interested actors runs deeply throughout the history of life. The great puzzles turn on how cooperative and efficient groups arise solely through self interest. The same basic forces seem to occur in both biology and in human affairs. I first recap the biology, then comment on the analogy to humans.

In biology, the tragedy of the commons rules in the absence of any special force that promotes cooperation. We see the tragedy of self interest everywhere: in sex ratios, in parasite virulence, in bacterial secretion of resource-acquiring molecules. In all cases, each member of a group would do better by promoting group cohesion and sharing in the benefits of greater group efficiency. But, without some force that curbs the free expression of self interest, competition within the group ultimately plays against everyone's interests.

Several forces in biology have overcome the tragedy to promote group cohesion. Correlated behaviors tie the success of actors together by matching behavior between partners and thus locking the success of the actor with its partner. If an actor's success is tied with its neighbors, then the value of group efficiency can dominate the disruptive force of self interested competition. In biology, correlated behavior most often arises by common genetics from shared ancestry – that is, by interactions between genetic kin. But any force that induces correlation can have the same effect.

Repression of competition within groups ties the interests of each individual to the group. With no opportunity to outcompete neighbors, each individual can gain only by promoting the efficiency and productivity of the group. Any process that randomizes the share of group resources provided to each individual effectively represses any opportunity for competition. The way that individuals divide their genetic material for transmission to offspring arose from a process of randomization, in which the probability that a particular gene passes to a child is determined randomly. We call this process fair meiosis – the basis throughout much of life for sex, reproduction, inheritance, and individuality.

Repression of competition may also be important at higher levels of social organization. I discussed one study of a primate, in which dominant males police conflict in the group. In the absence of those policing males, group cohesion deteriorated significantly. Effective policing of competition may be difficult to achieve in many biological settings. That difficulty explains why competition still rules much of life, and why in certain cases the ubiquitous force of competition may be overcome.

Self interested valuation of others' success arises naturally in biology through genetic kinship. Observed patterns of behavior ultimately depend on the rate at which competing behaviors are transmitted into the future. In this regard, an individual is shaped by natural selection to value the success of another in proportion to the correlation in their genetic tendency to pass the same behaviors on to future generations.

Regard for kin sets the foundation of social behavior. It is the reason parents care for offspring, sterile honeybees raise their siblings, and nonreproductive skin cells die to promote the success of sperm or egg. Kin correlations can shape behavior with great precision. I illustrated that precision by the relative valuation a social insect worker places on the queen's sons versus the other workers' sons. The class more highly valued switches depending on how many times the queen mates. The workers switch their treatment accordingly. When the workers are more closely correlated genetically with other workers' sons, they tolerate production of those sons without interference. By contrast, when workers are more closely correlated genetically with the queen's sons, they destroy the sons produced by other workers.

Finally, synergistic feedback between different aptitudes often provides benefit to both parties. I discussed how such synergism likely played a key role in the earliest evolutionary history of life. At some early stage, there must have been a consortium formed to produce the first cells. That consortium arose between separate molecules, each originally designed to replicate itself but not to interact cooperatively with other molecules. Some of those self interested replicators probably had synergistic biochemistry. The positive feedbacks combined with spatial associations imposed by cellular boundaries set the first great cooperative transition of life. I also discussed how such synergisms between complementary aptitudes shaped the historical trends to greater specialization and complexity in social insect colonies.

Correlated behaviors, repression of competition, other regarding valuation, and synergistic gains between different aptitudes rule conflict and cooperation throughout the history of life. The potential analogies with human behavior are clear.

But what is the value of such analogy between biology and human sociality? I see two related benefits. First, study of biology has greatly clarified the logic of self interest. I have, in this paper, outlined a rich theory of conflict and cooperation that has succeeded well in explaining and in predicting diverse behaviors. Many of those ideas from biology have arisen independently in the theory of games or in studies of human behavior. But the biological theory has a firmer conceptual foundation and greater connection with observed phenomena.

Second, analogies from nature suggest hypotheses about the forces that have shaped human societies. For example, Alexander (1979, 1987) has argued that many aspects of human morality turn on reducing competition between neighbors to promote group cohesion. Alexander developed his hypothesis from close study of biology followed by analogy to human self interest.

How useful are such analogies in forming hypotheses about human sociality? Certainly, both large mistakes and great insights may follow from analogy. Thus, one can reasonably defend both caution and boldness. But caution cannot solve puzzles. And many puzzles remain with regard to the forces that shape human cooperation and competition.

6 Historical Consequence

Historical analogy simply provides a source of ideas about how self interest plays out in human societies. By contrast, historical consequence means that humans have been shaped by the same forces that have operated throughout biological history. Such historical consequence still allows that humans are unique. From a strictly biological perspective, humans have particular attributes that set us apart. For example, other animals have culture and specialized cognitive abilities, but the great development of human culture and cognition define qualitative distinctions of human sociality.

Thinking about historical consequence leads to obvious questions. How strongly does genetic kinship shape behavior? How much does learning and culture alter the dynamics of behavior? How much bias has biological history built into the way we learn and transmit aspects of culture? How much does a history of group against group competition align self interested tendencies with those of the group?

Biological history has had at least some consequential effects along these lines. It simply does not make sense to suppose that history does not matter. But we still do not know how to weigh various factors. How should we proceed to learn more?

One approach is to consider the following question (Alexander 1990): What was the most important challenge to survival and reproduction that caused evolution to transform our ancestors from something like a chimpanzee into a modern human?

To start, consider how biologists think about evolutionary transformation in response to challenges of survival and reproduction. It is easier to see the structure of the argument if we begin with a nonhuman example. I use kangaroos. I describe the example in detail, because it is essential to understand the biological approach to analyzing evolutionary transformations. After I finish with the kangaroo example, I apply the same logic to the evolutionary transformation of humans from their ape ancestors.

Most of the 63 species of kangaroo make their living on the ground. Those ground dwelling species include the large well-known hoppers. However, at least 10 species of kangaroo belong to a group that has become specialized for life in the trees (Flannery et al. 1996). Adaptation for tree life led to several specialized characteristics.

The key here is that a single broad challenge – moving from the ground into the trees – explains a wide array of evolutionary changes to deal with the challenge. We will want to discuss what sort of challenge and what sort of changes characterize the evolutionary transformation of humans from ape ancestors. But first, let use consider the kangaroos. Diamond (1997) has described the characteristic changes so well that I quote directly:

> The lifestyle of arboreal tree-kangaroos required them to reverse millions of years of kangaroo evolution in many respects: saving weight by a 25 per cent reduction in muscle mass; developing long, strong, curved claws; big, powerful grasping forearms, and a rotator cuff in the shoulder (shared with humans but not with other kangaroos or most other mammals) to permit overhead use of the forearm; hind-feet that twist so that the soles can face each other to grasp a tree trunk; hair whorls to shed rain (also shared with humans); and a tail

> tufted with long fur in some species, used as a counterbalance in climbing and as a rudder in 'flight'.
>
> Those 'flights' are actually jumps to the ground from a height of 20 metres or more in the canopy. I know of no other big mammal that survives drops from such height... the animals' bones, muscles and ligaments must have become modified to withstand such shocks.
>
> ... Faced with a hard-to-digest, toxic, bulky leaf diet of low nutritional value, tree-kangaroos evolved a low metabolic rate. They decrease rather than increase their metabolic rate at low ambient temperature; spend 90 per cent of their time 'doing nothing' (that is, sitting and digesting); and have a complex stomach of several chambers, and regurgitate and rechew food, like cows chewing the cud.

The logic for tree kangaroos is simple. They moved from the ground to the trees: that move defined the central evolutionary challenge. Nearly all of the particular changes that separate tree kangaroos from those on the ground can be explained by evolutionary response to the key challenge.

We need to consider two steps in order to apply this same logic to humans. First, what are the particular characteristics that separate humans from their ape ancestors? This list of characteristics defines the changes that need to be explained. People argue over the exact limits of the differences, for example, how much symbolic processing chimpanzees and gorillas can accomplish. But in the end, these are more or less matters of fact that can be resolved by direct study.

Nonetheless, any direct statement about humans is likely to be controversial. Here is just one example of certain potentially unique traits of humans (Alexander 1990):

> Humans are the only mammal that lives in multi-male groups, in which confidence (likelihood) of paternity is high ... and the males are both extensively and complexly parental and also extensively and complexly cooperative with one another (and in which, I speculate, the males with the highest confidence of paternity also tend to be the most cooperative.)

In the second step, we formulate a hypothesis: What is the central evolutionary challenge that allows us to make sense of the particular evolutionary transformations that separate humans from ape ancestors?

Many theories have been proposed. Several emphasize the social environment. Again, I give just one brief account to indicate the way in which one may argue. Alexander (1990) gives a full scholarly discussion of past work and presents his own views as follows:

> ... At some point in their evolution humans obviously began to cooperate to compete... this intergroup competition becoming increasingly elaborate, direct, and continuous until it achieved the ubiquity with which it has been exhibited in modern humans throughout recorded history across the entire face of the earth... This unique kind of within-species balance-of-power race – involving, eventually, virtually all levels, or group sizes, within societies – would be a perpetual or unending one ... in which rapidly appearing differences in culture and technology could become significant unbalancers that could accelerate the process even further. It could be termed a case of "runaway social selection"... [calling] for adversarial and competing groups of humans to be central in creating the environment of brain and psyche selection. Unprecedented levels of cooperation within groups could thereby be generated, as well as unprecedented kinds of between-group adversarial relationships.

These points establish the problem of historical consequence. One may follow this problem in many directions. But, whatever direction, we can be sure that the concepts of self interested cooperation will play an important role in framing the key evolutionary challenge and the particular historical consequences for distinctly human characteristics.

It is, of course, possible that multiple factors explain the human transformation – different human characteristics may have different, unrelated explanations. But I follow Michael Ghiselin (1969) in his analysis of Darwin's greatness: above all, Darwin succeeded by his stubborn belief that a few simple processes explain much of the great complexity and variety of life. To discover a simple explanation, one must assume that a simple explanation is possible.

Acknowledgements National Institute of General Medical Sciences MIDAS Program grant U01-GM-76499 supports my research.

References

Alexander RD (1979) Darwinism and human affairs. University of Washington Press, Seattle

Alexander RD (1987) The biology of moral systems. Aldine de Gruyter, New York

Alexander RD (1990) How did humans evolve? Museum of Zoology, The University of Michigan 1:1–38

Alexander RD, Noonan KM, Crespi BJ (1991) The evolution of eusociality. In: Sherman PW, Jarvis JUM, Alexander RD (eds) The Biology of the Naked Mole Rat, Princeton University Press, Princeton, NJ, pp 3–44

Anderson RM, May RM (1982) Coevolution of hosts and parasites. Parasitology 85:411–426

Aumann RJ (1974) Subjectivity and correlation in randomized strategies. J Math Econ 1:67–96

Aumann RJ (1987) Correlated equilibrium as an expression of Bayesian rationality. Econometrica 55:1–18

Axelrod R, Hamilton WD (1981) The evolution of cooperation. Science 211:1390–1396

Binmore KG (1994) Game theory and the social contract. MIT, Cambridge, MA

Bot ANM, Rehner SA, Boomsma JJ (2001) Partial incompatibility between ants and symbiotic fungi in two sympatric species of *Acromyrmex* leaf-cutting ants. Evolution 55:1980–1991

Bourke AFG (1999) Colony size, social complexity and reproductive conflict in social insects. J Evol Biol 12:245–257

Bremermann HJ, Pickering J (1983) A game-theoretical model of parasite virulence. J Theor Biol 100:411–426

Colwell RK, Wilson DS (1981) Group selection is implicated in the evolution of female-biased sex ratios. Nature 290:401–404

Crow JF (1979) Genes that violate Mendel's rules. Sci Am 240:134–144

de Roode JC, Pansini R, Cheesman SJ, Helinski MEH, Huijben S, Wargo AR, Bell AS, Chan BHK, Walliker D, Read AF (2005) Virulence and competitive ability in genetically diverse malaria infections. Proc Natl Acad Sci USA 102:7624–7628

Diamond J (1997) Flying yellow kangaroos. Nature 385:692

Eigen M (1971) Self-organization of matter and the evolution of biological macromolecules. Naturwissenschaften 58:465–523

Eigen M, Schuster P (1979) The hypercycle: a principle of natural self-organization. Springer, New York

Fenner F, Day MF, Woodroofe GM (1956) Epidemiological consequences of mechanical transmission of myxomatosis by mosquitoes. J Hyg 54:173–194
Fisher RA (1983) Natural selection, heredity, and eugenics: including selected correspondence of RA Fisher with Leonard Darwin and Others, Edited by JH Bennett. Oxford University Press, New York
Flack JC, Krakauer DC, de Wall FBM (2005a) Robustness mechanisms in primate societies: a perturbation study. Proc R Soc Lond B 272
Flack JC, de Waal FBM, Krakauer DC (2005b) Social structure, robustness, and policing cost in a cognitively sophisticated species. Am Nat 165:E126–E139
Flack JC, Girvan M, de Waal FBM, Krakauer DC (2006) Policing stabilizes construction of social niches in primates. Nature 439:426–429
Flannery TF, Martin R, Szalay A (1996) Tree Kangaroos: a curious natural history. Reed Books, Australia
Frank SA (1985) Hierarchical selection theory and sex ratios. II. On applying the theory, and a test with fig wasps. Evolution 39:949–964
Frank SA (1992) A kin selection model for the evolution of virulence. Proc R Soc Lond B 250:195–197
Frank SA (1994a) Genetics of mutualism: the evolution of altruism between species. J Theor Biol 170:393–400
Frank SA (1994b) Kin selection and virulence in the evolution of protocells and parasites. Proc R Soc Lond B 258:153–161
Frank SA (1995a) Mutual policing and repression of competition in the evolution of cooperative groups. Nature 377:520–522
Frank SA (1995b) The origin of synergistic symbiosis. J Theor Biol 176:403–410
Frank SA (1996a) Host-symbiont conflict over the mixing of symbiotic lineages. Proc R Soc Lond B 263:339–344
Frank SA (1996b) Models of parasite virulence. Quart Rev Biol 71:37–78
Frank SA (1996c) Policing and group cohesion when resources vary. Anim Behav 52:1163–1169
Frank SA (1998) Foundations of social evolution. Princeton University Press, Princeton, NJ
Frank SA (2003) Repression of competition and the evolution of cooperation. Int J Org Evolution 57:693–705
Frank SA (2006) Social selection. In: Fox CW, Wolf JB (eds) Evolutionary genetics: concepts and case studies, Oxford University Press, New York, pp 350–363
Ghiselin MT (1969) The triumph of the Darwinian method. University of Chicago Press, Chicago
Godfray HCJ, Werren JH (1996) Recent developments in sex ratio studies. Trends Ecol Evol 11:59–63
Griffin AS, West SA, Buckling A (2004) Cooperation and competition in pathogenic bacteria. Nature 430:1024–1027
Hamilton WD (1967) Extraordinary sex ratios. Science 156:477–488
Hamilton WD (1970) Selfish and spiteful behaviour in an evolutionary model. Nature 228:1218–1220
Hamilton WD (1972) Altruism and related phenomena, mainly in social insects. Annu Rev Ecol Syst 3:193–232
Hamilton WD (1975) Innate social aptitudes of man: an approach from evolutionary genetics. In: Fox R (ed) Biosocial Anthropology, Wiley, New York, pp 133–155
Hamilton WD (1979) Wingless and fighting males in fig wasps and other insects. In: Blum MS, Blum NA (eds) Reproductive Competition and Sexual Selection in Insects, Academic, New York, pp 167–220
Hardin G (1968) The tragedy of the commons. Science 162:1243–1248
Hardy ICW (2002) Sex ratios: concepts and research methods. Cambridge University Press, Cambridge
Herre EA (1985) Sex ratio adjustment in fig wasps. Science 228:896–898
Herre EA (1993) Population structure and the evolution of virulence in nematode parasites of fig wasps. Science 259:1442–1446

Leigh EG (1971) Adaptation and diversity. Freeman, San Francisco
Leigh EG (1977) How does selection reconcile individual advantage with the good of the group? Proc Natl Acad Sci USA 74:4542–4546
Leigh EG (1991) Genes, bees and ecosystems: the evolution of a common interest among individuals. Trends Ecol Evol 6:257–262
Levin SA (1983) Some approaches to the modelling of coevolutionary interactions. In: Nitecki MH (ed) Coevolution, University of Chicago Press, Chicago, pp 21–65
Levin SA, Pimental D (1981) Selection of intermediate rates of increase in parasite-host systems. Am Nat 117:308–315
Lewontin RC (1970) The units of selection. Annu Rev Ecol Syst 1:1–18
Maynard Smith J (1979) Hypercycles and the origin of life. Nature 280:445–446
Maynard Smith J (1982) Evolution and the theory of games. Cambridge University Press, Cambridge
Maynard Smith J, Szathmáry E (1995) The major transitions in evolution. Freeman, San Francisco
Mueller UG (2002) Ant versus fungus versus mutualism: ant-cultivar conflict and the deconstruction of the attine ant-fungus symbiosis. Am Nat 160:S67–S98
Mueller UG, Poulin J, Adams RMM (2004) Symbiont choice in a fungus-growing ant (Attini, Formicidae). Behav Ecol 15:357–364
Nowak MA, Bonhoeffer S, May RM (1994) Spatial games and the maintenance of cooperation. Proc Natl Acad Sci USA 91:4877–4881
Pollack GB (1996) Kin selection, kin avoidance and correlated strategies. Evol Ecol 10:29–43
Queller DC (1994) Genetic relatedness in viscous populations. Evol Ecol 8:70–73
Rankin DJ, Bargum K, Kokko H (2007) The tragedy of the commons in evolutionary biology. Trends Ecol Evol 22:643–651
Ratnieks FLW, Reeve HK (1992) Conflict in single-queen Hymenopteran societies: the structure of conflict and processes that reduce conflict in advanced eusocial species. J Theor Biol 158:33–65
Skyrms B (1996) Evolution of the social contract. Cambridge University Press, Cambridge
Szathmáry E, Demeter L (1987) Group selection of early replicators and the origin of life. J Theor Biol 128:463–486
Taylor PD (1992a) Altruism in viscous poplations–an inclusive fitness approach. Evol Ecol 6:352–356
Taylor PD (1992b) Inclusive fitness in a heterogeneous environment. Proc R Soc Lond B 249:299–302
Taylor PD, Frank SA (1996) How to make a kin selection model. J Theor Biol 180:27–37
Trivers R (1971) The evolution of reciprocal altruism. Quart Rev Biol 46:35–57
Wenseleers T, Ratnieks FLW (2004) Tragedy of the commons in *Melipona* bees. Proc R Soc Lond B 271:S310–S312
Wenseleers T, Ratnieks FLW (2006) Comparative analysis of worker reproduction and policing in eusocial Hymenoptera supports relatedness theory. Am Nat 168:E163–E179
Wilson DS (1980) The natural selection of populations and communities. Benjamin-Cummings, Menlo Park, CA
Wilson DS, Pollock GB, Dugatkin LA (1992) Can altruism evolve in purely viscous populations? Evol Ecol 6:331–341
Wilson EO (1971) The insect societies. Harvard University Press, Cambridge, MA
Zhang MM, Poulsen M, Currie CR (2007) Symbiont recognition of mutualistic bacteria by *Acromyrmex* leaf-cutting ants. Int Soc Microb Ecol 1:313–320

How to Evolve Cooperation

Christine Taylor and Martin A. Nowak

Abstract Cooperation is needed for evolution to construct new levels of organization. The emergence of genomes, cells, multi-cellular organisms, social insects, and human society are all based on cooperation. Cooperation means that selfish replicators forgo some of their reproductive potential to help one another. But natural selection implies competition between individuals and therefore opposes cooperation unless a specific mechanism is at work. Five mechanisms for the evolution of cooperation are discussed: kin selection, direct reciprocity, indirect reciprocity, network reciprocity, and group selection. I will argue that cooperation is essential for evolvability.

1 Introduction

Evolution occurs in populations of reproducing individuals. Mutation, selection, and cooperation can be seen as the three fundamental principles of evolution. Cooperation is needed for evolution to construct new levels of organization. The origin of life, the emergence of the first cell, the arrival of eucarya, the rise of multi-cellular organisms, and the advent of human language are all based on cooperation. A higher level of organization emerges, whenever the competing units on the lower level begin to cooperate. Cooperation is always vulnerable to exploitation by defectors. Hence, the evolution of cooperation requires specific mechanisms, which allow natural selection to favor cooperation over defection. In this paper, we discuss five such mechanisms, and for each mechanism we derive the fundamental condition for the evolution of cooperation.

C. Taylor (✉)
Program for Evolutionary Dynamics, Department of Organismic and Evolutionary Biology, Department of Mathematics, Harvard University, Cambridge, MA 02138, USA
e-mail: taylor4@fas.harvard.edu

S.A. Levin (ed.), *Games, Groups, and the Global Good*, Springer Series in Game Theory, DOI: 10.1007/978-3-540-85436-4_2,

The payoff matrix of the prisoner's dilemma (Axelrod 1984; Rapoport and Chammah 1965) is given by

$$\begin{array}{c} \\ C \\ D \end{array} \begin{array}{c} \begin{array}{cc} C & D \end{array} \\ \begin{pmatrix} R & S \\ T & P \end{pmatrix} \end{array} \qquad (1)$$

The entries of the payoff matrix refer to the row player. If a cooperator, C, interacts with another cooperator, both get payoff R, which is the 'reward for mutual cooperation. If a cooperator, C, meets a defector, D, the cooperators gets the "sucker's payoff," S, while the defector gets the highest payoff of the game, T, which denotes the "temptation of defection." Two defectors obtain the payoff P, which stands for the "punishment" of mutual defection. The game is a prisoner's dilemma if $T > R > P > S$.

In the prisoner's dilemma, defectors dominate cooperators unless a mechanism for the evolution of cooperation is at work. We will discuss five mechanisms for the evolution of cooperation: direct reciprocity, indirect reciprocity, kin selection, group selection, and graph selection. Direct reciprocity is the idea that there are repeated encounters between the same two individuals: my action depends on what you have done to me in previous encounters. For indirect reciprocity there are repeated encounters in a population of individuals: my action depends on what you have done to me and to others. Kin selection studies games between genetic relatives. Group selection refers to the mechanism where competition not only occurs between individuals, but also between groups. Graph selection operates in structured populations, where cooperators can prevail over defectors by forming clusters.

All five approaches have led to different mathematical investigations, but here we show that the essential properties of each mechanism can be described by a transformation of the prisoner's dilemma payoff matrix. From these transformed matrices we can immediately derive the fundamental conditions for the evolution of cooperation. Our analysis reveals how each mechanism operates at a fundamental level.

This paper is an extension of earlier work (Nowak 2006b), which investigates a simplified prisoner's dilemma given by two parameters, b and c, denoting respectively the benefit and cost of cooperation. In this case, we have $R = b - c$, $S = -c$, $T = b$ and $P = 0$, which implies the restriction $R + P = T + S$. In contrast, the general prisoner's dilemma, which is studied here, has four parameters, displays a richer dynamical behavior and leads to generalizations of the fundamental rules presented in Nowak (2006b). Moreover, for kin selection, group selection and network reciprocity, the full Prisoner's Dilemma allows coexistence between cooperators and defectors if $R + P < T + S$ and bi-stability if $R + P > T + S$. Neither of these outcomes are possible for the simplified prisoner's dilemma with $R + P = T + S$.

We begin with some simple remarks on evolutionary game dynamics. Then we discuss each of the five mechanisms. Finally, we summarize the key findings.

2 Evolutionary Game Dynamics

Evolutionary game dynamics describe frequency dependent selection. The fitness of an individual is not constant, but depends on the relative abundance (= frequency) of various strategies (= phenotypes) in the population. The outcome of the game is related to reproductive success. Payoff determines fitness. Reproduction can be genetic or cultural. The first ideas of evolutionary game theory appeared in papers by Hamilton (1964); Maynard Smith and Price (1973); Trivers (1971). Books on evolutionary game theory include Cressman (2003); Gintis (2000); Hofbauer and Sigmund (1998); Maynard Smith (1982); Nowak (2006a); Samuelson (1997); Weibull (1995). For recent reviews see Hofbauer and Sigmund (2003); Nowak and Sigmund (2004). Evolutionary game theory is a general approach to evolutionary dynamics with constant selection being a special case.

Consider a game between two strategies, A and B, given by the payoff matrix

$$\begin{array}{c} \\ A \\ B \end{array} \begin{array}{c} \begin{array}{cc} A & B \end{array} \\ \begin{pmatrix} a & b \\ c & d \end{pmatrix} \end{array} \tag{2}$$

The entries denote the payoffs for the row player. Strategy A obtains payoff a when playing another A player, but payoff b when playing a B player. Srategy B obtains payoff c when playing an A player and payoff d when playing a B player.

If $a > c$ and $b > d$, then A dominates B. In this case, it is always better to use strategy A. The expected payoff of A players is greater than that of B players for any composition of a well-mixed population. If instead $a < c$ and $b < d$, then B dominates A and we have exactly the reverse situation Fig. 1.

If $a > c$ and $b < d$, then both strategies are best replies to themselves, which leads to a coordination game. In a population where most players use A, it is best to use A. In a population where most players use B, it is best to use B. There is bi-stability: both strategies are stable against invasion by the other strategy.

If $a < c$ and $b > d$, then both strategies are best replies to each other. In a population where most players use A, it is best to use B. In a population where most players use B, it is best to use A. There is stable co-existence between the two strategies.

If $a > c$ then A is a strict Nash equilibrium. Likewise, if $b < d$ then B is a strict Nash equilibrium. A strategy which is a strict Nash equilibrium is always an evolutionarily stable strategy (ESS). An ESS is stable against invasion by a small fraction of mutants using the other strategy in an infinitely large, well-mixed population (Maynard Smith 1982). The ESS condition does not imply protection by selection in a finite population (Nowak et al. 2004).

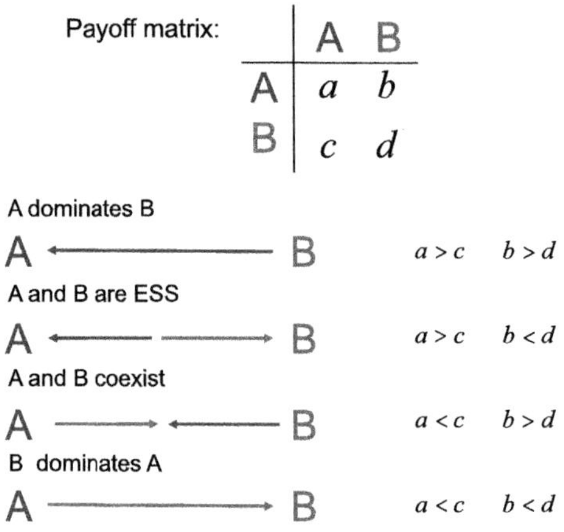

Fig. 1 Evolutionary dynamics of a two-player, two-strategy game. The entries of the matrix specify the payoff for the row player. There are four possibilities: (1) if $a > c$ and $b > d$, then strategy A dominates strategy B; (2) if $a > c$ and $b < d$, then both A and B are evolutionarily stable strategies (ESS); the game is bi-stable and is called a "coordination game"; (3) if $a < c$ and $b > d$, then both strategies are best replies to each other, and we have a hawk–dove game, where both strategies coexist; (4) if $a < c$ and $b < d$, then B dominates A

3 Direct Reciprocity

In a non-repeated prisoner's dilemma, it is best to defect no matter which strategy is adopted by the other player (because $R < T$ and $S < P$). But if there are repeated encounters between the same two individuals, then direct reciprocity can emerge and lead to the evolution of cooperation (Trivers 1971, 1985). Direct reciprocity is based on the idea "I help you and you help me." In each round the two players must choose between cooperation and defection. With probability w there is another round. With probability $1 - w$ the game is over. Hence, the average number of interactions between two individuals is $1/(1 - w)$.

There are many conceivable deterministic and stochastic strategies for the repeated prisoner's dilemma (Axelrod 1984; Axelrod and Hamilton 1981; Fudenberg and Maskin 1990; Imhof et al. 2005; Kraines and Kraines 1989; May 1987; Milinski 1984; Molander 1985; Nowak and Sigmund 1990, 1992, 1993), and the game can be played with simultaneous or alternating moves (Nowak and Sigmund 1994). But for the purpose of this paper we only need to consider two very simple strategies. Our defectors, D, defect in every move. Our cooperators, C, play tit-for-tat: they start with a cooperation and then do whatever the other player has done in the previous move.

If two cooperators meet, they cooperate all the time. If two defectors meet, they defect all the time. If a cooperator meets a defector, the cooperator cooperates in the first round and defects afterwards, while the defector defects in every round. The payoff matrix is given by

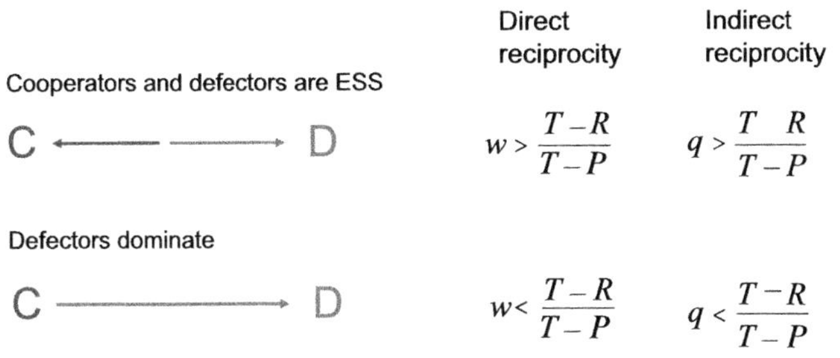

Fig. 2 Direct and indirect reciprocity can lead to the evolution of cooperation. While direct reciprocity is based on repeated encounters between the same two individuals, indirect reciprocity uses the experience of others. Defectors are always ESS. Cooperators are ESS if w or q exceed $(T-R)/(T-P)$, where w is the probability of another round in direct reciprocity, and q is the coefficient of social acquaintanceship under indirect reciprocity

$$\begin{array}{cc} & \begin{array}{cc} C & D \end{array} \\ \begin{array}{c} C \\ D \end{array} & \begin{pmatrix} \frac{R}{1-w} & S+\frac{wP}{1-w} \\ T+\frac{wP}{1-w} & \frac{P}{1-w} \end{pmatrix} \end{array} \tag{3}$$

Note that defectors, D, are always ESS (because $P > S$). Cooperators are ESS if

$$w > \frac{T-R}{T-P}. \tag{4}$$

Hence there are two possibilities: (1) if inequality (4) holds then both cooperation and defection are ESS, and the game is a coordination game; (2) if inequality (4) does not hold then defection dominates cooperation (Fig. 2). Therefore, inequality (4) represents a minimum requirement for the evolution of cooperation. If there are sufficiently many rounds, then direct reciprocity can lead to the evolution of cooperation. This argument is related to the Folk theorem (Binmore 1991; Fudenberg and Maskin 1986; Fudenberg and Tirole 1991).

4 Indirect Reciprocity

"Indirect reciprocity arises out of direct reciprocity in the presence of interested audiences" (Alexander 1987). For direct reciprocity, my decision is based on what you have done to me in previous encounters. For indirect reciprocity my decision is also based on what you have done to others. Indirect reciprocity represents the concept "I help you and somebody will help me." Indirect reciprocity is based on reputation (Nowak and Sigmund 1998a). Each event can be seen as an interaction between two people, according to a single prisoner's dilemma game given by payoff matrix (1). Each game is observed by others. Cooperation is costly, but leads to the reputation of being a helpful individual. Defection is more profitable in the short run, but leads to a bad reputation. Natural selection favors strategies that base their decision to cooperate or to defect on the reputation of oneself and of others (Brandt and Sigmund 2004, 2005; Fishman 2003; Fishman and Lotem 2001; Lotem et al. 1999; Nowak and Sigmund 1998a,b, 2005; Ohtsuki and Iwasa 2004, 2006, 2007; Panchanathan and Boyd 2004; Takahashi and Mashima 2003). Experimental studies confirm that helpful individuals are more likely to receive help in the future (Bolton et al. 2005; Dufwenberg et al. 2001; Engelmann and Fischbacher 2002; Milinski et al. 2002, 2006; Rockenbach and Milinski 2006; Wedekind and Braithwaite 2002; Wedekind and Milinski 2000).

In order to derive a necessary condition for the evolution of cooperation by indirect reciprocity, we study the interaction between two basic strategies: (1) defectors who always defect and (2) cooperators who cooperate unless they know the reputation of the other person to indicate a defector. The parameter q denotes the probability to know the reputation of another individual. A cooperator always cooperates with another cooperator, but cooperates with a defector only with probability $1-q$. Defectors never cooperate. We obtain the payoff matrix

$$\begin{array}{c} \\ C \\ D \end{array} \begin{array}{c} \begin{array}{cc} \quad C \quad & \quad D \quad \end{array} \\ \begin{pmatrix} R & (1-q)S+qP \\ (1-q)T+qP & P \end{pmatrix} \end{array} \tag{5}$$

This payoff matrix leads to exactly the same analysis as we have performed for direct reciprocity in the previous section. We obtain the same results, with q instead of w (Fig. 2). The probability to know the reputation of another player must exceed a certain threshold value,

$$q > \frac{T-R}{T-P}. \tag{6}$$

If this condition holds, then indirect reciprocity can lead to the evolution of cooperation.

5 Kin Selection

The concept of kin selection arose from the idea that evolutionary games are often played between individuals who are genetic relatives (Cavalli-Sforza and Feldman 1978; Frank 1998; Grafen 1979, 1985; Hamilton 1964, 1996; Maynard Smith 1964; Queller 1985, 1992; Rousset and Billiard 2000; Taylor 1996; Taylor and Frank 1996; Taylor et al. 2007). A gene which encodes altruistic behavior toward another individual promotes its own survival, if it is also present in the recipient of the altruistic act.

For the purpose of this paper, we use a method that was originally proposed by Maynard Smith (1982) for the hawk–dove game. Consider a population where the average relatedness between interacting individuals is given by r, which is a number between 0 and 1. The payoff received by the other player is added to your own payoff multiplied by r. The sum is divided by $1+r$ in order to keep the total payoff between two players constant. Therefore, we obtain the modified payoff matrix

$$\begin{array}{cc} & \begin{array}{cc} C & D \end{array} \\ \begin{array}{c} C \\ D \end{array} & \begin{pmatrix} R & \frac{S+rT}{1+r} \\ \frac{T+rS}{1+r} & P \end{pmatrix} \end{array} \tag{7}$$

Cooperators are ESS if

$$r > r_C = \frac{T-R}{R-S}. \tag{8}$$

Defectors are ESS if

$$r < r_D = \frac{P-S}{T-P}. \tag{9}$$

The evolutionary outcome depends on the relative ranking of r, r_C, and r_D. We must distinguish two parameter regions (Fig. 3).

I. If $R+P > T+S$ then $r_D > r_C$. There are three possibilities: (1) if $r_D > r_C > r$ then defectors dominate; (2) if $r_D > r > r_C$ then both cooperators and defectors are ESS; (3) if $r > r_D > r_C$ then cooperators dominate.

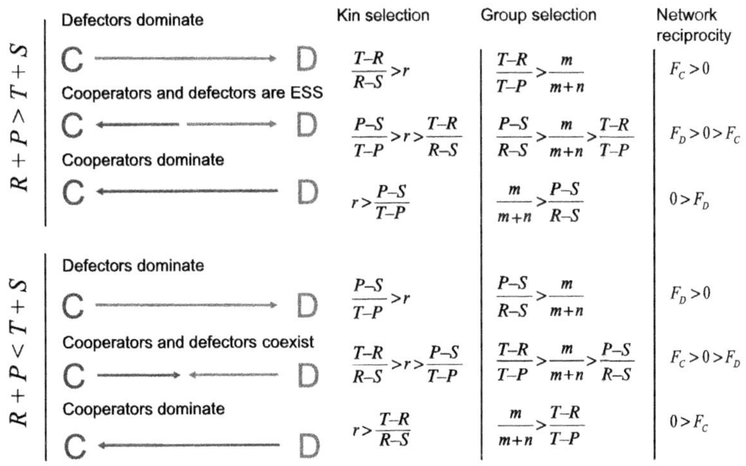

Fig. 3 Kin selection, group selection, and network reciprocity can lead to the evolution of cooperation. For kin selection, the parameter r is the coefficient of genetic relatedness between individuals. For group selection, the parameters m and n denote respectively the number of groups and the number of individuals per group (group size). For network reciprocity, we have $F_C = (T-R)k^2 - (T-P)k + (R+P-T-S)$ and $F_D = (P-S)k^2 - (R-S)k - (R+P-T-S)$, where k is the degree of the graph (that is the number of neighbors of each individual). For all three models we find: if $R+P > T+S$, then cooperators and defectors cannot co-exist; if $R+P < T+S$, then cooperators and defectors cannot simultaneously be ESS

II. If $R + P < T + S$ then $r_C > r_D$. Again there are three possibilities: (1) if $r_C > r_D > r$ then defectors dominate; (2) if $r_C > r > r_D$ then neither cooperators nor defectors are ESS; (3) if $r > r_C > r_D$ then cooperators dominate.

In the degenerate case, $R + P = T + S$, we have $r_C = r_D$. Then either C dominates (when $r > r_C$) or D dominates (when $r < r_D$). Neither coexistence, nor bi-stability between C and D are possible. The case $R + P = T + S$ is called "equal gains from switching" (Nowak and Sigmund 1990): when switching from C to D while playing against C the gain is $T - R$ and while playing against D the gain is $P - S$. The condition $T - R = P - S$ implies that the two gains are equal. The two-parameter prisoner's dilemma, where cooperators pay a cost, c, for the other person to receive a benefit, b, leads to equal gains from switching, because $T = b$, $R = b - c$, $P = 0$ and $S = -c$ (Nowak 2006b). In this special case, inequalities (8) and (9) lead to $r > c/b$, which is Hamilton's rule.

6 Group Selection

Group selection is based on the idea that competition occurs not only between individuals but also between groups. Many models of group selection have been proposed over the years (Bowles 2006; Boyd and Richerson 2002; Crow and Aoki 1982; Fletcher and Zwick 2004; Goodnight 1990a,b; Goodnight and Stevens 1997; Harpending and Rogers 1987; Harvey et al. 1985; Kerr and Godfrey-Smith 2002; Killingback et al. 2006; Leigh 1983; Maynard Smith 1976; Michod 1999; Nunney 1985; Paulsson 2002; Slatkin and Wade 1978; Szathmáry and Demeter 1987; Traulsen et al. 2005; Uyenoyama and Feldman 1980; Wade 1977, 1978; Williams 1966; Wilson 1975, 1983; Wilson and Hölldobler 2005; Wynne-Edwards 1962). Here we use an approach described by (Traulsen and Nowak, 2006). A population is subdivided into m groups. The maximum size of a group is n. Individuals interact with others in the same group according to a prisoner's dilemma. The fitness of an individual is $1 - \omega + \omega F$, where F is the payoff and ω the intensity of selection. At each time step, an individual from the entire population is chosen for reproduction proportional to fitness. The offspring is added to the same group. If the group reaches the maximum size, it can split into two groups with a certain probability, p. In this case, a randomly selected group dies to prevent the population from exploding. The maximum population size is mn. With probability $1 - p$ the group does not divide. In this case, a random individual of that group is chosen to die. For small p, the fixation probability of a single cooperator in the entire population is given by the fixation probability of a single cooperator in a group times the fixation probability of that group. The model can be extended to include migration (Traulsen and Nowak 2006).

The payoff matrix that describes the interactions between individuals of the same group is given by

$$\begin{array}{cc} & \begin{array}{cc} C & D \end{array} \\ \begin{array}{c} C \\ D \end{array} & \begin{pmatrix} R & S \\ T & P \end{pmatrix} \end{array} \tag{10}$$

Between groups there is no game dynamical interaction in our model, but groups divide at rates that are proportional to the average fitness of individuals in that group. The multi-level selection is an emerging property of the population structure. Therefore one can say that cooperator groups have a constant payoff R, while defector groups have a constant payoff P. Hence, in a sense the following "game" of constant selection describes the competition between groups

$$\begin{array}{cc} & \begin{array}{cc} C & D \end{array} \\ \begin{array}{c} C \\ D \end{array} & \begin{pmatrix} R & R \\ P & P \end{pmatrix} \end{array} \tag{11}$$

Again the "fitness" of a group is $1 - \omega + \omega F$ where F is its "payoff". Surprisingly, for weak selection ($w << 1$) and large n and m, the essence of the overall selection dynamics on two levels can be described by a single payoff matrix, which is the sum of matrix (10) multiplied by the group size, n, and matrix (11) multiplied by the number of groups, m (Nowak 2006b). The result is

$$\begin{array}{cc} & \begin{array}{cc} C & D \end{array} \\ \begin{array}{c} C \\ D \end{array} & \begin{pmatrix} (n+m)R & nS+mR \\ nT+mP & (n+m)P \end{pmatrix} \end{array} \tag{12}$$

The intuition for adding the two matrices multiplied with the respective population size is as follows. For fixation of a new strategy in a homogeneous population using the other strategy, first the game dynamics within one group of size n have to be won and then the game dynamics between m groups have to be won. For weak selection and large m and n, the overall fixation probability is the same as the fixation probability in the single game using the combined matrix 12 and population size, mn. Therefore, some aspects of the stochastic selection process on two levels can be studied by a standard replicator equation using the combined matrix.

From matrix (12) we see that cooperators are ESS if

$$\frac{m}{m+n} > \frac{T-R}{T-P}. \tag{13}$$

Defectors are ESS if

$$\frac{m}{m+n} < \frac{P-S}{R-S}. \tag{14}$$

Again there are two parameter regions defined by $R + P > T + S$ and $R + P < T+S$. The same six cases apply as for kin selection, but the thresholds have different values (Fig. 3).

7 Graph Selection

Spatial games can lead to cooperation in the absence of any strategic complexity: cooperators can coexist with defectors and sometimes even outcompete them (Nowak and May 1992). Frank (1998) points out that spatial structure or kin selection may be related to Aumann's notion of correlated games (Aumann, 1974, 1987). Spatial games are usually played on regular lattices such as square, triangular, or hexagonal lattices (Herz, 1994; Nowak and May, 1992; Rand and Wilson, 1995). Evolutionary graph theory (Lieberman et al. 2005) is a general approach to study the effect of population structure or social networks on evolutionary or ecological dynamics (Abramson and Kuperman 2001, Durrett and Levin 1994, Ebel and Bornholdt 2002, Hassell et al. 1994, Hauert and Doebeli 2004, May 2006; Nakamaru et al. 1997, Nakamaru et al. 1998, Rousset 2004, Santos and Pacheco 2005, Santos et al. 2006, Skyrms and Pemantle 2000, Szabó et al. 2005, Vukov and Szabó 2005, Vukov et al. 2006, Wu et al. 2006). The individuals of a population occupy the vertices of a graph. The edges denote who interacts with whom. In principle, there can be two different graphs (Ohtsuki et al. 2007): the "interaction graph" determines who plays with whom; the "replacement graph" determines who competes with whom for reproduction. Here we assume that the interaction and replacement graphs are identical. Network reciprocity is a generalization of spatial reciprocity to graphs (Ohtsuki et al. 2006): on graphs cooperators form clusters which can enable them to outcompete defectors.

We consider a "two coloring" of the graph: each vertex is either a cooperator or a defector. Each individual interacts with all of its neighbors according to the standard payoff matrix (1). The payoffs are added up. The fitness of an individual is given by $1 - \omega + \omega F$ where $\omega \in [0, 1]$ denotes the intensity of selection and F denotes the payoff for this individual. Here we consider evolutionary dynamics according to death–birth updating (Ohtsuki et al. 2006): in each time step a random individual is chosen to die; then the neighbors compete for the empty site proportional to their fitness.

A calculation using pair approximation on regular graphs (where each vertex has k edges) leads to a deterministic differential equation which describes how the expected frequency of cooperators (and defectors) changes over time (Ohtsuki and Nowak 2006a). This differential equation turns out to be a standard replicator equation (Hofbauer et al. 1979; Taylor and Jonker 1978; Zeeman 1980) with a modified payoff matrix. For the interaction between cooperators and defectors on a graph with degree $k > 2$, the modified payoff matrix is of the form

$$\begin{array}{c} \\ C \\ D \end{array} \begin{array}{c} \begin{array}{cc} C & D \end{array} \\ \begin{pmatrix} R & S+H \\ T-H & P \end{pmatrix} \end{array} \tag{15}$$

where

$$H = \frac{(k+1)(R-P) - T + S}{(k+1)(k-2)}. \tag{16}$$

For a derivation of this transformation see (Ohtsuki and Nowak 2006a). Cooperators are ESS if $R > T - H$ or

$$F_C = k^2(T - R) - k(T - P) + (R + P - T - S) < 0. \tag{17}$$

Defectors are ESS if $S + H < P$ or

$$F_D = k^2(P - S) - k(R - S) + (T + S - R - P) > 0. \tag{18}$$

Note that

$$F_C - F_D = (T + S - P - R)(k^2 - k - 2).$$

We have two parameter regions when $k > 2$ (Fig. 3):

I. If $R + P > T + S$ then $F_D > F_C$. There are three possibilities: (1) if $F_D > F_C > 0$ then defectors dominate; (2) if $F_D > 0 > F_C$ then both cooperators and defectors are ESS; (3) if $0 > F_D > F_C$ then cooperators dominate.
II. If $R + P < T + S$ then $F_C > F_D$. Again there are three possibilities: (1) if $0 > F_C > F_D$ then cooperators dominate; (2) if $F_C > 0 > F_D$ then neither cooperators nor defectors are ESS; (3) if $F_C > F_D > 0$ then defectors dominate.

In the degenerate case, $R + P = T + S$, we have $F_C = F_D$ and therefore either cooperators or defectors dominate; neither bi-stability nor coexistence are possible. The same is true for graphs with $k = 2$ (Ohtsuki and Nowak 2006b).

8 Conclusion

We have studied five mechanisms for the evolution of cooperation. Each mechanism leads to a transformation of the prisoner's dilemma payoff matrix. From the transformed matrices, we have derived the fundamental condition for each mechanism to facilitate the evolution of cooperation. The transformed matrices can also be used to study evolutionary success in finite populations (Nowak 2006a,b).

Direct reciprocity can lead to evolution of cooperation if

$$w > \frac{T - R}{T - P}. \tag{19}$$

The parameter w denotes the probability of playing another round in the repeated Prisoner's Dilemma game.

Indirect reciprocity can lead to evolution of cooperation if

$$q > \frac{T - R}{T - P}. \tag{20}$$

The parameter q denotes the probability of knowing the reputation of the other person. It is a measure for the degree of social acquaintanceship in the population.

Kin selection can lead to evolution of cooperation if

$$r > \min\left\{\frac{T-R}{R-S}, \frac{P-S}{T-P}\right\}. \tag{21}$$

The parameter r denotes the fraction of the payoff of the other person that is added to my payoff. It can be seen as a measure for genetic relatedness in the population. The notation "min$\{x, y\}$" means to take the smaller value of either x or y. In our case, this notation is a simple way to cover the two cases $R + P > T + S$ and $R + P < T + S$.

Group selection can lead to evolution of cooperation if

$$\frac{m}{m+n} > \min\left\{\frac{T-R}{T-P}, \frac{P-S}{R-S}\right\}. \tag{22}$$

The parameters m and n denote respectively the number of groups and the group size.

Graph selection can lead to evolution of cooperation if

$$\min\{F_C, F_D\} < 0, \tag{23}$$

where

$$F_C = k^2(T-R) - k(T-P) + (R+P-T-S) \tag{24}$$
$$F_D = k^2(P-S) - k(R-S) + (T+S-R-P). \tag{25}$$

The parameter k is the degree of the regular graph.

For kin, group, and graph selection, the original prisoner's dilemma matrix is tranformed into a new matrix

$$\begin{array}{c} \\ C \\ D \end{array}\!\!\begin{array}{c} \begin{array}{cc} C & D \end{array} \\ \begin{pmatrix} R & S \\ T & P \end{pmatrix} \end{array} \longrightarrow \begin{array}{c} \\ C \\ D \end{array}\!\!\begin{array}{c} \begin{array}{cc} C & D \end{array} \\ \begin{pmatrix} R' & S' \\ T' & P' \end{pmatrix} \end{array} \tag{26}$$

given by 7, 12, and 15, respectively. We note that for each of the transformed matrices, the sign of $R' + P' - S' - T'$ is the same as the sign of $R + P - S - T$. Therefore, the sign of $R + P - S - T$ determines which evolutionary outcomes are possible for the transformed payoff matrix. If $R + P - T - S < 0$, then cooperators and defectors cannot simultaneously be ESS in the transformed game. If $R + P - T - S > 0$, then cooperators and defectors cannot coexist in the transformed game. If $R + P - T - S = 0$, then either cooperators or defectors must dominate.

Acknowledgements Support from the John Templeton foundation and the NSF/NIH joint program in mathematical biology (NIH grant R01GM078986) is gratefully acknowledged. The Program for Evolutionary Dynamics at Harvard University is sponsored by Jeffrey Epstein.

References

Abramson G, Kuperman M (2001) Social games in a social network. Phys Rev E 63:030901
Alexander R (1987) The biology of moral systems. Aldine De Gruyter, New York
Aumann R (1974) Subjectivity and correlation in randomized strategies. J Math Econ 1:67C96
Aumann R (1987) Correlated equilibrium as an expression of Bayesian rationality. Econometrica 55:1C18
Axelrod R (1984) The evolution of cooperation. Basic Books, New York
Axelrod R, Hamilton WD (1981) The evolution of cooperation. Science 211:1390–1396
Binmore K (1991) Fun and Games. Heath, Lexington, MA
Bolton GE, Katok E, Ockenfels A (2005) Cooperation among strangers with limited information about reputation. J Public Econ 89:1457–1468
Boyd R, Richerson PJ (2002) Group beneficial norms spread rapidly in a structured population. J Theor Biol 215:287–296
Bowles S (2006) Group competition, reproductive leveling, and the evolution of human altruism. Science 314:1569–1572
Brandt H, Sigmund K (2004) The logic of reprobation: assessment and action rules for indirect reciprocity. J Theor Biol 231:475–486
Brandt H, Sigmund K (2005) Indirect reciprocity, image scoring, and moral hazard. PNAS 102:2666–2670
Cavalli-Sforza LL, Feldman MW (1978) Darwinian selection and "altruism". Theor Popul Biol 14(2):268–280
Cressman R (2003) Evolutionary Dynamics and Extensive Form Games. MIT Press, Cambridge
Crow JF, Aoki K (1982) Group selection for a polygenic behavioral trait: a different proliferation model. Proc Natl Acad Sci USA 79:2628–2631
Dufwenberg M, Gneezy U, Güth W, van Damme E (2001) Direct vs indirect reciprocity: an experiment. Homo Oecon 18:19–30
Durrett R, Levin S (1994) The importance of being discrete (and spatial). Theor Pop Biol 46:363–394
Ebel H, Bornholdt S (2002) Coevolutionary games on networks. Phys Rev E 66:056118
Engelmann D, Fischbacher U (2002) Indirect reciprocity and strategic reputation building in an experimental helping game. Univ Zürich working paper no. 132
Fishman MA (2003) Indirect reciprocity among imperfect individuals. J Theor Biol 225:285–292
Fishman MA, Lotem A, Stone L (2001) Heterogeneity stabilises reciprocal altruism interaction. J Theor Biol 209:87–95
Fletcher JA, Zwick M (2004) Strong altruism can evolve in randomly formed groups. J Theor Biol 228:303–313
Frank SA (1998) Foundations of Social Evolution. Princeton University Press, Princeton
Fudenberg D, Maskin E (1986) Folk Theorem for repeated games with discounting or with incomplete information. Econometrica 54:533–554
Fudenberg D, Maskin E (1990) Evolution and Cooperation in Noisy Repeated Games. Am Econ Rev 80:274–279
Fudenberg D, Tirole J (1991) Game Theory. MIT Press, Cambridge
Gintis H (2000) Game Theory Evolving. Princeton University Press, Princeton
Goodnight CJ (1990a) Experimental studies of community evolution. I. The response to selection at the community level. Evolution 44(6):1614–1624
Goodnight CJ (1990b) Experimental studies of community evolution. II. The ecological basis of the response to community selection. Evolution 44(6):1625–1636
Goodnight CJ, Stevens L (1997) Experimental Studies of Group selection: what do they tell us about group selection in nature. Am Nat 150:S59–S79
Grafen A (1979) The hawk-dove game played between relatives. Anim Behav 27:905–907
Grafen A (1985) A geometric view of relatedness. Oxford Surveys Evol Biol 2:28–89
Hamilton WD (1964) The genetical evolution of social behaviour. J Theor Biol 7:1–16

Hamilton WD (1996) Narrow Roads of Gene Land: I Evolution of Social Behaviour WH Freeman, Oxford
Harpending H, Rogers A (1987) On Wright's mechanism for intergroup selection. J Theor Biol 127:51–61
Harvey PH, Partridge L, Nunney L (1985) Group selection and the sex ratio. Nature 313:10–11
Hassell MP, Comins HN, May RM (1994) Species coexistence and self-organizing spatial dynamics. Nature 370:290–292
Hauert C, Doebeli M (2004) Spatial structure often inhibits the evolution of cooperation in the Snowdrift game. Nature 428:643–646
Herz AVM (1994) Collective phenomena in spatially extended evolutionary games. J Theor Biol 169:65–87
Hofbauer J, Sigmund K (1998) Evolutionary Games and Population Dynamics. Cambridge University Press, Cambridge, UK
Hofbauer J, Sigmund K (2003) Evolutionary Game Dynamics. Bull Am Math Soc 40:479–519
Hofbauer J, Schuster P, Sigmund K (1979) A note on evolutionary stable strategies and game dynamics. J Theor Biol 81:609–612
Imhof LA, Fudenberg D, Nowak MA (2005) Evolutionary cycles of cooperation and defection. P Natl Acad Sci USA 102:10797–10800
Kerr B, Godfrey-Smith P (2002) Individualist and multi-level perspectives on selection in structured populations. Biol Philos 17:477–517
Killingback T, Bieri J, Flatt T (2006) Evolution in group-structured populations can resolve the tragedy of the commons. Proc R Soc B 273:1477–1481
Kraines D, Kraines V (1989) Pavlov and the prisoner's dilemma. Theory and Decision 26:47–49
Leigh EG (1983) When does the good of the group override the advantage of the individual? Proc Natl Acad Sci USA 80:2985–2989
Lieberman E, Hauert C, Nowak MA (2005) Evolutionary dynamics on graphs. Nature 433:312–316
Lotem A, Fishman MA, Stone L (1999) Evolution of cooperation between individuals. Nature 400:226–227
May RM (1987) More evolution of cooperation. Nature 327:15–17
May RM (2006) Network structure and the biology of populations. Trends in Ecol & Evol 21:394–399
Maynard Smith J (1964) Group selection and kin selection. Nature 200:1145–1147
Maynard Smith J (1976) Group selection. Quart Rev Biol 51:277–283
Maynard Smith J (1982) Evolution and the Theory of Games. Cambridge University Press, Cambridge, UK
Maynard Smith J, Price GR (1973) The logic of animal conflict. Nature 264:15–18
Michod RE (1999) Darwinian dynamics. Princeton University Press, Princeton
Milinski M (1984) A predator's costs of overcoming the confusion-effect of swarming prey. Animal Behav 32:1157–1162
Milinski M, Semmann D, Krambeck H-J (2002) Reputation helps solve the 'tragedy of the commons.' Nature 415:424–426
Milinski M, Semmann D, Krambeck H-J, Marotzke J (2006) Stabilizing the Earth's climate is not a losing game: supporting evidence from public goods experiments. PNAS 103:3994–3998
Molander P (1985) The optimal level of generosity in a selfish, uncertain environment. J Conflict Resolut 29:611–618
Nakamaru M, Matsuda H, Iwasa Y (1997) The evolution of cooperation in a lattice-structure population. J Theor Biol 184:65–81
Nakamaru M, Nogami H, Iwasa Y (1998) Score-dependent fertility model for the evolution of cooperation in a lattice. J Theor Biol 194:101–124
Nowak MA (2006a) Evolutionary dynamics. Harvard University Press, Cambridge
Nowak MA (2006b) Five rules for the evolution of cooperation. Science 314:1560–1563
Nowak MA, May RM (1992) Evolutionary games and spatial chaos. Nature 359:826–829

Nowak MA, Sasaki A, Taylor C, Fudenberg D (2004) Emergence of cooperation and evolutionary stability in finite populations. Nature 428:646–650
Nowak MA, Sigmund K (1990) The evolution of stochastic strategies in the prisoner's dilemma. Acta Appl Math 20:247–265
Nowak MA, Sigmund K (1992) Tit for tat in heterogenous populations. Nature 355:250–253
Nowak MA, Sigmund K (1993) A strategy of win-stay, lose-shift that outperforms tit for tat in prisoner's dilemma. Nature 364:56–58
Nowak MA, Sigmund K (1994) The alternating prisoner's dilemma. J Theor Biol 168:219–226
Nowak MA, Sigmund K (1998a) Evolution of indirect reciprocity by image scoring. Nature 393:573–577
Nowak MA, Sigmund K (1998b) The dynamics of indirect reciprocity. J Theor Biol 194:561–574
Nowak MA, Sigmund K (2004) Evolutionary dynamics of biological games. Science 303:793–799
Nowak MA, Sigmund K (2005) Evolution of indirect reciprocity. Nature 437:1291–1298
Nunney L (1985) Group selection, altruism, and structure-deme models. Am Nat 126:212–230
Ohtsuki H, Iwasa Y (2004) How should we define goodness? Reputation dynamics in indirect reciprocity. J Theor Biol 231:107–120
Ohtsuki H, Iwasa Y (2006) The leading eight: social norms that can maintain cooperation by indirect reciprocity. J Theor Biol 239:435–444
Ohtsuki H, Iwasa Y (2007) Global analyses of evolutionary dynamics and exhaustive search for social norms that maintain cooperation by reputation. J Theor Biol 244:518–531
Ohtsuki H, Nowak MA (2006a) The replicator equation on graphs. J Theor Biol 243:86–97
Ohtsuki H, Nowak MA (2006b) Evolutionary games on cycles. Proc R Soc B 273:2249–2256
Ohtsuki H, Hauert C, Lieberman E, Nowak MA (2006) A simple rule for the evolution of cooperation on graphs and social networks. Nature 441:502–505
Ohtsuki H, Pacheco J, Nowak MA (2007) Evolutionary graph theory: breaking the symmetry between interaction and replacement. J Theor Biol 246:681–694
Panchanathan K, Boyd R (2004) Indirect reciprocity can stabilize cooperation without the second-order free rider problem. Nature 432:499–502
Paulsson J (2002) Multileveled selection on plasmid replication. Genetics 161:1373–1384
Queller DC (1985) Kinship, reciprocity and synergism in the evolution of social behaviour. Nature 318:366–367
Queller DC (1992) A general model for kin selection. Evolution 46:376–380
Rand DA, Wilson HB (1995) Using spatio-temporal chaos and intermediate-scale determinism to quantify spatially extended ecosystems. Proc Biol Sci 259:111–117
Rapoport A, Chammah AM (1965) Prisoner's dilemma: a study in conflict and cooperation. University of Michigan Press, Ann Arbor
Rockenbach B, Milinski M (2006) The efficient interaction of indirect reciprocity and costly punishment. Nature 444:718–723
Rousset F (2004) Genetic structure and selection in subdivided populations. Princeton University Press, Princeton
Rousset F, Billiard S (2000) A theoretical basis for measures of kin selection in subdivided populations: finite populations and localized dispersal. J Evol Biol 13:814–825
Samuelson L (1997) Evolutionary games and equilibrium selection. MIT, Cambridge
Santos FC, Pacheco JM (2005) Scale-free networks provide a unifying framework for the emergence of cooperation. Phys Rev Lett 95:098104
Santos FC, Pacheco JM, Lenaerts T (2006) Cooperation prevails when indviduals adjust their social ties. PLoS Comput Biol 2:1284–1291
Skyrms B, Pemantle R (2000) A dynamic model of social network formation. Proc Natl Acad Sci USA 97:9340–9346
Slatkin M, Wade MJ (1978) Group selection on a quantitative character. Proc Natl Acad Sci USA 75:3531–3534
Szabó G, Vukov J, Szolnoki A (2005) Phase diagrams for an evolutionary prisoner's dilemma game on two-dimensional lattices. Phys Rev E 72:047107

Szathmáry E, Demeter L (1987) Group selection of early replictors and the origin of life. J Theor Biol 128:463–486
Takahashi N, Mashima R (2003) The emergence of indirect reciprocity: is the standing strategy the answer? Center for the study of cultural and ecological foundations of the mind, Hokkaido University, Japan, Working paper series no. 29
Taylor PD (1996) Inclusive fitness arguments in genetic models of behaviour. J Math Biol 34:654–674
Taylor PD, Frank S (1996) How to make a kin selection model. J Theor Biol 180:27–37
Taylor PD, Jonker LB (1978) Evolutionarily stable strategies and game dynamics. Math Bio Sci 40:145–156
Taylor PD, Wild G, Gardner A (2007) Direct fitness or inclusive fitness: how should we model kin selection. J Evol Biol 20:296–304
Traulsen A, Nowak MA (2006) Evolution of cooperation by multilevel selection. PNAS 103:10952–10955
Traulsen A, Sengupta AM, Nowak MA (2005) Stochastic evolutionary dynamics on two levels. J Theor Biol 235:393–401
Trivers R (1971) The evolution of reciprocal altruism. Q Rev Biol 46:35–37
Trivers R (1985) Social evolution. Benjamin/Cummings, Menlo Park
Uyenoyama MK, Feldman MW (1980) Theories of kin and group selection: a population genetic perspective. Theor Pop Biol 17:380–414
Vukov J, Szabó G (2005) Evolutionary prisoner's dilemma game on hierarchical lattices. Phys Rev E 71:036133
Vukov J, Szabó G, Szolnoki A (2006) Evolutionary prisoner's dilemma game on hierarchical lattices. Cooperation in the noisy case: Prisoner's dilemma game on two types of regular random graphs. Phys Rev E 74:067103
Wade MJ (1977) An experimental study of group selection. Evolution 31:134–153
Wade MJ (1978) A critical review of the models of group selection. Qrt Rev Biol 53:101–114
Wedekind C, Braithwaite VA (2002) The long-term benefits of human generosity in indirect reciprocity. Curr Biol 12:1012–1015
Wedekind C, Milinski M (2000) Cooperation through image scoring in humans. Science 288:850–852
Weibull J (1995) Evolutionary game theory. MIT, Cambridge
Williams GC (1966) Adaption and natural selection: a critique of some current evolutionary thought. Princeton University Press, Princeton
Wilson DS (1975) A theory of group selection. Proc Nat Acad Sci USA 72:143–146
Wilson DS (1983) The group selection controversy and current status. Annu Rev Ecol Syst 14:159–187
Wilson EO, Hölldobler B (2005) Eusociality: origin and consequences. Proc Natl Acad Sci USA 102:13367–13371
Wu Z, Xu X, Huang Z, Wang S, Wang Y (2006) Evolutionary prisoner's dilemma game with dynamic preferential selection. Phys Rev E 74:021107
Wynne-Edwards VC (1962) Animal dispersion in relation to social behavior. Oliver and Boyd, Edinburgh
Zeeman EC (1980) Population dynamics from game theory. In: Nitecki A, Robinson C (eds) Proceedings of an International Conference on Global Theory of Dynamics Systems. Lecture Notes in Mathematics 819. Springer, Berlin

Beyond Enlightened Self-Interest: Social Norms, Other-Regarding Preferences, and Cooperative Behavior

Samuel Bowles and Herbert Gintis

Abstract Both economists and biologists have developed repeated interaction models of cooperation in social dilemmas with groups of self-regarding individuals. Repeated interactions do provide opportunities for cooperative individuals to discipline defectors, and may be effective in groups of two individuals. However, these models are inadequate for groups of larger size, making plausible assumptions about the information available to each individual. Moreover, even presupposing extraordinary cognitive capacities and levels of patience among the cooperating individuals, it is unlikely that a group of more than two individuals would ever adopt the cooperative equilibria that the models have identified, and almost certainly, if it were to adopt one, its members would abandon it in short order. Though intended as models of decentralized interaction, the models by which self-regarding *Homo economicus* is said to cooperate implicitly presume implausible levels of coordination such as might in the real world be provided by social norms. The inadequacy of these models, coupled with extensive experimental and other empirical evidence of human cooperation suggests that other-regarding preferences in the context of social norms that facilitate and direct human cooperation must be part of an adequate explanation.

1 Introduction

While various forms of cooperation are found among many animals, humans are distinct in the scope and variety of kinds of cooperation in which we engage. In contrast to biology, where cooperative behaviors have become a central research focus only in recent decades, a major goal of economic theory since its inception has been to explain wide-scale cooperation. Since early in the twentieth century this endeavor involved the development of the so-called Walrasian model of

---

S. Bowles (✉)
Santa Fe Institute, 1399 Hyde Park Road, Santa Fe, NM 87501, USA and University of Siena, Siena, Italy

S.A. Levin (ed.), *Games, Groups, and the Global Good*, Springer Series in Game Theory,
DOI: 10.1007/978-3-540-85436-4_3, © Springer Physica-Verlag Berlin Heidelberg 2009

economy-wide competitive exchange and the affirmation of Adam Smith's "invisible hand" conjecture. The success of this endeavor culminated in the Fundamental Theorem of Welfare Economics (Arrow and Debreu 1954; Debreu 1959; Arrow and Hahn 1971), sustaining Smith's insight that self-regarding behaviors might support socially valued economic outcomes. But the theorem's essential assumption that all relevant aspects of all exchanges could be completely specified in contracts enforceable at zero cost to the exchanging parties is widely recognized as not applicable to any real world economy (Arrow 1971; Bowles and Gintis 1993; Gintis 2002; Bowles 2004).

A second major thrust of economic theory eschewed this widely implausible assumption and developed sophisticated repeated game models in which the outcomes of exchanges are determined by bargaining, collusion, and other forms of strategic interaction. These models refine and extend the insights of Shubik (1959), Trivers (1971), Taylor (1976), and Axelrod and Hamilton (1981) that retaliation against defectors by withdrawal of cooperation may enforce cooperation among self-regarding individuals. This literature culminates in the folk theorems of Fudenberg and Maskin (1986), Fudenberg et al. (1994), and others. A great virtue of these models, in contrast to the Walrasian paradigm in economics, is that in recognizing the ubiquity of incomplete or unenforceable contracts, they describe the real world of interactions among most animals, including humans (Blau 1964; Gintis 1976; Stiglitz 1987; Tirole 1988; Laffont 2000; Bowles and Hammerstein 2003).

The folk theorems were not developed for the purpose of evolutionary explanation and have not be extensively used in this way. The most ambitious attempts in this direction, applied towards understanding the broad historical and anthropological sweep of human experience, are in the work of Robert Sugden (1986) and Binmore (1993, 1998, 2005). These works offers an evolutionary approach to morality, in which moral rules form a cultural system that developed historically with the emergence of *Homo sapiens*. A society's moral rules, in this view, are instructions for behavior in conformity with one of the myriad of Nash equilibria of a repeated n-player social interaction. Because the interactions are repeated, the self-regarding individuals who comprise the social order will conform to the moral rules of their society as a type of self-fulfilling prophecy (if all other individuals play their part in this Nash equilibrium, an individual has no incentive to deviate from playing his part as well).

The evolutionary solutions of Sugden and Binmore belong to a broad class of models developed to explain cooperation among self-regarding individuals as a results of repeated interactions. In this essay, we will show that while the insight that repeated interactions provide opportunities for cooperative individuals to discipline defectors is correct, none of these models adequately explains human cooperation. The reason is that even presupposing extraordinary cognitive capacities and levels of patience among the cooperating individuals, it is unlikely that a group of size greater that two would ever adopt the cooperative equilibria that the models have identified, and almost certainly, if it were to adopt one, its members would abandon it in short order, unless some elements of other-regarding preferences, in the form of altruistic cooperation and punishment are present.

2 Folk Theorems and Evolutionary Dynamics

The folk theorem is based on a stage game played an indefinite number of times, with a constant, strictly positive, probability that in each period the game will continue for an additional period. The restrictions on the stage game tend to be minimal and rather technical (Fudenberg and Maskin 1986; Fudenberg et al. 1994). Player strategies in the repeated game are conditioned on the pattern of behavior, usually interpreted as cooperation and defection, in previous periods. The information concerning this pattern of behavior is a signal that may be *perfect* (completely accurate and received by all individuals) or *imperfect* (inaccurate with positive probability and/or received only by a subset of individuals). An imperfect signal can be *public* (all players receive the same signal) or *private* (different players receive difference signals, and some may receive no signal at all). We may think of imperfect public signals as caused by execution errors, which are then seen by all other players, while private signals are caused by *limited scope*, in which players observe the behaviors of only a subset of their group members, or perceptual error, in which specific individuals incorrectly interpret cooperation as defection, or vice-versa. However execution errors can be private because they are seen only by a subset of individuals, and hence private signals need not involve perceptual error.

With either perfect or public imperfect signals, a folk theorem can be proved, asserting that any feasible allocation of payoffs to the players that dominates the minimax payoff of each player can be achieved, or approximated as closely as desired, as the equilibrium per-period payoff to the repeated game, for some discount factor strictly less than unity (Fudenberg and Maskin 1986; Fudenberg et al. 1994). A similar folk theorem can be proved for certain types of private signals. Significant contributions to this literature include Sekiguchi (1997), Piccione (2002), Ely and Välimäki (2002), Bhaskar and Obara (2002), and Mailath and Morris (2006). Private signaling creates especially grave problems. First, the sequential equilibrium requires strictly mixed strategies on the part of all players in all periods. Yet, individuals have no incentive to play these mixed strategies. Second, the equilibria require that private signals be sufficiently close to being public, so all individuals receive nearly the same signal concerning the behavior of any given group member. When this is not the case, the equilibrium will not exist. Thus, these models apply only to forms of cooperation where all members observe the actions of (nearly) all others with a high level of accuracy. The equilibrium concept employed is that of sequential equilibrium, which is a Nash equilibrium in which players choose best response and use Bayesian updating of beliefs at all information sets, whether on or off the path of play (Kreps and Wilson 1982).

In Sect. 3 we show that repeated game models and their associated folk theorems are merely a first step in understanding cooperation. Proving the existence of a sequential Nash equilibrium must be followed by an analysis of the dynamical out-of-equilibrium behavior of the system, with the goal of showing that the equilibrium is asymptotically stable (i.e., has a basin of attraction) and the system is highly likely eventually to enter the basin of attraction of the equilibrium and remain there. When this is unlikely to be true, we say that the result is an "dynamically irrelevant Nash equilibrium."

Recent advances in the epistemological foundations of equilibrium concepts in game theory provide a possible way forward in dealing with out-of-equilibrium dynamics (Aumann and Brandenburger 1995). Common knowledge of rationality and even common priors do not ensure that player beliefs are sufficiently aligned to produce Nash equilibrium in all but the smallest and simplest games. Nor does game theory provide an explanation of how individual beliefs can be aligned in a manner allowing a group to coordinate on the kinds of complex behaviors required by the folk theorems.

However, sociologists (1933 [1902]; Parsons and Shils 1951) and anthropologists (Benedict 1934; Boyd and Richerson 1985; Brown 1991) have found that virtually every society has such processes, and that they are key to understanding strategic interaction. Borrowing a page from sociological theory, we posit that groups have *social norms* specifying how a game ought to be played. Learning a social norm includes learning that the norm is common knowledge among those who know it, learning what behavior is suggested by the norm, and learning that a large fraction of group members knows the norm and follows it.

Social norms do not ensure equilibrium, because error, mutation, migration, and other dynamical forces may lead individuals to reject beliefs or behavior fostered by the norm, because the moral commitments associated with the norm might conflict with an individual's personal ethics, or its suggested behavior may be rejected as not it the individual's best interest; i.e., the action fostered by a social norm must be a best response to the behaviors of the other group members, given the beliefs engendered by the social norm and the individual's Bayesian updating. Moreover, social norms cannot be introduced as a deus ex machina, as if laid down by a centralized authority, without violating the objective to provide a "bottom up" theory of cooperation that does not presuppose preexisting institutional forms of cooperation. Social norms are thus discretionary, because any institution that is posited to enforce behavior should itself be modeled within the dynamical system, unless plausible reasons are given for taking a macro-level institution as unproblematically given. Nor are social norms fixed in stone. A group's social norms must themselves be subject to change, those groups producing better outcomes for their members displacing groups with less effective norms, and changing social and demographic conditions leading to the evolutionary transformation of norms within groups.

The idea of social norms is akin to Binmore's notion of moral rules that choose among Nash equilibria, except that in our model, social norms facilitate the attainment of a Nash equilibrium by appropriately aligning beliefs, rather than selecting among Nash equilibria, as is the case in Binmore's model. Employing the terminology of interactive epistemology (Aumann and Brandenburger 1995), a social norm leads agents to align their Bayesian priors, and generates a correlated equilibrium with the potential to coordinate cooperative activity and provide incentives for individuals to play their part in this activity.

Beginning with Sect. 4, we restrict consideration to the n-player public goods game, which is the appropriate model for many social dilemmas in which contemporary humans exhibit a high level of cooperation, including team production, voting, common pool resource management, and collective action, as well as in common defense, information sharing, and hunting in Pleistocene ancestral communities.

Fudenberg et al. (1994) proved the folk theorem for stage games with imperfect public signals. Gintis (2009a) has shown that their argument applies to the public goods game, deriving expressions linking the error rate ϵ, the group size n, and the discount factor δ, and showed that for any given discount factor δ, there is a maximum $n\epsilon$, order of magnitude unity, that supports cooperation. This means that cooperation may be sustained in groups experiencing less than one error per period, and otherwise not . Thus, either large groups or large error rates are incompatible with cooperation, despite the folk theorem, unless the discount factor is permitted to approach arbitrarily close to unity, which is ruled out by such demographic realities as the probability of mortality, as well as subjective time preference.

In Sect. 4 we use an agent-based simulation to the public goods game with imperfect public signaling to show that without social norms, a high level of cooperation can be attained only with very small group size ($n \leq 4$) or near zero error rates. When we introduce social norms reflecting the game-theoretic strategy of punishing defections but ignoring defections by others for whom defecting is a punishment of a third party, we can attain quite high levels of cooperation as long as we do not allow the social norms themselves to evolve. However, when social norms are subject to competitive pressure, they collapse, leading to the exceedingly low levels of cooperation characteristic of models without social norms.

The reason for this unraveling is straightforward. When the error rate is low, the optimal social norm is to tolerate zero defections. However, when all groups follow this social norm, a group that tolerates a single defection has higher average payoff than the zero-tolerance groups. Thus, by adopting a norm that is very intolerant of defections, a group is providing a public good to the rest of the population at a cost to itself. Hence, other groups copy the less stringent social norm until all allow a single defector. But, in this situation, a group that tolerates two defectors has higher average payoff than the groups that tolerate a single defector, so tolerating two defectors eventually becomes the universal social norm. At some point a within-group selection process takes over: there are now so many defectors that individuals who ignore the social norms altogether and merely tolerate zero defectors have higher payoffs than group members who conform to social norms. Because defections are now present in all groups these zero-tolerance individuals defect at a high rate. This leads quickly to the abandoning of the social norm and hence the unraveling of cooperation.

This exercise shows both the value of the social norm approach, and the weakness of the repeated game solution in the context of the public goods game in an evolutionary setting. Our negative assessment of the folk theorem is due in part to the particular game we have studied. There is a serious problem with the public goods game as a model of cooperation: the only incentive mechanism is the threat of withdrawal of cooperation in response to an observed defection. If the public good in question is a vital service to the group, the idea of executing a coordinated failure to provide the service in response to an infraction is implausible. This form of punishment cannot be directed at the miscreant, but rather is shared by all. Thus, in large groups, or groups with imperfect signals, the efficiency costs of incentives can completely offset any gains from cooperation. For instance, (1) fishers cooperating to maintain a common pool resource cannot possibly respond to overfishing on the part of one member by all members' intentionally overfishing; (2) a band of hunters cannot respond to an observed shirking incident by shirking in response; (3) in time of conflict, warriors are unlikely to punish cowards within their midst by refusing to fight.

Section 5 suggests that a general alternative in such cases is to use *directed punishment*, whereby a miscreant must pay a fine a cost to the individual imposing the fine. However, implementing a truly decentralized directed punishment mechanism is challenging. If punishment is costly, self-regarding individuals must have adequate incentives for carrying it out, and the signaling mechanism must include information on punishing activity. If punishing is rewarding to the punisher (e.g., failing to help a miscreant in need), then we must have a mechanism that limits punishing acts to miscreants alone. This problem is difficult when signals are public, but approaches being insurmountable when signals are private, so that a punishing act that is justified according to one observer may not be justified according to another. In short, directed punishment merely shifts the problem of cooperation from the stage game to the directed punishment game.

As we show in Sect. 6, altruistic punishment can sustain cooperation in the public goods game by dropping the requirement that the payoff to punishing must be sufficient to motivate punishing behavior on the part of self-regarding individuals. Not surprisingly, then, the public goods game induces cooperation only when accompanied by altruistic participants.

Of course, this raises the question as to how altruistic cooperation and punishment may have arisen through an evolutionary dynamic. We have dealt with this in several papers (Gintis 2000, 2003; Bowles et al. 2003; Bowles and Gintis 2004; Bowles 2006; Choi and Bowles 2007), but provide a new and relatively concise and plausible model in Sect. 7.

3 Dynamically Irrelevant Equilibria

John Nash (1950) developed the equilibrium concept that bears his name, the idea was elaborated upon by Kuhn (1953) and promoted in the influential volume by Luce and Raiffa (1957). John Harsany (1967) extended the notion of Nash equilibrium to games of incomplete information, and Reinhard Selten (1975) offered the first, and most important so-called equilibrium refinement, the *subgame perfect* equilibrium, which ruled out Nash equilibria involving incredible threats; i.e., actions that a player registers the intention of choosing, but are not best responses, so will not in fact be chosen should the occasion arise. There followed a decade of research into equilibrium refinements, including the well-known sequential equilibrium that we have already encountered. Jean Tirole (1990) documented a revolution in the economic field of Industrial Organization accomplished by searching for appropriately Nash equilibrium refinements. By the time Kreps (1990) and Fudenberg and Tirole (1991) published their influential textbooks, it had become accepted wisdom that "solving" a game meant finding its subgame perfect or sequential equilibria.

But, these "solutions" do not explain human behavior because there is no reason to believe that individuals would ever adopt the behaviors making up the equilibrium, or if they did, would not swiftly abandon them. While there are conditions under which individuals can "learn" to play a Nash equilibrium (Fudenberg and Levine 1997; Young 2006), these conditions do not obtain for repeated games, which are much more complex entities than their stage games. The strongest arguments in favor of the assertion that a Nash equilibrium will be played come from evolutionary game theory, where it is shown that every stable equilibrium of a dynamical system governed by a monotone dynamic, such as the replicator dynamic (Taylor and Jonker 1978), is a Nash equilibrium of the underlying game (Nachbar 1990; Samuelson and Zhang 1992). However, if there are multiple equilibria, as in the case in repeated game theory, this argument does not yield any prediction about behavior and does not imply that a high level of cooperation will occur even if full cooperation equilibria exist. Indeed, this argument does not even imply that an evolutionary system has a stable equilibrium, the alternative being a limit cycle or other non-equilibrium behavior.

It is surprising how little can be said about strategic interaction even assuming that individuals are predisposed and able to best respond given full knowledge of the game. We have known since (Pearce 1984; Bernheim 1984) that the assumption that it is common knowledge that individuals maximize their payoffs implies only that players will use only strategies that survive the iterated elimination of strictly dominated strategies, and there are many examples of games that cast doubt on the adequacy of even this assumption (Milgrom and Roberts 1990; Carlsson and van Damme 1993; Basu 1994; Vives 2005). Moreover, the equilibria in the repeated game models of cooperation are not achievable by the iterated elimination of strictly dominated strategies.

Recent research in interactive epistemology suggests that the conditions for achieving Nash equilibrium are quite stringent and rarely satisfied, except in the simplest of cases (Aumann and Brandenburger 1995). The problem with achieving a Nash equilibrium is that individuals may have heterogeneous and incompatible *beliefs* concerning how other players will behave, and indeed what other players believe concerning one's own behavior. It is clear from this research that the epistemological requirements for Nash equilibrium in all but the simplest games cannot be deduced from the assumption of rationality alone. This is because when there are multiple Nash equilibria, even the assumption that other players will choose a Nash strategy (an assumption that is itself difficult to justify) is insufficient to ensure a Nash equilibrium. Rather, there must be a social process leading to the alignment of conjectures and the constitution of common priors. This idea has a long history in the context of pure coordination games (Lewis 1969), but it in fact applies quite generally (Aumann 1987).

4 Social Norms in the Public Goods Game

The concept of social norms provides at least a partial solution. A cooperative equilibrium with social norms as one in which not only is the equilibrium strategy mix evolutionarily stable, but social norms are themselves an evolutionary adaptation, stable against invasion by competing social norms. To see this, we develop an agent-based model of the public goods game.

Suppose a large population forms N groups of n members each, and each group plays a public goods game repeatedly d times. We will call this series of d rounds an *encounter*. At the end of each encounter, players re-assort randomly into new groups of size n and carry out another encounter. By cooperating, a player confers a benefit of b on the other members (i.e., a benefit of $b/(n-1)$ per other member) and a cost c to himself. We assume that there are $n+1$ possible types of players, called *t-Cooperators*, for $t = 0, \ldots, n$. A t-Cooperator cooperates in the current round provided at least t other players cooperated in the previous round. We call an n-Cooperator, who never cooperates, a *Defector*, and we call a 0-Cooperator an *unconditional Cooperator*. On the first round, all players apply the t-criterion as though everyone cooperated in the previous round (i.e., all types cooperate except the n-Cooperator, who defects on all rounds). Finally we assume that a player who attempts to cooperate will accidentally defect with probability $\epsilon > 0$. We call this an *execution error*, and we call ϵ the *execution error rate*. Note that when $n = 2$, our model reduces to standard conditional cooperation, where the Conditional Cooperators is a 1-Cooperator, universal defect is a 2-Cooperator, and an unconditional cooperator is a 0-Cooperator.

Our central question will be the frequency of cooperation that can be sustained in the long run of this system, for different choices of the benefit b, error rate ϵ, duration of the encounter d, and the group size n. We created a population with 25 groups of size $n = 2, 4, 6, 8, 10, 12, 14$ playing a public goods game repeatedly for 25 periods ($d = 25$), in which by cooperating, an individual contributes b to the other players at a cost of $c = 1$ to himself, where $b = 2$ and $b = 4$. We begin (initialize) the simulation by assigning to each individual an t in the range $t = 0, \ldots, n-1$ with equal probability. The long-run behavior of our model does not depend on the initialization procedure, because it is a finite Markov chain with no absorbing states. However we use a highly randomized initialization procedure that leads the simulation to attain its long-run dynamic in a very short time (several hundred periods).

At the end of each encounter, 5% of individuals are replaced by new individuals, using a Darwinian fitness criterion according to which the probability of reproduction (respectively, death) is proportional to the individual's payoff relative to others in the population. We assume a mutation rate of 2% per newly-created individual, which means there is one mutation about every 50 encounters. A mutant is assigned an t in the range $t = 0, \ldots, n$ with equal probability. Also, we assume an execution error rate varying from $\epsilon = 0\%$ to $\epsilon = 10.0\%$. To promote cooperation in the face of errors, we assume that an individual who accidentally defected (we assume this fact is known to all group members, so it is public information) cooperates unconditionally on the next two rounds, thus allowing cooperation to be restored (this behavior is known as contrite in the literature).

To implement the concept of social norms, suppose each group G promotes a minimum cooperate level t_g, such that members cooperate as long as at least t_g members cooperated on the previous round. We assume that the group is in one of the states {Cooperate, Punish}. The game starts in state Cooperate, and remains there until fewer than t_g members signal Cooperate, whereupon the state Punish is entered. In state Punish, the social norm is for the Defectors to Cooperate and all

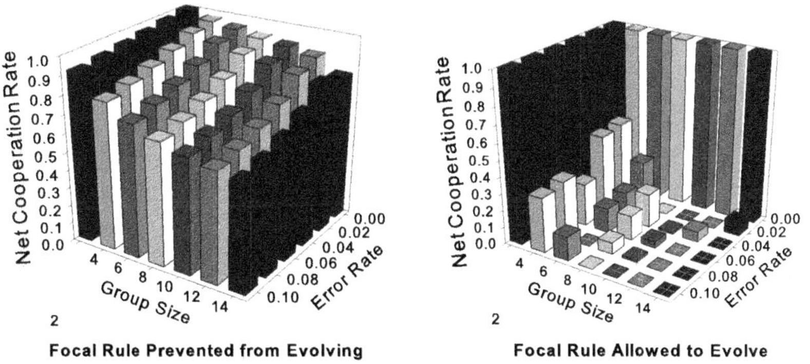

Fig. 1 The rate of cooperation in the public goods game for various group sizes and error rates, $b/c = 2$. The *left pane* assumes an exogenously fixed social norm $t_g = n$. The *right pane* assumes that the social norm is subject to evolutionary pressures

others to Defect. If the Defector succeed in cooperating, the system moves back into state Cooperate, where everyone cooperates, and otherwise remains in state Punish.

Suppose each individual is either a "Conformist" or an "Independent." A Conformist follows the normative minimum cooperate level t_g of whatever group he is in, while the Independent follows his own strategy, given by his being an t-Cooperator for a given t. We begin by assuming that all groups have the same t_g, which is exogenously given and not subject to change during the simulation. The fraction of Conformists in the population and the fraction of each type of t-Cooperator among the Independents, however, evolve endogenously according to a payoff-based replicator dynamic. The results are exhibited in Fig. 1 for $t_g = n$. Results are similar for values of t_g as small as 70% of group size. For smaller values of t_g, the fraction of Conformists declines to low levels and little cooperation can be sustained. In general, if a group's focus does not provide the proper incentives, Independents will have higher payoffs than Conformists, and the social norm will be abandoned as the frequency of Independents increases and that of Conformists decreases.

It is clear from the left pane in Fig. 1 that the addition of an appropriate social norm entails high efficiency of cooperation, attaining 70% even for groups of size 14 with a 10% error rate. However, by maintaining a stringent defection threshold, a group is bearing a cost to provide a benefit to the rest of the population, because defectors punished as a result of the stringent threshold will then have low fitness, and hence will tend to be replaced by Cooperators, thus benefitting all members of the population. Thus, members of each group will do better if their group raises the number of defections it permits and if other groups maintain zero or low tolerance for defectors. If the defection rate that each group permits before triggering retaliation is permitted to evolve, it will therefore fall. The results of this dynamic are depicted in the right pane of Fig. 1.

5 Directed Punishment

The analysis to this point suggests that even when we have the luxury of public signaling, the inability to direct punishment explicitly to the offending party renders the repeated game model of cooperation an infeasible or inefficient instrument in many cases. The obvious alternative is to allow some form of punishment directed specifically at the miscreant.

Suppose defectors can be identified, and are punished an amount p by other members of the group. We must have $p \geq c$ for punishment to deter defection, which means each other group members must punish an offender an amount at least $c/(n-1)$. If there are execution errors occurring with probability ϵ, then in a fully cooperative equilibrium each member each period will punish others an expected amount $\epsilon(n-1)c/(n-1) = \epsilon c$, and will himself receive an amount of punishment $\epsilon p = \epsilon c$ if punishment is set to its minimum effective level $p = c$. Suppose the cost c_p of meting out punishment p is αp where $-1 < \alpha$ (i.e., the cost may be negative, providing an incentive to punish). Then, assuming full cooperation, the cost of punishing and being punished per individual is $\epsilon c(1+\alpha)$, and the net payoff per period is

$$b(1-\epsilon) - c(1+\epsilon(1+\alpha)). \tag{1}$$

Note that the cost of punishing is just $\epsilon(1+\alpha)$ per period per group member, which is independent of group size. Thus, directed punishment appears to solve the problem of cooperation in large groups. Moreover, there is nothing in principle preventing $-1 < \alpha < 0$, so the directed punishment solution has the potential of being extremely efficient.

There is a catch, however: if $\alpha > 0$, players have no incentive to carry out the punishment, and if $\alpha < 0$, players have no incentive to limit their extractions to shirkers. Thus, an equilibrium of this type cannot be sustained by self-regarding individuals. To create incentives for individuals to punish defectors in the $\alpha > 0$ case, suppose members agree that any individual who is detected not punishing a defector is himself subject to punishment by the other players. Suppose with probability ϵ an individual who intends to punish fails to do so, or is perceived publicly by the other members to have failed. For simplicity, we choose ϵ to be the same as the error rate of cooperation. If all individuals cooperate and punish, the number of observed defections will be ϵn. Suppose all members must punish defectors equally. Then, the mean number of punishment of defector events per period will be ϵn^2. But, of course, $\epsilon^2 n^2$ of these events will erroneously be viewed as non-punishing, so we must have $\epsilon^2 n^3$ punishing of non-punisher events (let us call this second order punishment). Similarly, we must have $\epsilon^3 n^4$ third-order punishment to enforce second-order punishment. Assuming we have punishment on all levels, the total amount of punishment per individual per period will be

$$\epsilon n(1 + \epsilon n + \epsilon^2 n^2 + \ldots) = \frac{\epsilon n}{1 - \epsilon n},$$

provided $\epsilon < 1/n$. If the reverse inequality holds, this mechanism cannot work because each order of punishment involves greater numbers than the previous. Assuming $\epsilon < 1/n$, the expected payoff to a Cooperator under conditions of complete cooperation (assuming one engages in one's own punishment) is given by the recursion equation

$$v = b(1-\epsilon) - c - \epsilon n \frac{p + c_p}{1 - \epsilon n} + \delta v,$$

so

$$v(1-\delta) = (b(1-\epsilon) - c)(1 - \epsilon n) - \epsilon n(p + c_p). \qquad (2)$$

which becomes negative when ϵ is sufficiently close to $1/n$. Thus, cooperation will not be sustainable for large n unless error rates are quite low.

6 Altruistic Punishing in the Public Goods Game

The critical problem in the above model, that of motivating the punishment of free-riders, appears to be solved in both natural and experimental public goods games by the fact that many people appear to enjoy punishing those who violate social norms and happily do this without the expectation of any material reward.

In this section we therefore develop a model of cooperation in the Public Goods Game in which each agent is motivated by self-interest, unconditional altruism, and altruistic punishing, based on Carpenter et al. (2009). We investigate the conditions for a cooperative equilibrium, as well as how the efficiency of cooperation depends on the level of altruism and reciprocity. We show that if there is a stable equilibrium including both cooperation and shirking, an increase in either altruism or reciprocity motives will generate higher efficiency.

We consider a group of size $n > 2$, where member i works with probability e_i, so i contributes an amount qe_i to the other members of the group where $q > 1$, while the cost of working is a quadratic function $s(e_i) = e_i^2/2$. We assume the members of the group share their output equally. We assume member i can impose a cost s on member j at cost $c_i(s)$ to himself. The cost s results from public criticism, shunning, ostracism, physical violence, exclusion from desirable side deals, or another form of harm. We assume the social norm is $e = 1/2$, in the sense that other members judge j as prosocial or anti-social according as $e > 1/2$ or $e < 1/2$.

To model cooperative behavior with social preferences, we say that individual i's utility depends on his own material payoff π_i and the material payoff π_j to each other individual j, weighted by a factor $w_{ij} = a_i + l_i(2e_j - 1)$, where $a_i > 0$ means i is unconditionally altruistic and $a_i < 0$ means i is unconditionally spiteful. Also, $l_i \geq 0$ is a measure of i's strength of strong reciprocity motive. Note that if $l_i > 0$, then i places positive value on j's payoff if $e_j > 0$, and negative value if $e_j < 0$; i.e. i values helping or hurting j according as j's behavior is pro- or anti-social (Rabin 1993; Levine 1998).

If l_i and a_i are both positive, the individual is termed a strong reciprocator, motivated to reduce the payoffs of an individual who shirks even at a cost to himself. Players maximize the sum of own material payoff and the payoffs of the other players, where j's payoff is weighted by w_{ij}. It follows that when j increases his shirking level, i may respond by increasing his own shirking level (because he values j's benefit from his effort less), and/or by punishing j (because that lowers j's payoff, and hence raises i's utility).

This model requires only that a certain fraction of group members be strong reciprocators. This is in line with the evidence from behavioral game theory that in most experimental settings a certain fraction of the subjects do not act reciprocally, either because they are self-regarding or they are purely altruistic (Gintis 2009b). Note also that the punishment system could elicit a high level of cooperation, yet a low level of net material payoff. This is because punishment is not strategic in this model. In real societies, the amount of punishment of shirkers is generally socially regulated, and punishment beyond the level needed to secure compliance is actively discouraged (Wiessner 2005).

It is possible to show, using this model with plausible parameters, that the level of punishment by i imposed on j, is decreasing in i's unconditional altruism a_i, increasing in i's reciprocity motive, l_i, increasing in the level e_j of j's shirking, increasing in the harm that j inflicts upon i by shirking; and decreasing in group size. Moreover, we show in Carpenter et al. (2009) that an increase in i's unconditional altruism a_i leads i to shirk less, and an increase in i's reciprocity motive l_i leads i to shirk more when i's partners shirk on balance and to shirk less when i's partners work on balance.

Carpenter et al. (2009) also conducted an experimental public goods game that provides evidence for the behavioral relevance of altruistic rewarding and punishing in teams. In a treatment approximating a one-shot interaction, and on the terminal round of the game, shirkers are punished. They also find that shirkers respond to punishment by increasing their contributions to the public good. Neither self-regarding nor unconditionally altruistic motivations can account for these results.

7 The Evolutionary Emergence and Stability of Altruistic Punishment

The unsurprising fact that cooperation may be sustained by the presence of members ready to punish free-riding fellow group members without the expectation of a material reward solves the problem of motivating punishment but it raises another: how could individuals motivated to punish free-riders have initially emerged and proliferated, even when this behavior is altruistic so that the punisher would enhance his fitness by declining to punish? To provide an answer we will develop an agent-based model that illustrates a plausible dynamic through which altruistic punishment might have emerged from the strategic interaction of self-regarding agents, and then persist because it is a locally stable equilibrium.

Suppose members of a group may cooperate by contributing an amount of effort e to a public project, with $0 \leq e \leq 1$, thereby producing a benefit be that is shared equally by the other members of the group, at a cost ce to himself. We assume that $0 < c < b$, so the maximum benefit to the group occurs when each member is compliant, supplying effort $e = 1$, and each receives net payoff $b - c$. However, in the absence of punishment, non-altruistic individuals maximize their payoff by setting $e = 0$, and each member's payoff will be zero.

We also assume that there is a collective punishment process such that at a cost c, group members can impose a cost s on another group member who is discovered defecting (i.e., supplying effort $e < 1$). There are three types of agents: Cooperators who are unconditionally compliant (i.e., they always supply effort level $e = 1$) and never punish, Selfish types, who are compliant when the expect cost of non-compliance exceeds the cost of compliance and also never punish, and Punishers, who are compliant under the same conditions as the Selfish types, but who punish non-compliance when detected, provided that the fraction f of Punishers in the group exceeds a collectively agreed-upon threshold q, which we call the quorum level.

Individuals have one genetic locus at which there is a single copy of one of three alleles (i.e., individuals are haploid), corresponding to the three behavioral types, Cooperator, Selfish, and Punisher. Individuals have two parents (i.e., reproduction is diploid) and inherit the genetic type of each parent with probability 1/2. The couple's share of the population's total number of offspring surviving to reproductive age is increasing in the couple's fraction of the total payoffs of the population.

We assume that the cost of punishing non-compliance is shared among Punishers, so as the fraction f of Punishers increases, the individual cost of punishing declines. Moreover, when the fraction of Punishers is low, the probability of being detected defecting will also be low, so all non-Cooperators will defect. In this situation, in the absence of a quorum, not only would Punishers incur the costs of punishing, but also would forego the benefits of cooperation. This is the obstacle that in most models prevents the proliferation of a punishment strategy when rare. With a quorum, however, Punishers can invade a population of non-punishers.

Quorum sensing requires Punishers to know the number of other Punishers in their group. At the outset, individuals know their own type, but not the types of other group members. The type of an unknown member can be determined by paying a quorum-sensing cost c, which is shared by all known Punishers in the group. If the fraction of known Punishers in the group is at least q, when non-compliance is detected, it will be punished by all Punishers, and thus the types of all Punishers will become known. If the fraction of group members who are either known Punishers or are of unknown type is below q, punishing will not occur and the unknown types will remain unknown. This is reasonable because in this case, even if all unknown types are Punishers, the quorum threshold cannot be attained. When neither of these is the case and an act of non-compliance is detected. Punishers share the quorum-sensing cost of determining the type of each unknown member. All immigrants to the group, and all newborns (who may be mutants) are initially unknown.

We assume that groups are initially composed of 18 members of a single generation, representing two bands of the size thought to be typical among our Pleistocene ancestors (Marlowe 2005). There are 250 groups, giving a total population of 4,500. Because of birth, death, and migration, group size is variable over time. We assume that 50% of newborns are located in the parental group, and 50% are relocated randomly in the population. This implies a migration rate of 50% per generation (25 years), allowing for complete exogamy. Moreover, we assume that when a group becomes sufficiently small, it is repopulation by additional migration from larger groups. This repopulation is a form of migration that supplements the exogamy migration of 50% of newborns. Consistent with Pleistocene demographics (Hassan 1973), we study a population with constant size, so the death rate is 4% per year, and new individuals are born at the rate of 4% per year as well. To avoid the artificial situation in which punishment is costless because all individuals comply, we assume that with 3% probability, an individual who attempts to cooperate fails, and in fact defects. This means that in a group of 18 at least one defection will occur in each year with a probability of 42%.

We start each simulation with only Selfish types in the population, so Cooperators and Punishers only enter a group through the random mutation of newborns. We assume offspring mutate with probability 0.001 from the parentally supplied genotype. Whatever the parental type, the mutant is Selfish with probability 0.40, and is a Cooperator or a Punisher, each with probability 0.3. Thus, there are 4.5 mutants per generation, of which $4.5 \times 0.30 = 1.35$ are Punishers. We normalize payoffs by a linear transformation so that the thereby normalized payoffs π vary from zero to one. We assume a baseline fitness $w < 1$ and a selection coefficient $1 - w$ on π, so that individual fitness is given by $w + (1 - w)\pi$, normalized so as to maintain a constant population size. We set $w = 0.8$ in our simulations. We have found that weaker selection ($w = 0.95$) shortens the expected takeoff time, but does not otherwise materially affect our results.

Figure 2 shows a single simulation of the evolution of this model over 4,000 generations (100,000 years). The Punisher allele does not go to fixation because Selfish types have higher fitness than Punishers in groups in which f is far above the quorum threshold $q = 50\%$. The long-run average rate of non-compliance after the emergence and proliferation of Punishers is 6%, half of which is due to behavioral error, and the remainder due to agents' misjudging f, due to noise in the quorum sensing process, or f falling below the quorum level.

To better understand the process of takeoff and the subsequent evolutionary success of the Punisher strategy, we define two terms, both derived from the decomposition of evolutionary processes into additive within-group and between-group selection effects due to Price (1970). The first, β_G, is the effect of the fraction of a group that are Punishers on the group average expected fitness, while the second, β_I, is the effect of an individual's own type (Punisher or not) on individual fitness. If β_I is negative then punishing is altruistic, while β_G may be positive if the presence

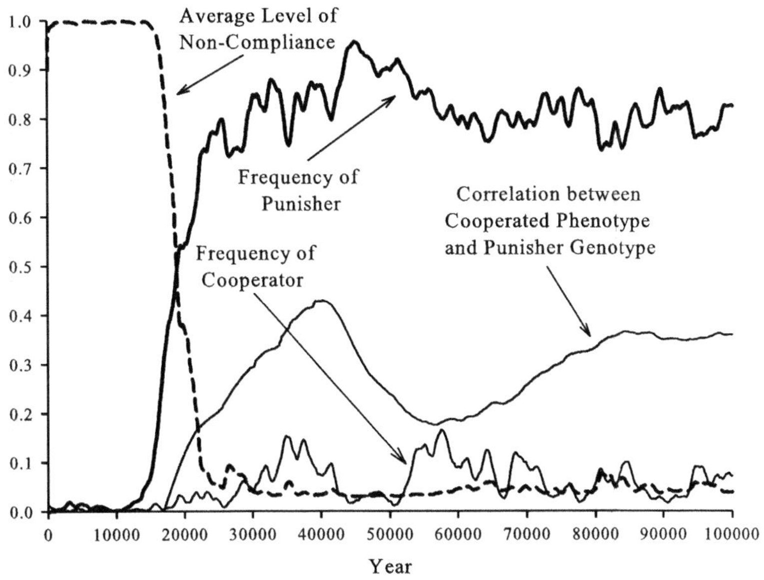

Fig. 2 The successful invasion of punishers and stabilization of cooperation. The timing of the takeoff to cooperation, which we define as 70% or more punishers and 10% or less non-compliance depends on both the assumed rate of production of punishers by the mutation process. In the simulations shown, $\mu = 1/1000$ and 30% of mutants are punishers. The strength of the selection process is $1 - w$, and the quorum level is $q = 0.5$ for these simulations, although any value of q between 0.25 and 0.50 produces similar results

of a large number of Punishers in group supports a high level of cooperation and hence high fitness for members of the group on average.

Figure 3 shows the average movement of these two coefficients in 25 replications of the process of taking off from zero to sustained cooperation. It is clear that in the long run, Punishers are altruists: after 30,000 years, the within-group term is negative, indicating that a Punisher's fitness would be increased by switching to non-Punisher. But the between-group term is positive, indicating that on average, members of groups with a high frequency of Punishers have a fitness advantage over members of groups with few Punishers. Because most punishers are in such groups their within group disadvantage is counteracted.

What accounts for the eventual success of altruistic punishment in this model? Those bearing the Punish allele are only modestly more likely to be in groups with other Punishers. The fraction of the variance of the Punish allele that is between groups measured by the $F_{ST} = 0.11$ (average of 25 runs, with differences only in the third decimal place across runs), confirming that the degree of genetic assortment implied our model and the simulation parameters is not at variance with observed levels (Bowles 2006). The success of Punishers is due to the fact that quorum sensing creates a strong relationship between compliance and punishing behavior that

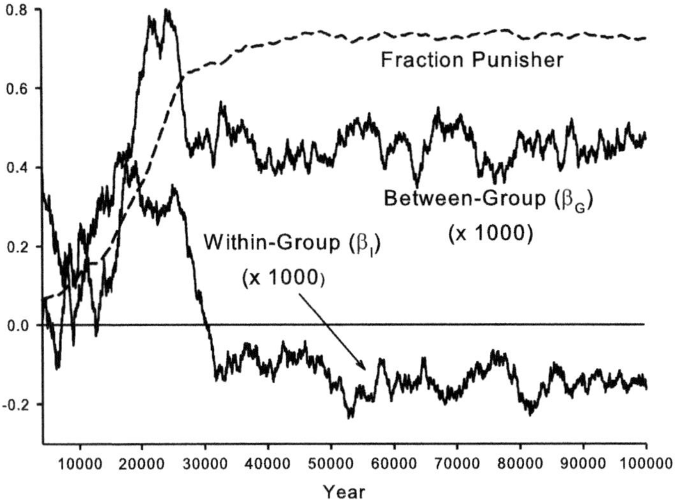

Fig. 3 Between-group (β_G) and within-group (β_I) selection for punishers. This is an average of 25 runs of the model

is only operative when the fraction of Punishers is sufficiently large. This entails a positive correlation between the Punisher genotype and the cooperating phenotype. As is well known (Queller 1992; Fletcher and Zwick 2006), such a correlation is necessary for natural selection to favor an altruistic genotype. Figure 2 shows that this situation is indeed the case: the correlation among groups between the phenotypic trait "fraction cooperated" and the genetic trait "fraction Punisher" becomes strongly positive and exceeds $r = 0.20$ when a state of high-level cooperation is attained. Experiments in dyads have suggested that punishment of defectors may lower group average payoffs (Dreber et al. 2008). This is not surprising, because in very small groups cooperation may be sustained by means of reciprocal altruist strategies such as tit-for-tat without resort to a punishment option. In the much larger groups studied here, the punishment option is essential to sustaining cooperation and high group average payoffs. In fact, average fitness is much higher in groups with substantial fractions of Punishers than in groups where the frequency of Punishers is just above the threshold.

Figure 3 also shows that in the takeoff-period, Punishers are not altruistic; were they to eschew punishing their payoffs, and hence fitness, would fall. The only groups in which punishing could be individually fitness enhancing are those in which the shift of a single individual from Punisher to non-Punisher would move the group below q, or a shift from non-Punisher to Punisher would move the group above q. We call such groups tipping groups, and hypothesize that the takeoff of Punishers occurs precisely when most Punishers, fortuitously, are located in tipping groups. This hypothesis is illustrated by the simulation shown in Fig. 4, which charts

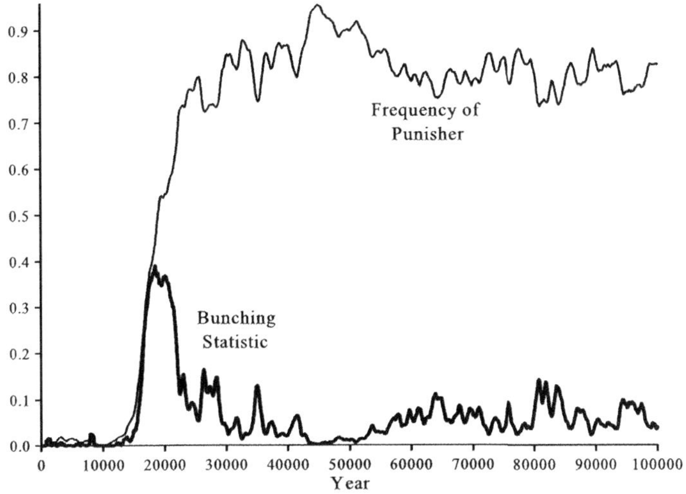

Fig. 4 Most punishers are in tipping groups when punishing proliferates. The bunching statistic is the probability that a punisher is in a group where one individual moving from selfish to punishing pushes the group from below to above the quorum level, or moving from punishing to selfish pushes the group from above to below the quorum level

the probability that a Punisher is in a tipping group, which we term the "bunching statistic." For the simulation shown, this statistic shows a rapid, sustained increase during the takeoff period, after which it falls back to a low level when the frequency of Punishers is above 70%.

Figure 5 illustrates the importance of quorum sensing for the operation of the model. In this figure, all parameters are at their benchmark levels except that we assume there is no quorum; i.e., Punishers always punish. Starting with a high level of Punishers (75%) and a low level of Cooperators (5%), the cooperators invade so that by year 40,000 Punishers are driven to low levels. This permits the Selfish types to invade, reducing Cooperators to very low levels by year 50,000. After year 50,000, the rate of non-compliance (not shown in the figure) is above 85%.

8 Conclusion: The Missing Choreographer

The economic theory of cooperation based on repeated games proves the existence of equilibria with desirable properties, while leaving the question of how such equilibria are achieved as an afterthought. The folk theorem on cooperation in repeated games shares this defect with the even more celebrated Fundamental ("Invisible Hand") Theorem, demonstrating the Pareto-efficiency of competitive market allocations. The latter purports to model decentralized market interactions, but on close inspection requires a level of coordination that is not explained, but rather posited

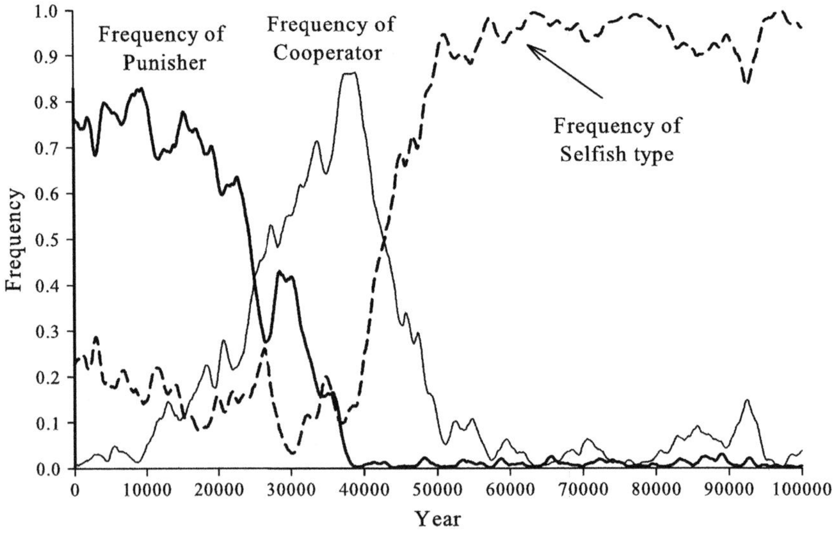

Fig. 5 Cooperation fails without quorum sensing. Starting with a high level of punishers the selfish types eventually invade

as a deus ex machina (Sonnenschein 1972; Kirman 1989; Ingrao and Israel 1990). We have shown, similarly, for the case of cooperation supported by retaliation as in the Folk Theorem, that the social norms on which the coordination must be based will not evolve spontaneously. Yet, highly choreographed coordination on complex strategies capable of deterring defection are supposed to materialize quite without the need for a choreographer. As in the case of the Fundamental Theorem, the dynamics are thus unspecified and, if we are correct, impossible to provide without a fundamental change in the underlying theory.

The failure of the models underlying both the Folk Theorem and the Fundamental Theorem is hardly surprising, for the task we set for them, that of explaining the emergence and persistence of cooperation among large numbers of self regarding strangers without recourse to pre-existing cooperative institutions, is not only formidable; it most likely never occurred in the history of our species. Humans are indeed unique among living creatures in the degree and range of cooperation among large numbers of substantially unrelated individuals. The global division of labor and exchange, the modern democratic welfare state, and contemporary warfare alike evidence our distinctiveness. These forms of cooperation emerged historically and are today sustained as a result of the interplay of self-regarding and social preferences operating under the influence of group-level institutions of governance and socialization that favor Cooperators, in part by helping to coordinate their actions so as to target transgressions for punishment and thus protecting them from exploitation by defectors.

The norms and institutions that have accomplished this evolved over millennia through trial and error. Consider how real world institutions addressed two of the shoals on which the economic models foundered. First, the private nature of information, as we have seen, makes it virtually impossible to coordinate the targeted punishment of miscreants. In many small-scale societies this problem is attenuated by such cooperative customs as eating in public so that violations of sharing norms can be easily detected. Cooperative shrimp fishermen in Japan, who pool their fishing returns, deliberately land their catch at the same time of day for the same reason (Platteau and Seki 2001).

But, where larger numbers are involved, converting private information about transgressions to public information that can provide the basis of legitimate punishment often involves civil or criminal trials, elaborate processes that rely on commonly agreed upon rules of evidence and ethical norms of appropriate behavior. Even these complex institutions frequently fail to transform the private protestations of innocence and accusations of guilt into common knowledge.

Second, cooperation often unravels when the withdrawal of cooperation by the civic-minded intending to punish a defector is mistaken by others as itself a violation of a cooperative norm, inviting a spiral of further defections. A similar dynamic is observed in some experiments with public goods games in which subjects have the option of punishing other members; while the punishment option sustains cooperation, it often also leads to vendetta-like cycles of punishment and counter punishment, which may more than offset the gains to cooperation (Herrmann et al. 2008). In virtually all surviving societies, this problem is addressed by the creation of a corps of specialists entrusted with carrying out the more severe of society's punishments. Their uniforms convey the civic purpose of the punishments they mete out, and their professional norms, seek to ensure that the power to punish was not used for personal gain. Like court proceedings as a way to transform private into public information, this institution works imperfectly.

Modeling the complex process by which we became a cooperative species is a major challenge of contemporary science. Economic theory, favoring parsimony over realism, has instead sought to explain cooperation without reference to other-regarding preferences and with a minimalist or fictive description of social institutions. This research trajectory, as we have seen, has produced significant insights. But it may have run its course.

For students of human cooperation, the challenge thus shifts from that favored by biologists and economists over the last half century – showing why self-regarding individuals would nonetheless cooperate – to explaining how the other-regarding preferences and group-level institutions that sustain cooperation could have emerged and proliferated in an empirically plausible evolutionary setting, a task that we have touched on briefly in the previous section and addressed at some length in our forthcoming book, *A Cooperative Species*, and in a series of related papers (Gintis 2000, 2003; Bowles et al. 2003; Bowles 2006; Choi and Bowles 2007).

Acknowledgments We would like to thank E. Somanathan for initial conversations that stimulated this research and Robert Boyd, Jessica Flack, Eric Maskin, and Robert Sugden for comments on an earlier version. This chapter draws extensively on material in our book, *A Cooperative Species*. We would also like to thank the European Science Foundation and the Behavioral Sciences Program of the Santa Fe Institute for research support.

References

Arrow KJ (1971) Political and economic evaluation of social effects and externalities. In: Intriligator MD (ed) Frontiers of quantitative economics. North Holland, Amsterdam, pp 3–23

Arrow KJ, Debreu G (1954) Existence of an equilibrium for a competitive economy. Econometrica 22(3):265–290

Arrow KJ, Hahn F (1971) General competitive analysis. Holden-Day, San Francisco

Aumann RJ (1987) Correlated equilibrium and an expression of Bayesian rationality. Econometrica 55:1–18

Aumann RJ, Brandenburger A (1995) Epistemic conditions for Nash equilibrium. Econometrica 65(5):1161–1180

Axelrod R, Hamilton WD (1981) The evolution of cooperation. Science 211:1390–1396

Basu K (1994) The traveler's dilemma: paradoxes of rationality in game theory. Am Econ Rev 84(2):391–395

Benedict R (1934) Patterns of culture. Houghton Mifflin, Boston

Bernheim BD (1984) Rationalizable strategic behavior. Econometrica 52(4):1007–1028

Bhaskar V, Obara I (2002) Belief-based equilibria: the repeated prisoner's dilemma with private monitoring. J Econ Theory 102:40–69

Binmore KG (1993) Game theory and the social contract: playing fair. MIT Cambridge, MA

Binmore KG (1998) Game theory and the social contract: just playing. MIT Cambridge, MA

Binmore KG (2005) Natural justice. Oxford University Press, Oxford

Blau P (1964) Exchange and power in social life. John Wiley, New York

Bowles S (2004) Microeconomics: behavior, institutions, and evolution. Princeton University Press, Princeton

Bowles S (2006) Group competition, reproductive leveling, and the evolution of human altruism. Science 314:1669–1672

Bowles S, Gintis H (1993) The revenge of Homo economicus: contested exchange and the revival of political economy. J Econ Perspect 7(1):83–102

Bowles S, Gintis H (2004) The evolution of strong reciprocity: cooperation in heterogeneous populations. Theor Popul Biol 65:17–28

Bowles S, Hammerstein P (2003) Does market theory apply to biology? In: Hammerstein P (ed) Genetic and cultural evolution of cooperation. MIT, Cambridge, MA, pp 153–165

Bowles S, Choi Jung-kyoo, Hopfensitz A (2003) The co-evolution of individual behaviors and social institutions. J Theor Biol 223:135–147

Boyd R, Richerson PJ (1985) Culture and the evolutionary process. University of Chicago Press, Chicago

Brown DE (1991) Human universals. McGraw-Hill, New York

Carlsson H, van Damme E (1993) Global games and equilibrium selection. Econometrica 61(5):989–1018

Carpenter J, Bowles S, Gintis H, Hwang SH (2009) Strong reciprocity and team production. J Econ Behav Organ

Choi J-K, Bowles S (2007) The coevolution of parochial altruism and war Science 318(26): 636–640

Debreu G (1959) Theory of value. Wiley, New York

Dreber A, Rand DG, Fudenberg D, Nowak MA (2008) Winners don't punish. Nature 452:348–351

Durkheim E (1933 [1902]) The division of labor in society. The Free Press, New York

Ely JC, Välimäki J (2002) A robust folk theorem for the prisoner's dilemma. J Econ Theory 102:84–105

Fletcher JA, Zwick M (2006) Unifying the theories of inclusive fitness and reciprocal altruism. Am Nat 168(2):252–262

Fudenberg D, Maskin E (1986) The folk theorem in repeated games with discounting or with incomplete information. Econometrica 54(3):533–544

Fudenberg D, Tirole J (1991) Game theory. MIT Cambridge, MA

Fudenberg D, Levine K (1997) The theory of learning in games. MIT Cambridge, MA
Fudenberg D, Levine K, Maskin E (1994) The folk theorem with imperfect public information. Econometrica 62:997–1039
Gintis H (1976) The nature of the labor exchange and the theory of capitalist production. Rev Radic Polit Econ 8(2):36–54
Gintis H (2000) Strong reciprocity and human sociality. J Theor Biol 206:169–179
Gintis H (2002) Some implications of endogenous contract enforcement for general equilibrium theory. In: Petri F, Hahn F (eds) General equilibrium: problems and prospects. Routledge, London, pp 176–205
Gintis H (2003) The hitchhiker's guide to altruism: genes, culture, and the internalization of norms. J Theor Biol 220(4):407–418
Gintis H (2009) The bounds of reason: game theory and the unification of the behavioral sciences. Princeton University Press, Princeton, NJ
Gintis H (2009) Game theory evolving, 2nd edn. Princeton University Press, Princeton, NJ
Hassan FA (1973) Determination of the size, density, and growth rate of hunting-gathering populations. In: Polgar S (ed) Population, ecology, and social evolution. Mouton, The Hague, pp 27–52
Harsany JC (1967) Games with incomplete information played by bayesian players, Parts I, II, and III. Behav Sci 14:159–182, 320–334, 486–502
Herrmann B, Thöni C, Gächter S (2008) Anti-social punishment across societies. Science 319:1362–1367
Ingrao B, Israel Giorgio (1990) The invisible hand: economic equilibrium in the history of science. MIT, Cambridge
Kirman A (1989) The intrinsic limits of modern economic theory: the emperor has no clothes. Econ J 99(395):126–139
Kreps DM (1990) A course in microecoomic theory. Princeton University Press, Princeton, NJ
Kreps DM, Wilson R (1982) Sequential equilibria. Econometrica 50(4):863–894
Kuhn HW (1953) Extensive games and the problem of information. In Kuhn HW, Tucker AW (eds) Contributions to the theory of games, Vol 2 of Annals of mathematics studies. Princeton University Press, Princeton, NJ pp 193–216
Laffont JJ (2000) Incentives and political economy. Oxford University Press, Oxford
Levine DK (1998) Modeling altruism and spitefulness in experiments. Rev Econ Dyn 1(3): 593–622
Lewis D (1969) Conventions: a philosophical study. Harvard University Press, Cambridge, MA
Luce R (1957) Raiffa HD Games and decisions. Wiley New York
Mailath GJ, Morris S (2006) Coordination failure in repated games with almost-public monitoring. Theor Econ 1:311–340
Marlowe F (2005) Hunter-gatherers and human evolution. Evol Anthropol 14:54–67
Milgrom PR, Roberts J (1990) Rationalizability, learning, and equilibrium in games with strategic complementarities. Econometrica 58(6):1255–1277
Nachbar JH (1990) Evolutionary selection dynamics in games: convergence and limit properties. Int J Game Theory 19:59–89
Nash JF (1950) Equilibrium points in n-Person games. Proc Nat Acad Sci 36:48–49
Parsons T, Shils E (1951) Toward a general theory of action. Harvard University Press, Cambridge, MA
Pearce D (1984) Rationalizable strategic behavior and the problem of perfection. Econometrica 52:1029–1050
Piccione M (2002) The repeated prisoner's dilemma with imperfect private monitoring. J Econ Theory 102:70–83
Platteau J-P, Seki E (2001) Community arrangements to overcome market failure: pooling groups in Japanese fisheries. In: Hayami M, Hayami Y (eds) Communities and markets in economic development. Oxford University Press, Oxford pp 344–402
Price GR (1970) Selection and covariance. Nature 227:520–521
Queller DC (1992) A general model for kin selection. Evolution 42(2):376–380

Rabin M (1993) Incorporating fairness into game theory and econometrics. Am Econ Rev 83(5):1281–1302
Samuelson L, Zhang J (1992) Evolutionary stability in asymmetric games. J econ Theory 57(2):363–391
Sekiguchi T (1997) Efficiency in repated prisoner's dilemma with private monitoring. J Econ Theory 76:345–361
Selten R (1975) Re-examination of the perfectness concept for equilibrium points in extensive games. Int J Game Theory 4:25–55
Shubik M (1959) Strategy and market structure: competition, oligopoly, and the theory of games. Wiley New York
Sonnenschein H (1972) Market excess demand functions. Econometrica 40:549–563
Stiglitz J (1987) The causes and consequences of the dependence of quality on price. J Econ Lit 25:1–48
Sugden R (1986) The economics of rights, co-operation and welfare. Basil Blackwell, Oxford
Taylor M (1976) Anarchy and cooperation. Wiley, London
Taylor P, Jonker L (1978) Evolutionarily stable strategies and game dynamics. Math Biosci 40:145–156
Tirole J (1988) The theory of industrial organization. MIT, Cambridge, MA
Tirole J (1990) The theory of industrial organization. MIT Cambridge, MA
Trivers RL (1971) The evolution of reciprocal altruism. Q Rev Biol 46:35–57
Vives X (2005) Complementarities and games: new developments. J Econ Lit 43:437–479
Wiessner P (2005) Norm enforcement among the Juh́oansi Bushmen: a case of strong reciprocity? Hum Nat 16(2):115–145
Young HP (2006) Strategic learning and its limits. Oxford Uniersity Press, Oxford

Evolution, Cooperation, and Repeated Games[1]

E. Maskin

Abstract I discuss recent work that characterizes what outcomes correspond to evolutionarily stable strategies in two-player symmetric repeated games when players have a positive probability of making a mistake.

The theory of repeated games has been an important tool in the behavioral and biological sciences. Indeed, it provides the central model for explaining how agents with selfish objectives might nevertheless behave cooperatively and efficiently in a long-term relationship. For that reason, it has been invoked often by economists, political scientists, ecologists, anthropologists and others interested in human cooperation.

Repeated-game theory offers a beautifully simple answer to the question of why selfish agents should cooperate: namely, they should do so to ensure continued cooperation in the future. For an illustration of this point, consider the classic embodiment of the tension between self-interest and cooperation, the Prisoner's Dilemma.

	C	D
Cooperate	2,2	−1,3
Defect	3,−1	0,0

Prisoner's Dilemma

In this game there are two players – a row player and a column player – each of whom has two possible actions: to "cooperate" (C) or to "defect" (D). The payoffs to each combination of player's actions are given by the table above, where in each box the first number is the row player's payoff and the second is the column player's payoff. Notice that if the game is played just once, then regardless of what the other

[1] This chapter is based on work with D. Fudenberg.

E. Maskin
Institute for Advanced Study, School of Social Science, Einstein Drive, Princeton, NJ 08540, USA

S.A. Levin (ed.), *Games, Groups, and the Global Good*, Springer Series in Game Theory, 79
DOI: 10.1007/978-3-540-85436-4_4, © Springer Physica-Verlag Berlin Heidelberg 2009

player does, it is optimal (i.e., a *dominant strategy*) for each player to play D: by doing so he gets 3 rather than 2 if the other player plays C, and 0 rather than -1 if the other player plays D. Thus, the predicted outcome of the one-shot game is for each player to play D (and thereby get a zero payoff), even though both would be better off if they played C (they would then each get 2).

Now imagine that the game is repeated many times (formally, infinitely many times), and that a player cares about his payoff on average in the long run rather than in any particular iteration. Unlike before, playing cooperatively may now be in the player's best interest. In fact, the strategy in which a player (1) plays C in the first iteration, (2) continues to play C as long as both players have cooperated in the past, and (3) plays D otherwise (call this strategy CC for "conditionally cooperate") generates an *equilibrium* in the sense that if both players adopt it, neither has the incentive to deviate to any other strategy unilaterally. To see this, note that when players both adhere to CC, a never-ending stream of cooperation results, giving each player an average payoff of 2 per period. If, however, a player unilaterally deviates to some other strategy, then either (1) in some iteration, he plays D – in which case his opponent always plays D thereafter, so that he (the deviator) obtains a zero average payoff at best, or else (2) despite the deviation, he continues to play C in every iteration – in which case, he still gets an average of 2, and so doesn't gain from deviating. Thus, repetition makes cooperation plausible as a possible outcome.

Unfortunately, cooperation is not the *only* plausible outcome. For example, the strategy in which a player *always plays D* (AD) also generates an equilibrium if both players adopt it. And these two extremes – CC, which leads to cooperation in every iteration, and AD, which leads to no cooperation at all – aren't the only possibilities: the Folk Theorem of repeated games (cf. Fudenberg and Maskin 1986) tells us that every intermediate possibility between full cooperation and full defection can occur in equilibrium as well. Moreover, the theory does not favor any particular equilibrium over the others. That is, the theory makes scarcely any clear-cut prediction about behavior at all. It is practically unfalsifiable.

Evolution (either biological or cultural) might be expected to help resolve this predictive murkiness. Evolutionary forces often foster efficiency, and so we could hope that uncooperative behavior will be rooted out and cooperative behavior promoted by selective pressure. Indeed, this view was set out forcefully in Axelrod and Hamilton (1981) and later in the popular book Axelrod (1984). The basic idea is easy to convey. Imagine that we start with a population in which all players use the strategy AD. Now, suppose that a small group of "mutants" are introduced who use the strategy CC. Notice that CC earns the same average payoff (namely, zero) against AD that AD does against itself. Moreover, CC earns an average payoff of 2 against itself, whereas AD gets only 0 on average against CC. Thus, in expectation (assuming that pairs of strategies are picked at random from the entire population), CC performs strictly better than AD. And so, according to standard evolutionary dynamics, CC will replicate faster than AD and eventually take over the population. That is, uncooperative behavior in the guise of AD will eventually be driven out.

Unfortunately, among uncooperative strategies, AD is unrepresentatively easy to drive out. Consider instead ALT, the strategy that alternates between C and D until someone breaks the alternating pattern, after which point it always plays D. Unlike AD, a population of ALTs *cannot* be invaded by a mutant strategy. To see this, notice that a mutant would have to conform to the alternating pattern: otherwise, it would do strictly worse (an average payoff of 0) against ALT than ALT itself does against (average payoff 1), and so could not grow relative to ALT. But a mutant conforming to the alternating pattern would also fail to perform better in expectation than ALT. Thus, ALT is *evolutionarily stable*[2] (ES) despite being quite inefficient and uncooperative.[3]

Yet there is a sense in which ALT is too inflexible. It relies, after all, on perfect alternation: any deviation from the pattern $C, D, C, D, C\ldots$ is "punished" by an infinite succession of Ds. This suggests that it might not fare so well in an environment in which strategies are not always executed correctly, i.e., in which there is a small but positive probability that a strategy makes a mistake. Indeed, I claim that in such an environment, ALT is no longer ES.

Specifically, consider mutant strategy s' that is identical to ALT except when the alternating pattern has been broken. In the iteration just after the pattern is broken, s' plays C (unlike ALT, which plays D) for one period – to "signal" its willingness to cooperate. If the other player also plays C in that iteration, then s' plays C from then on. But if the other player plays D, s' (like ALT) plays D thereafter.

I claim that s', so constructed, will successfully invade a population of ALTs if strategies are subject to a small probability of inaccurate execution (so that, with positive probability, the alternating pattern will be broken). To see this, notice that (1) s' is identical to ALT before the alternating pattern is broken; (2) s' and ALT each get a payoff of 0 per period against ALT after the pattern is broken, (3) s' gets a payoff of almost 2 but ALT gets only 0 against s' after the pattern is broken. Thus, in expectation, s' performs better than ALT, and so will successfully invade.

In Fudenberg and Maskin (1990, 2007), Drew Fudenberg and I characterize the payoffs corresponding to evolutionarily stable strategies in two-player symmetric[4] repeated games when (1) there is a positive probability p in each iteration that a strategy is misexecuted, and (2) players discount future payoffs at a positive rate r, so that instead of maximizing the long-run average payoff, a player maximizes the *discounted* average payoff

---

[2] Roughly speaking, a strategy s is ES if no mutant strategy s' performs better than s in expectation against a population consisting mostly of s but with a small proportion of s'.

[3] In fact, the situation is even worse than ALT would suggest. Consider the strategy that repeatedly follows the pattern C followed by *two* Ds until the pattern is broken, at which point it thereafter plays D. For the same reason as ALT, this more elaborate strategy is ES, yet it attains an average payoff of only $\frac{2}{3}$. In fact, by continuing to add more Ds to the repeated pattern, we can obtain an ES strategy that is arbitrarily close in average payoff to the fully uncooperative strategy AD.

[4] A two-player game is symmetric if the two players have the same set of actions and if interchanging actions between the players causes the corresponding payoffs to be interchanged.

$$\frac{r}{1+r}\sum_{t=1}^{\infty}\left(\frac{1}{1+r}\right)^{t-1}\pi_t,$$

where π_t is his payoff in iteration t. So that I can state these characterization results reasonably precisely, let us define a pair of payoffs (v_R, v_C) (where v_R and v_C are the row and column players' payoffs respectively) in a two-player symmetric game g to be *strongly efficient* if (1) (v_R, v_C) are feasible payoffs for g and (2) $v_R + v_C$ maximizes the sum of the two players' payoffs among all feasible payoffs for g. Thus in the Prisoner's Dilemma (see the table on p. 79) the unique strongly efficient payoffs are (2, 2). But in the Battle of the Sexes, as given in the following table,

	Ballet	Boxing
Boxing	0,0	1,2
Ballet	2,1	0,0

Battle of the Sexes

any convex combination of (1, 2) and (2, 1) is strongly efficient (assuming that players can randomize between the two points). Let $\underline{v}$ be the smallest payoff consistent with strong efficiency, i.e.,

$$\underline{v} = \min\{v_R \mid (v_R, v_C) \text{ is strongly efficient}\}.$$

Thus, $\underline{v} = 2$ in the Prisoner's Dilemma, but $\underline{v} = 1$ in the Battle of the Sexes.

The first result asserts that if the mistake probability and discount rate are small (but positive), the payoff resulting from an evolutionarily stable strategy cannot be much less that $\underline{v}$. More formally we have

Theorem 1 *Given $\varepsilon > 0$, then if $p > 0$ and $r > 0$ are sufficiently small, the payoffs generated when both player use an ES strategy s cannot be less than $\underline{v} - \varepsilon$.*

Note that Theorem 1 implies that, for the Prisoner's Dilemma, evolutionarily stable behavior must be almost fully cooperative. This is not the case, however, for the Battle of the Sexes. Full cooperation would entail playing (Ballet, Ballet) or (Boxing, Boxing), so that, on average, players would be getting a payoff of $1\frac{1}{2}$ each. Evolutionary stability, by contrast, ensures only that average payoffs will not be (much) below 1.

The proof of Theorem 1 is similar to the argument above showing that ALT is not ES when p is positive. If s fails to attain the payoff $\underline{v}$, then we can construct a mutant strategy s' that mimics s most of the time but exploits s's inefficiency by "signaling" its willingness to play more efficiently. It the other player reciprocates this signal, then s' will thereafter play strongly efficiently. If not, then s' will revert to playing like s.

Theorem 1, by itself, is an incomplete characterization because it doesn't deal with the issue of whether ES strategies actually exist. Accordingly, the second main result establishes that if v exceeds $\underline{v}$, there exists an ES strategy s attaining the payoff v (approximately), provided that r and p are small. More formally, we can state

Theorem 2 *Let (v, v) be feasible payoffs with $v \geq \underline{v}$. For all $\varepsilon > 0$, there exist r and p sufficiently small so that there exists an ES s for which, if both players use s, the corresponding payoffs are within ε of (v, v).*

For the Prisoner's Dilemma, Theorem 2 asserts the existence of an ES strategy that (approximately) attains full cooperation. What form might such a strategy take? Like ALT, CC is too inflexible when p is positive: one mistake leads to D forever. The strategy Tit-for-tat (play C in the first iteration and thereafter do whatever the other player did in the previous iteration), emphasized by Axelrod and Hamilton (1981), is similarly too readily thrown off track by mistakes. If say, the row player (by mistake) plays D in the first iteration, Tit-for-tat will have the column player follow with D in the second iteration, and then will induce the row player to play D again in the third iteration, etc. That is, there will be a stream of Ds that ends only when someone makes another mistake. A more robust strategy against mistakes takes the form: play C in the first iteration, and thereafter play C if *both* players played C the previous iteration or if *neither* did. Indeed, this strategy is ES for the payoffs given the Prisoner's Dilemma table above (see Fudenberg and Maskin 1990).

To see how ES strategies can generate payoffs quite short of full cooperation in games resembling the Battle of the Sexes, consider the game given by the following table

	a	*b*	*c*	*d*
a	0,0	4,1	0,0	0,0
b	1,4	0,0	0,0	0,0
c	0,0	0,0	0,0	0,0
d	0,0	0,0	0,0	2,2

Expanded Battle of the Sexes

Consider the strategy s that (1) plays d as long as in every past iteration either both players played d or neither d, (2) plays a forever if the other player was the first to deviate from d, and (3) plays b forever if it itself was the first to deviate from d. It can be shown that s is ES for r and p sufficiently small. Nevertheless, the combination (d, d) generates payoffs (2, 2) that are not strongly efficient (strongly efficient payoffs sum to 5 in this example). The reason why, despite this inefficiency, s is invulnerable to invasion by a mutant is that the "punishment" for deviating from d is itself strongly efficient – i.e., play generates payoffs (1, 4) or (4, 1) – and so there is no way that a mutant can get a toehold against s.

Bibliographical note: Theorems 1 and 2 are established by Fudenberg and Maskin (1990) for the case in which r and p are *infinitesimal*. Fudenberg and Maskin (2007) treat the case of finite values of r and p.

References

Axelrod R (1984) The evolution of cooperation. Basic Books

Axelrod R, Hamilton W (1981) The evolution of cooperation. Science 211:1390–1396

Fudenberg D, Maskin E (1986) The folk theorem in repeated games with discounting or with incomplete information. Econometrica 54(3):533–554 (Reprinted in Rubinstein A (ed) Game theory in economics, London, Edward Elgar, 1995)

Fudenberg D, Maskin E (1990) Evolution and cooperation in noisy repeated games. Am Econ Rev 80:274–279

Fudenberg D, Maskin E (2007) Evolution and repeated games. Mimeo

Public Good Games with Incentives: The Role of Reputation

Hannelore De Silva and Karl Sigmund

Abstract Both the Trust Game and the Ultimatum Game reduce, in their most simplified versions, to a Public Good Game with an added incentive: namely a reward in the first case, and a sanction in the other. In this paper, the evolutionary game dynamics of these games is analyzed by means of the replicator equation. Positive and negative incentives have very different but complementary effects. We investigate the role of reputation, and show how occasional failures to contribute can lead to stabilizing cooperation.

1 A Philosophical Entente Cordiale

In *Leviathan* (1651), the English philosopher Thomas Hobbes described life in the absence of a central authority as "solitary, poore, nasty, brutish, and short." Selfish urges lead to "such a war as is every man against every man." The contemporary French philosopher Blaise Pascal held an equally dim view: "Nous naissons injustes; car chacun tend à soi... La pente vers soi est le commencement de tout désordre en guerre, en police, en économie etc." (We are born unfair; for everyone inclines towards himself... The tendency towards oneself is the origin of every disorder in war, polity, economy etc.)

In the following century, views on selfishness underwent a remarkable turnabout. The Scottish philosopher Adam Smith held that the selfish person works inadvertently for the public benefit. "By pursuing his own interest he frequently promotes that of the society more effectually than when he really intends to promote it." Greed promotes behavior beneficial to others. And most famously: "It is not from the benevolence of the butcher, the brewer, or the baker, that we expect our dinner, but from their regard to their own self-interest. We address ourselves, not to their humanity but to their self-love, and never talk to them of our own necessities but of their advantages."

H. De Silva (✉)
WU Vienna, Department Finance and Accounting, Heiligenstädter Strasse 46–48, 1190 Vienna, Austria

S.A. Levin (ed.), *Games, Groups, and the Global Good*, Springer Series in Game Theory,
DOI: 10.1007/978-3-540-85436-4_5, © Springer Physica-Verlag Berlin Heidelberg 2009

An intriguingly similar view had been expressed, well before Smith, by Voltaire in his *Lettres philosophiques* (also known as *Lettres anglaises*): "Il est bien vrai que Dieu aurait pu faire des créatures uniquement attentives au bien d'autrui. Dans ce cas, les marchands auraient été aux Indes par charité et le macon eut scié de la pierre pour faire plaisir à son prochain. Mais Dieu a établi les choses autrement.... C'est par nos besoins mutuels que nous sommes utiles au genre humain; c'est le fondement de tout commerce; c'est l'éternel lien des hommes." ("Assuredly, God could have created beings uniquely interested in the welfare of others. In that case, traders would have been to India by charity, and the mason would saw stones to please his neighbor. But God designed things otherwise... It is through our mutual needs that we are useful to the human species; this is the grounding of every trade; it is the eternal link between men.")

The French term "amour propre" certainly sounds a lot better than "self-love." Voltaire boldly claimed: "Il est aussi impossible qu'une société puisse se former et subsister sans amour propre, qu'il serait impossible de faire des enfants sans concupiscence, de songer à se nourrir sans appetit, etc. C'est l'amour de nous-même qui assiste l'amour des autres." ("It is as impossible that a society could emerge and subsist without self-love than that people could produce children without lust, feed themselves without appetite, etc. Love for oneself assists the love for others.")

It is unknown whether Hobbes, during his time in Paris, ever met Pascal; but Smith most certainly had associated with Voltaire.

2 Public Goods and Private Incentives

So much for philosophical views on selfishness. They vary. But economic models make it clear that self-interested individuals will not act to achieve their group interest, except when prodded by incentives directed selectively towards individuals in the group, i.e., punishing exploiters or rewarding contributors (Olson 1965; Hardin 1968; Henrich and Boyd 2001; Sigmund 2007). Self-love is not always beneficial: it needs help to escape from the traps of social dilemmas. In this chapter, we investigate the role of reputation to promote an "enlightened self-interest." The importance of reputation as a kind of second (non-monetary) currency is well-established in economics literature, of course. Here we present a treatment based on evolutionary game dynamics (Hofbauer and Sigmund 1998; Nowak 2006). If players simply imitate what is successful in the long run, with nothing but self-interest in their mind, populations can evolve towards economically beneficially behavior. We analyze a few basic models, starting with two scenarios which at first glance seem quite different, and which are well-known in behavioral game theory as trust game and ultimatum game (Kagel and Roth 1995; Camerer 2003; Camerer and Fehr 2006).

Both are one-shot, two-person games. In both, a coin toss first decides who of the two players is the proposer and who is the responder. The proposer is then endowed with a certain amount of money. In the trust game (Berg et al. 1995), the proposer can decide to donate part of this endowment to the responder, knowing that it will be multiplied by a factor $r > 1$ by the experimenter. The responder can then decide whether or not to return a part of this donation to the proposer. This concludes the trust game. The ultimatum game (Güth et al. 1982) does not take much time either. The endowment, in this case, is conditional. The proposer has to offer a percentage p of it to the responder, and if the responder accepts, the proposer keeps the rest; but if the responder declines, the experimenter withdraws the whole sum, so that both players gain nothing.

In the trust game, a purely selfish responder will never return anything, and a purely selfish proposer, anticipating this, should offer nothing. In the ultimatum game, a responder's self-interest will accept any positive sum, since it is better than nothing. Accordingly, the proposer should offer only a very small sum. In real experiments, the observed behavior differs considerably from these predictions of what a card-board "homo economicus" ought to do. Indeed, in the trust game, responders often return a large part of their gift, and in the ultimatum game, responders often reject offers which they deem too small (Camerer 2003; Henrich 2006). Accordingly, proposers in both types of games tend to transfer substantial proportions of their endowment, to both players' mutual benefit.

Both trust and ultimatum games are used to study norms of behavior, such as fairness and concern for one another. We shall study the evolutionary dynamics of simplified versions of these games, and then apply these results to address the issue of public goods with positive or negative incentives. Our main claim is that the concern for one's own reputation plays an essential role in causing us to deviate from what is prescribed for "homo economicus," and hence to turn to economically more profitable behavior.

3 The Mini-Trust Game

In a minimal variant of the trust game, we assume that the proposer has only to decide whether or not to donate a fixed amount c. Thus a proposer has the choice between two moves $\mathbf{e}_1$ (donate) and $\mathbf{e}_2$ (defect). A responder who receives a donation (i.e., the amount $b = rc$) has a choice between two moves, namely to return a certain amount β or not: these two moves will be denoted by $\mathbf{f}_1$ and $\mathbf{f}_2$. To make the game interesting, we will assume that $c < \beta < b$. In this case, if both players cooperate, both can make a gain. The payoff matrix is

$$\begin{array}{c|cc} & \mathbf{f}_1 & \mathbf{f}_2 \\ \hline \mathbf{e}_1 & (\beta - c, b - \beta) & (-c, b) \\ \mathbf{e}_2 & (0,0) & (0,0) \end{array} \qquad (1)$$

Since the players are with equal probability in the role of proposer and responder, they are involved in a symmetric game. There exist four strategies, namely (a) the "pro-social" strategy $G_1 = \mathbf{e}_1\mathbf{f}_1$ (donate, return); (b) the strategy $G_2 = \mathbf{e}_2\mathbf{f}_1$ (such a player does not donate, but returns a donation; (c) the asocial strategy $G_3 = \mathbf{e}_2\mathbf{f}_2$ (neither donate nor return); and finally (d) the strategy $G_4 = \mathbf{e}_1\mathbf{f}_2$ (such a player donates, but does not return). It is easy to compute the expected payoff values. But before doing this, we interpolate two brief sections on the replicator dynamics of two-role, two-strategy games (Hofbauer and Sigmund 1988; Sigmund et al. 2001), in order to make this chapter self-contained.

4 The Dynamics of Two-Role Games

Let us consider a game with two roles I and II, and with two strategies for each role, which we denote by $\mathbf{e}_i$ and $\mathbf{f}_j$. The payoff matrix is

$$\begin{array}{c|cc} & \mathbf{f}_1 & \mathbf{f}_2 \\ \hline \mathbf{e}_1 & (A,a) & (B,b) \\ \mathbf{e}_2 & (C,c) & (D,d) \end{array} \tag{2}$$

Let us assume that a coin toss decides which role to assign to which player. The strategies for the resulting symmetric game will be denoted by $G_1 = \mathbf{e}_1\mathbf{f}_1$, $G_2 = \mathbf{e}_2\mathbf{f}_1$, $G_3 = \mathbf{e}_2\mathbf{f}_2$ and $G_4 = \mathbf{e}_1\mathbf{f}_2$. The payoff for a player using G_i against a player using G_j is given, up to the factor 1/2 which we shall henceforth omit, by the (i, j)-entry of the matrix

$$M = \begin{pmatrix} A+a & A+c & B+c & B+a \\ C+a & C+c & D+c & D+a \\ C+b & C+d & D+d & D+b \\ A+b & A+d & B+d & B+b \end{pmatrix}. \tag{3}$$

Let us assume that players tend to imitate successful individuals, and hence occasionally switch from one strategy to another. They compare their average payoff with that of another player and adopt that player's strategy with a probability proportional to the payoff difference, if it is positive (if not, they do not switch). Since the payoffs depend on the state of the (well-mixed) population, given by the frequencies $x_i(t)$ of the strategies G_i, this yields an evolutionary dynamics in the state space $S_4 = \{(x_1, x_2, x_3, x_4) \in R^4_+ : x_1 + \ldots + x_4 = 1\}$. It is given by the replicator equation

$$\dot{x}_1 = x_i[(M\mathbf{x})_i - \bar{M}], \tag{4}$$

where $\bar{M} = x_1(M\mathbf{x})_1 + \ldots + x_4(M\mathbf{x})_4$ is the average payoff in the population. Since the dynamics are unaffected if one modifies the payoff matrix M by replacing m_{ij} by $m_{ij} - m_{1j}$, we can use the matrix

$$\begin{pmatrix} 0 & 0 & 0 & 0 \\ R & R & S & S \\ R+r & R+s & S+s & S+r \\ r & s & s & r \end{pmatrix}. \tag{5}$$

with $R := C - A$, $r := b - a$, $S := D - B$ and $s := d - c$.

5 Staying in the Saddle

We shall denote matrix (5) again by M. It has the property that

$$m_{1j} + m_{3j} = m_{2j} + m_{4j} \tag{6}$$

for $j = 1, 2, 3, 4$. Hence

$$(M\mathbf{x})_1 + (M\mathbf{x})_3 = (M\mathbf{x})_2 + (M\mathbf{x})_4 \tag{7}$$

holds for all $\mathbf{x}$. From this follows easily that the function $V = x_1x_3/x_2x_4$ satisfies

$$\dot{V} = V[(M\mathbf{x})_1 + (M\mathbf{x})_3 - (M\mathbf{x})_2 - (M\mathbf{x})_4] = 0 \tag{8}$$

in the interior of S_4, and hence the value of V remains unchanged along every orbit.

Hence the interior of the state simplex S_4 is foliated by the surfaces

$$W_K := \{\mathbf{x} \in S_4 : x_1x_3 = Kx_2x_4\}, \tag{9}$$

with $0 < K < \infty$. These are saddle-like surfaces which are spanned by the quadrangle of edges G_1G_2, G_2G_3, G_3G_4 and G_4G_1 joining the vertices of the simplex S_4 (see Fig. 1).

The orientation of the flow on the edges can easily be obtained from the previous matrix. For instance, if $R = 0$, then the edge G_1G_2 consists of fixed points. If $R > 0$, the flow along the edge points from G_1 towards G_2 (in the absence of the strategies G_3 and G_4, the strategy G_2 dominates G_1), and conversely, if $R < 0$, the flow points from G_2 to G_1.

Generically, the parameters R, S, r and s are nonzero. This corresponds to 16 orientations of the quadrangle $G_1G_2G_3G_4$, which by symmetry can be reduced to 4

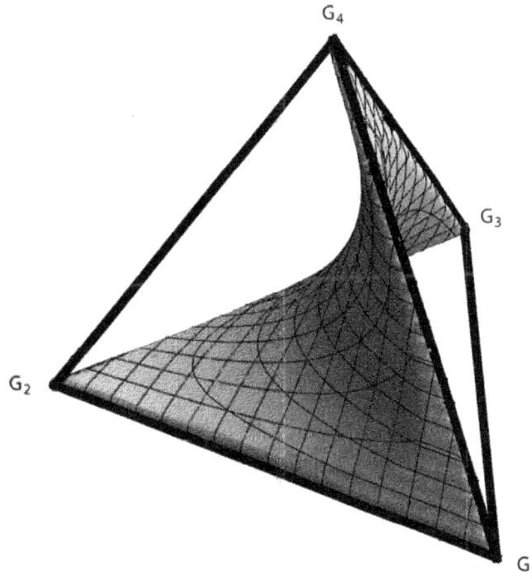

Fig. 1 The state space S_4 (a simplex with four corners $G_i, i = 1, 2, 3, 4$, corresponding to the four strategies of a symmetrized two-roles, two-strategies game), and a saddle-like surface W_K spanned by the edges $G_1 \to G_2 \to G_3 \to G_4 \to G_1$ (see text). The evolving states remain on their initial surface W_K. If there exist fixed points in the interior of the state space, they form a line intersecting each W_K

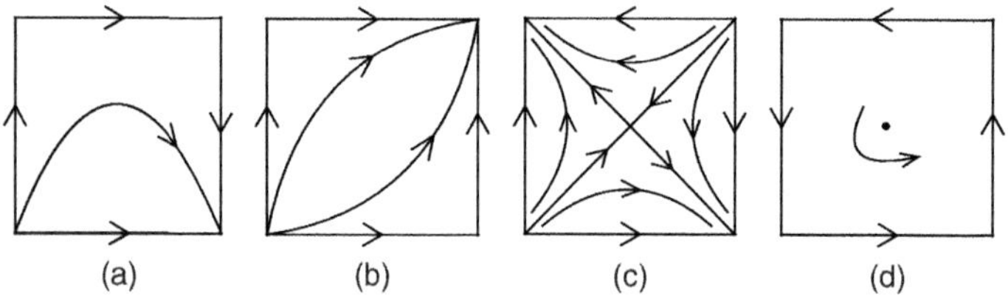

Fig. 2 The four generic orientations of the quadrangles spanning the saddle-like surfaces. The orientations depend on the signs of R, S, r and s (see text). In cases (c) and (d), there exists a fixed point in the interior of W_K

(see Fig. 2). Fixed points in the interior of the simplex S_4 must satisfy $(M\mathbf{x})_i = 0$ for $i = 2, 3, 4$ (since $(M\mathbf{x})_1$ trivially vanishes). This implies for $S \neq R$

$$x_1 + x_2 = \frac{S}{S - R}, \tag{10}$$

and for $s \neq r$

$$x_1 + x_4 = \frac{s}{s - r}. \tag{11}$$

Such solutions lie in the simplex if and only if $RS < 0$ and $rs < 0$, which corresponds to the orientations (c) and (d) of the quadrangle spanning the saddle-like surfaces W_K. If this is the case, one obtains a line of fixed points which intersects each W_K in exactly one point (see Fig. 1). The solutions can be written as

$$x_i = m_i + \xi \tag{12}$$

for $i = 1, 3$ and

$$x_i = m_i - \xi \tag{13}$$

for $i = 2, 4$, with ξ as parameter and

$$\mathbf{m} = \frac{1}{(S - R)(s - r)}(Ss, -Sr, +Rr, -Rs) \in W_1. \tag{14}$$

6 Farewell to Trust

For the corresponding payoff matrix, we obtain $R = c - \beta < 0$, $r = \beta > 0$, $S = c > 0$ and $s = 0$ (see Fig. 3). If $x_3 = x_4 = 0$, i.e., if everyone in the population is ready to return a donation, it is best to donate, i.e., G_1 dominates G_2. If $x_2 = x_3 = 0$, i.e., if donations can be taken for granted, then it is best not to return it, i.e., G_4 dominates G_1. If $x_1 = x_2 = 0$, i.e., if no one ever returns a donation, then G_3 dominates G_4, i.e., it is best not to donate. Finally, if $x_1 = x_4 = 0$, i.e., if nobody ever donates, then it does not matter whether one is willing to return a

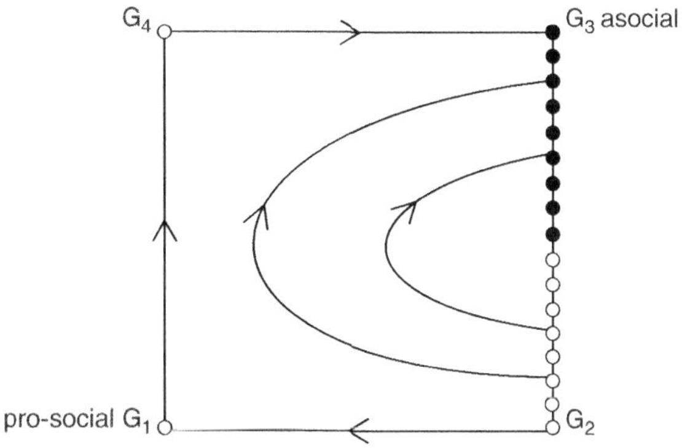

Fig. 3 Dynamics on a saddle-like surface for the trust game (or for a public good game with reward). The edge G_2G_3 consists of fixed points, the segment G_3Q of stable fixed points which are Nash equilibria

donation or not. In this case, the state of the population is a fixed point. Neither G_2 nor G_3 has an advantage.

It is easy to see that the segment QG_3, where

$$Q = \left(0, \frac{c}{\beta}, \frac{\beta - c}{\beta}, 0\right), \tag{15}$$

consists of saturated fixed points, i.e., of Nash equilibria. Indeed, for $x_1 = x_4 = 0$, both $(M\mathbf{x})_1$ (which is normalized to 0) and $(M\mathbf{x})_4$ are smaller than the average payoff $\bar{M} = (M\mathbf{x})_2 = (M\mathbf{x})_3 = c - \beta x_2$. The flow along the edges leads from G_2 to G_1, from there to G_4, and then to G_3. All orbits in the interior converge to the segment QG_3 for $t \to +\infty$ and to the segment QG_2 for $t \to -\infty$. Thus the population will, in the long run, consist only of players who, as proposers, never donate (and consequently, as responders, never return anything). From the economic viewpoint, the minimal version of the trust game does not take off: no donations, no paybacks.

7 Ultimate Offers

We now turn to the ultimatum game. It is simple enough, but we shall simplify it even further (cf. Nowak et al. 2000), and assume that the proposer has only a choice between offering a high percentage h (for instance, 45%) or a low percentage l (for instance 15%), with $0 < l < h < 1$. The responder could, in principle, accept both offers, one of them, or none. Again, we simplify by assuming that he has to choose between two strategies only: the strategy denoted by h, which consists in accepting the high offer only, or the strategy denoted by l, which consists in accepting both possible offers.

In this reduced version of the ultimatum game, the two strategies for role I, namely $\mathbf{e}_1$ and $\mathbf{e}_2$, are given by the offers h and l; and the two strategies $\mathbf{f}_1$ and $\mathbf{f}_2$ for role II will again denoted by h and l, for convenience; these strategies correspond now the responder's aspiration levels. The payoff matrix is given by

$$\begin{array}{c|cc} & \mathbf{f}_1 & \mathbf{f}_2 \\ \hline \mathbf{e}_1 & (1-h,h) & (1-h,h) \\ \mathbf{e}_2 & (0,0) & (1-l,l) \end{array} \tag{16}$$

The strategy G_1 corresponds to (h,h): high offers, and a high aspiration level. We may term it as the fair strategy. By contrast, $G_3 = (l,l)$ epitomizes the selfish strategy. It leads to the acceptance of any positive offer, and aims to part with as little as possible. The strategy $G_2 = (l,h)$ is paradoxical: it offers little, but insists on a high offer. G_4, finally, makes a good offer, but accepts a low offer. For want of a better term, we call it the mild strategy. The payoff parameters are $R = h - 1 < 0$, $r = 0$, $S = h - l > 0$ and $s = l > 0$. The selfish strategy dominates the mild strategy and the paradoxical strategy, which itself is dominated by the fair strategy; but the mild and the fair strategies are equivalent, in the absence of the other two strategies, one does as well as the other: all offers are fair, and the average payoff is 1/2.

There exist no fixed points in the interior of S_4. Indeed, whenever $x_2 > 0$ or $x_3 > 0$, we have $(M\mathbf{x})_4 > (M\mathbf{x})_1$ and hence both ratios x_4/x_1 and x_3/x_2 are increasing. On each surface W_K, the flow is as shown in Fig. 4. On the edge $x_2 = x_3 = 0$, all points are fixed points. If $x_1 < \frac{h-l}{1-l}$, then both $(M\mathbf{x})_2$ and $(M\mathbf{x})_3$ are larger than $\bar{M}$. Let us denote by $\mathbf{Q}$ the point $(\frac{h-l}{1-l}, 0, 0, \frac{1-h}{1-l})$. Then the symmetric Nash equilibria of the game are those on the segment G_1Q, and the vertex G_3. We note that on the edge $x_2 = x_4 = 0$, there exists another fixed point P, with coordinates $(h, 0, 1-h, 0)$. In a population with selfish and fair players only, we have a bistable competition. The fair strategy is risk-dominant (i.e., a population

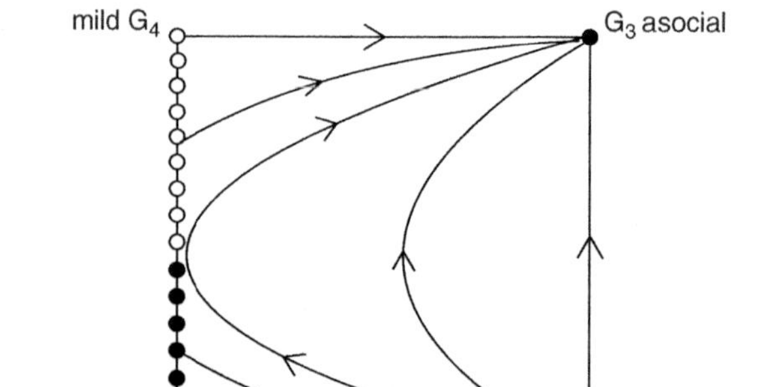

Fig. 4 Dynamics on a saddle-like surface for the ultimatum game (and public good game with punishment). The edge G_1G_4 consists of fixed points, the segment G_1Q of stable fixed points which are Nash equilibria

consisting in equal numbers of selfish and fair players will see fair players win) if $h < 1/2$.

The orbits in the interior of S_4 either converge to G_3, or else to the set of Nash equilibria, as shown in Fig. 4. If we assume that random shocks occasionally perturb the state of the population, we will expect that they lead to neutral drift along the edge $x_2 = x_3 = 0$. As soon as $x_1 < \frac{h-l}{1-l}$, a random perturbation sending the state into *int* S_4 will cause the fixation of G_3. This implies that eventually, the population consists of selfish players only. Thus evolutionary game theory leads to the same prediction as classical game theory; both are in contrast to experimental evidence.

8 Bifurcation Through Reputation

So far, we have considered conditions of strict anonymity. Let us now assume that with some (possibly small) probability, players may know their co-player by reputation, and in particular may know about the offers previously accepted by that co-player. Let us furthermore assume that occasionally, players offer less than they usually would, if they have reason to believe that they can get away with it; more precisely, if they know that their co-player has previously accepted low offers. The two assumptions seem reasonable enough: they only require some information about other players in the group, and a touch of opportunistic selfishness. In that case, accepting a low offer can have the regrettable consequence that one is offered less, in future games.

In order to analyze this situation, let us assume that $\mu > 0$ is the probability that a "fair" (h, h) proposer encountering a mild (h, l) responder knows that this player accepts a low offer, and consequently offers l instead of h. This yields the payoff matrix

	$\mathbf{f}_1$	$\mathbf{f}_2$
$\mathbf{e}_1$	$(1-h, h)$	$(1-h+\mu(h-l), h-\mu(h-l))$
$\mathbf{e}_2$	$(0, 0)$	$(1-l, l)$

(17)

which differs from (1) in one position only, by the term $\mu(h - l)$ which can be arbitrarily small. It can be viewed as a perturbation of the previous game, due to the effect of reputation. The corresponding symmetrized game (5) is now given by $R = h - 1$, $r = -\mu(h - l)$, $S = (h - l)(1 - \mu)$ and $s = l$. For $\mu < 1$, we have $R < 0$, $S > 0$, $s > 0$ (as before) and $r > 0$ (while we had $r = 0$ in the unperturbed case). This yields now a generic case, corresponding to case (c) in Fig. 2. There exists a line of fixed points in the interior of the state space S_4. Each of the surfaces W_K (for $K > 0$) intersects this line in a saddle point. For $\mu \to 0$, the point $\mathbf{m}$, and with it all interior fixed points, converge to the point Q on the edge G_1G_4. The dynamics on each surface W_K is bistable, the vertices $\mathbf{e}_1$ and $\mathbf{e}_3$ are the attractors (see Fig. 5). Hence, depending on the initial condition, the population will either converge to the fair or to the selfish strategy.

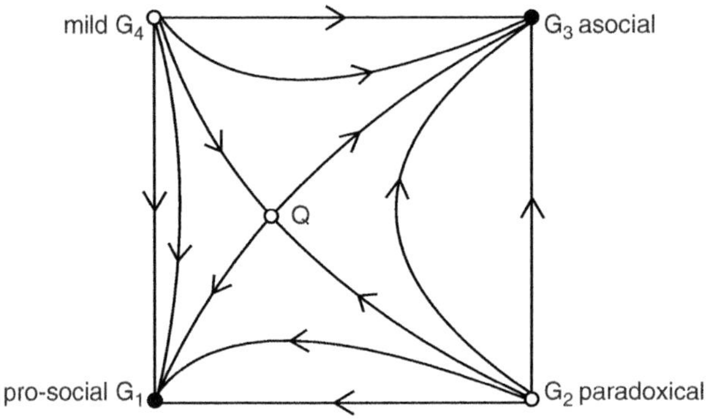

Fig. 5 Dynamics on a saddle-like surface for an ultimatum game with reputation (or for a public good game with reputation). The dynamics is bistable, the pro-social state G_1 and the asocial state G_3 are attractors

9 Public Goods with Punishment

In a simple form of the public goods game, each of the N players participating in the game has the possibility of contributing a fixed amount c to the common pool. The experimenter multiplies each player's contribution by a factor $r > 1$, and divides the resulting amount equally among all other $N - 1$ players participating in the game.

For $N = 2$, this is a prisoner's dilemma game: both players can decide whether or not to send a donation $b = rc$ to the other player, at a cost c for themselves. The dominant solution is to defect. But let us now introduce a second stage to this game, by allowing the players to punish defectors. We shall assume that the sanction consists in imposing a fine of size β. This fine is not collected by the punishing player. On the contrary, the punisher has to pay a fee, which costs him an amount γ. The first stage of the game offers scope for altruism (helping another player at a cost for oneself), and the second stage scope for spite (harming the other player at a cost for oneself). Obviously, in both stages, the dominating solution is to avoid the cost. A selfish player should defect in the first stage, and refuse to punish in the second stage.

If we assume that players can impose their fine conditionally, fining only those who have failed to help them, the long-term outcome will be, as before, that no pro-social behavior emerges (see Sigmund et al. 2001). Indeed, let us label with $\mathbf{e}_1$ those players who cooperate by sending a donation to their co-player, and with $\mathbf{e}_2$ those who do not, i.e., who defect; similarly, let $\mathbf{f}_1$ denote those who punish defectors, and $\mathbf{f}_2$ those who do not. The payoff matrix is given by

	$\mathbf{f}_1$	$\mathbf{f}_2$
$\mathbf{e}_1$	$(-c, b)$	$(-c, b)$
$\mathbf{e}_2$	$(-\beta, -\gamma)$	$(0, 0)$

(18)

Here, the first number in each entry is the payoff for the corresponding row player, and the second number for the column player. We have used the same notation as for two-role games, although the situation is completely symmetric: instead of being either in one role or in the other, a player is first in one role and then in the other. Despite this difference, we can apply the same method as before. Indeed, each strategy for this two-stage game must specify what to do in the first stage, and what to do in the second. Hence, it is given by a pair $\mathbf{e}_i\mathbf{f}_j$ (with $i, j \in \{1, 2\}$). As in section 3, we denote the resulting four strategies with $G_1 = \mathbf{e}_1\mathbf{f}_1$, $G_2 = \mathbf{e}_2\mathbf{f}_1$, $G_3 = \mathbf{e}_2\mathbf{f}_2$ and $G_4 = \mathbf{e}_1\mathbf{f}_2$. The strategy G_1 corresponds to the "pro-social" behavior: to give help, and to punish those who don't. G_3 is the selfish strategy which avoids any costs: a player using it does not help the co-player, and expects no help. G_2 can again be viewed as paradoxical: a G_2-player defects, but punishes a co-player who defects. Finally, G_4 can again be viewed as a "mild" strategy: a G_4 player sends a donation to the co-player but does not react if this is not reciprocated.

10 Dynamics with Reputation

We can follow the same approach as before, and obtain $R = c - \beta$, $S = c$, $r = 0$ and $s = \gamma$. Again, the manifolds $W_K = \{\mathbf{x} \in S_4 : x_1x_3 = Kx_2x_4$ are invariant (for $K > 0$) and the dynamics is as in Fig. 3. In fact, the ultimatum mini-game can be viewed as a special case, with $\gamma = l$, $\beta = 1 - l$, and $b = c = h - l$. Intuitively, this simply means that in the ultimatum game, the donation consists of making the high offer instead of the low offer. The benefit to the recipient (i.e., the responder) $h - l$ is equal to the cost to the donor (i.e., the proposer). The punishment consists of refusing the offer. This costs the responder l (the amount offered) and punishes the proposer by the amount $1 - l$, which is large if the offer is low.

The fixed points in W_K are the corners $\mathbf{G}_i$ and the points on the edge $\mathbf{G}_1\mathbf{G}_4$. $\mathbf{G}_3$ is a Nash equilibrium, $\mathbf{G}_2$ is not. A point $\mathbf{x}$ on the edge $\mathbf{G}_1\mathbf{G}_4$ is a Nash equilibrium whenever $x_1 \geq c/\beta$. Thus if $c > \beta$, $\mathbf{G}_3$ is the only Nash equilibrium. This case is of little interest. From now on, we restrict our attention to the case $c < \beta$: the fine costs more than the donation. We denote the point $(c/\beta, 0, 0, (\beta - c)/\beta)$ with $\mathbf{Q}$ and see that the closed segment $\mathbf{QG}_1$ consists of Nash equilibria. In the long run, in spite of the segment of Nash equilibria, random shocks will ultimately establish the asocial state $\mathbf{G}_3$.

Still following the parallel with the ultimatum game, let us assume that with a probability μ, cooperators (i.e., $\mathbf{e}_1$-players) defect against nonpunishers, i.e., $\mathbf{f}_2$-players. (Hence μ is the probability that (1) the $\mathbf{f}_2$-type becomes known and (2) the $\mathbf{e}_1$-type decides to defect). The payoff matrix becomes

	$\mathbf{f}_1$	$\mathbf{f}_2$
$\mathbf{e}_1$	$(-c, b)$	$(-c(1-\mu), b(1-\mu))$
$\mathbf{e}_2$	$(-\beta, -\gamma)$	$(0, 0)$

(19)

We obtain $R = (c - \beta) < 0$, $S = c(1 - \mu) > 0$, $s = \gamma > 0$ and $r = -b\mu < 0$. Thus the edge $\mathbf{G}_1\mathbf{G}_4$ consists no longer of fixed points, but of an orbit converging to $\mathbf{G}_1$. The dynamics is as in Fig. 5. On each saddle-like surface W_K, and therefore in the whole interior of the state space S_4, the dynamics is bistable, with attractors $\mathbf{G}_1$ and $\mathbf{G}_3$. Depending on the initial condition, every orbit converges to one of these two attractors, namely the asocial state $\mathbf{G}_3$ (no contributions, no punishment) and the pro-social regime $\mathbf{G}_1$ (cooperate, punish defectors).

11 Revealing Errors

The previous model is, in a certain sense, incomplete. Indeed, it essentially depends on altering the dynamics on the edge $\mathbf{G}_1\mathbf{G}_4$ by introducing the reputation effect. But on that edge, the population consists of two types only, both contributing to the public good. How should players learn whether the co-player is of type $\mathbf{f}_1$ or $\mathbf{f}_2$, i.e., willing to punish a defector, or not? Even if each player plays many rounds of the game, no defection ever arises.

There are several ways to deal with this question. One possibility would be to assume that players learn about their co-players' propensity to punish from other sources. It seems not unlikely that we can get a good idea about the irascibility or meekness of our co-players by watching their interactions with noisy children or their reactions to the daily news, rather than merely from observing how they act in the donation game. But it is probably better to complete the model without appealing to other interactions.

The simplest approach is to introduce the possibility of errors. Let us assume that player play the game repeatedly, and that players intending to donate will, with a certain probability ϵ, fail to implement their intention. (This could be due to a mistake, or to a lack of resources.) In the absence of reputation, this yields the following payoff structure:

	$\mathbf{f}_1$	$\mathbf{f}_2$
$\mathbf{e}_1$	$(-(1-\epsilon)c-\epsilon\beta,(1-\epsilon)b-\epsilon\gamma)$	$(-(1-\epsilon)c,(1-\epsilon)b)$
$\mathbf{e}_2$	$(-\beta,-\gamma)$	$(0,0)$

(20)

Compared with the situation in the previous section, s remains unchanged, whereas R and S are multiplied by $(1-\epsilon)$, which does not affect the sign, and hence conserves the dynamics on the corresponding edge. But r is now equal to $\epsilon\gamma$, and hence positive. This means that on the edge $\mathbf{G}_1\mathbf{G}_4$, the flow points towards $\mathbf{G}_4$: punishment is dominated. As a result, we obtain a dynamics as in case (b) of Fig. 2. All orbits in the interior of the simplex S_4 converge to the vertex $\mathbf{G}_3$. The asocial type wins.

Now let us introduce reputation. For simplicity, we will assume that players who know that their co-player is not of the punishing type never donate. (It would suffice to assume that they defect with a small probability). The parameter μ, then, is simply the probability to learn that the co-player has, once in the past, failed to punish a defector. If we assume perfect information, this reduces to the probability that the co-player has encountered a defection. On the edge $x_2 = x_3 = 0$, all players are willing to donate, and a defection occurs only by mistake. The probability that the co-player, in his k previous rounds, never faced a mistaken defection is $(1-\epsilon)^k$. If the number of rounds is distributed geometrically, with a constant probability $w<1$ for a further round, then $w^k(1-w)$ is the probability that the co-player has experienced k rounds. This means that

$$\mu = \frac{w\epsilon}{1-w(1-\epsilon)}. \tag{21}$$

If we assume that a player does not donate if he knows that he can get away with it (or if he commits an error), this yields

$$\begin{array}{c|cc} & \mathbf{f}_1 & \mathbf{f}_2 \\ \hline \mathbf{e}_1 & (-(1-\epsilon)c-\epsilon\beta, (1-\epsilon)b-\epsilon\gamma) & (-(1-\epsilon)(1-\mu)c, (1-\epsilon)(1-\mu)b) \\ \mathbf{e}_2 & (-\beta,-\gamma) & (0,0) \end{array} \tag{22}$$

We see that $r=\epsilon\gamma-\mu(1-\epsilon)b$ is negative if

$$\gamma < \frac{w(1-\epsilon)b}{1-w(1-\epsilon)}, \tag{23}$$

i.e., if the fee for punishing the defector is not too high.

Of course this can also be applied to the ultimatum game. In that case, $r=\epsilon\gamma-\mu(1-\epsilon)b$ is negative if

$$l < w(1-\epsilon)h, \tag{24}$$

i.e., if the low offer is sufficiently smaller than the high offer.

12 Public Goods with Rewards

Let us now consider a public good game (still with $N=2$ players only), but assume that the players have, in a second phase of the game, the option of rewarding contributors. Thus we consider a positive rather than a negative incentive. We shall assume that players who reward their donors have to pay a cost γ, and that the rewarded player receives an amount β (if $\beta=\gamma$ this is simply a payback). We assume $0<c<\beta$ and $0<\gamma<b$. If $\mathbf{e}_1$ and $\mathbf{e}_2$ are the two options for the first stage (to contribute or not), and $\mathbf{f}_1$ and $\mathbf{f}_2$ for the second stage (to reward donors or not), then the payoff structure is given by

$$\begin{array}{c|cc} & \mathbf{f}_1 & \mathbf{f}_2 \\ \hline \mathbf{e}_1 & (\beta-c, b-\gamma) & (-c,b) \\ \mathbf{e}_2 & (0,0) & (0,0) \end{array}. \tag{25}$$

The minimal variant of the trust game, introduced in Sect. 3, can be viewed as a special case (making the usual analogy between a two-role game and a two-stages game). There exist four strategies, namely (a) the "pro-social" strategy $G_1 = \mathbf{e}_1\mathbf{f}_1$ (donate, reward); (b) the strategy $G_2 = \mathbf{e}_2\mathbf{f}_1$ (such a player does not donate, but rewards a donor); (c) the asocial strategy $G_3 = \mathbf{e}_2\mathbf{f}_2$ (which neither donates nor rewards); and finally (d) the strategy $G_4 = \mathbf{e}_1\mathbf{f}_2$ (such a player donates, but does not reward). For the corresponding payoff matrix (5), we obtain $R = c - \beta < 0$, $r = \gamma > 0$, $S = c > 0$ and $s = 0$ (see Fig. 3). The outcome is exactly the same as for the trust game. Thus the population will, in the long run, consist only of players who always defect (and consequently never reward).

Let us now introduce reputation effects into the public goods game with rewards. We shall assume that with a small likelihood μ, cooperators defect if they know that the other player is not going to reward them, i.e., is of type $\mathbf{f}_2$. (μ is the probability that (1) the $\mathbf{f}_2$-type becomes known and (2) the $\mathbf{e}_1$-type decides to defect). Similarly, we denote by ν the likelihood that defectors cooperate if they know that they will be rewarded. (ν is the probability that (1) the $\mathbf{f}_1$-type becomes known and (2) the $\mathbf{e}_2$-type reacts and donates). This yields the payoff matrix

$$\begin{array}{c|cc} & \mathbf{f}_1 & \mathbf{f}_2 \\ \hline \mathbf{e}_1 & (\beta - c, b - \gamma) & (-c(1-\mu), b(1-\mu)) \\ \mathbf{e}_2 & ((\beta - c)\nu, (b - \gamma)\nu) & (0,0) \end{array} \tag{26}$$

Now $R = (c-\beta)(1-\nu) < 0$, $S = c(1-\mu) > 0$, $r = \gamma - b\mu$ which is positive if μ is small, and $s = (\gamma - b)\nu$, which is negative. It is this last condition that differs from the unperturbed system. The edge G_2G_3 no longer consists of fixed points. Instead, G_3 is dominated by G_2: if players can acquire a reputation for rewarding donations, this can motivate co-players to donate. The essential parameter, therefore, is ν.

Let us begin by assuming that μ is small, so that r is positive. For $\nu > 0$, the flow on the edge $\mathbf{G}_2\mathbf{G}_3$ leads towards $\mathbf{G}_3$, so that the frame spanning the saddle-like surfaces W_K is cyclically oriented (see Fig. 6). As before, there exists now a line of fixed points in the interior of S_4. On each saddle-like surface W_K, the orbits rotate around this fixed point; they spiral towards it for $0 < K < 1$ and away from it for $K > 1$. The surface W_1 consists of periodic orbits.

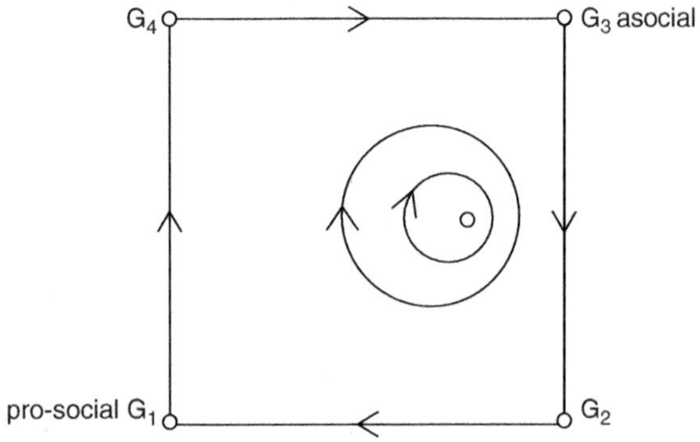

Fig. 6 Dynamics on a saddle-like surface for the trust game with reputation (μ small, $\nu > 0$). The edges are cyclically oriented. For W_1, the orbits are periodic. The orbits on W_K converge either to the inner fixed point or to the boundary, depending on whether $0 < K < 1$ or $K > 1$

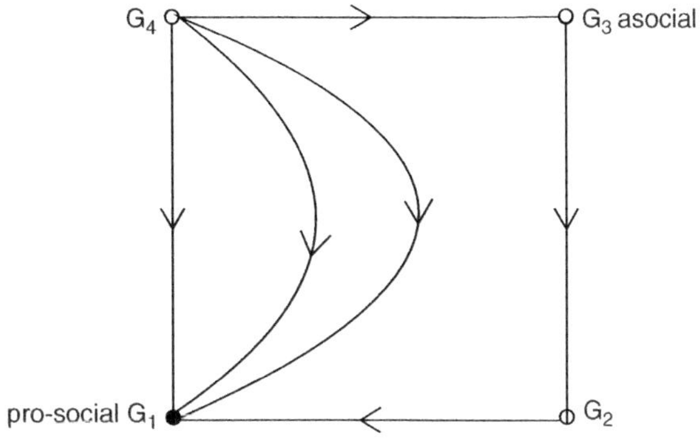

Fig. 7 Dynamics on the saddle-like surface for the trust game if $\mu > \beta/b$. In that case, all orbits converge to G_1

We stress the highly unpredictable dynamics if $\nu > 0$ and μ small. For one half of the initial conditions, the replicator dynamics sends the state towards the line of fixed points. But there, random fluctuations will eventually lead to the other half of the simplex, where the replicator dynamics leads to the heteroclinic cycle $\mathbf{G}_1\mathbf{G}_4\mathbf{G}_3\mathbf{G}_2$. The population seems glued for a long time to one strategy, then suddenly switches to the next, remains there for a still longer time etc. However, an arbitrarily small random shock will send the state back into the half-simplex where the state converges again to the line of fixed points, etc. Not even the time averages of the frequencies of strategies converge. One can only say that the most probable state of the population is either monomorphic (i.e., close to one corner of S_4) or else close to the attracting part of the line of fixed points (all four types present, the proportion of cooperators larger among rewarders than among nonrewarders, and – if the value ν is small – a frequency of rewarders close to c/β, and a frequency of donors which is small).

Let us note that we encounter the same problem as for the public good with punishment. If $x_1 = x_4 = 0$, then nobody ever donates. In this case, how should the $\mathbf{f}_1$-trait (rewarding donors) ever reveal itself? The assumption that occasionally players commit errors is far less plausible as in the previous case, since donating inadvertently is far less likely than failing in the intention to donate.

Finally, let us briefly consider the case when the fact that a player does not reward has a high probability to become publicly known. In that case, it is unlikely that such a player receives a donation. This means that μ is close to 1, and hence that the parameter $r = \gamma - b\mu$ is negative. In that case, all orbits in the saddle-like surface W_K converge to G_1 (see Fig. 7): the social strategy wins.

13 Larger Groups

So far we have only considered games with two players. Many economic interactions, and in particular many joint enterprises, involve more than two actors. In Sect. 9 we have introduced a so-called *others only* version of the public good game with N players. Each player's contribution is multiplied by $r > 1$ and divided equally among all $N - 1$ other players. In another version, we can assume that it is divided among all N players, so that if a player contributes c, then a part $\frac{r}{N}$ is returned to the donor. In the simplest case, when each contribution is of the same value c and if N_c players contribute, then the total amount cN_c is multiplied by $r > 1$ and divided equally among all N participants. A social dilemma holds if $r < N$. In alternative models, the total amount is a nonlinear function of the number of contributors.

Similarly, there are many ways of modeling punishment. In the simplest approach, each punisher pays a fee γ to inflict a fine β upon each defector. The resulting game dynamics is like that with two players (Hauert et al. 2004). If random shocks occasionally perturb the system, then in the long run, the asocial strategy (no contribution, no punishment) dominates the population. Again, the situation can be redressed if we assume that players can obtain information about the type of their co-players, and that contributors occasionally yield to the temptation of exploiting their co-players if they know that they can get away with it (i.e., that there are few or no punishers in their group).

With positive rewards, the situation is again similar to that of a two-person game, at least for a large set of parameter values.

14 Discussion

It is unlikely that one-shot interactions between anonymous players, such as the ultimatum game or the trust game, play a prominent role in human economy. In fact, their artificiality is an advantage for experiments. From early on, most experiments in physics or physiology are similarly based on artificial situations, such as a feather in a vacuum tube etc.

On the other hand, some everyday parallels to trust and ultimatum games exist. For instance, sellers who fix a (non-negotiable) price tag to an object displayed in their shopwindow are proposing an offer to the passersby, who can reject it or not. This has similarities with the ultimatum game. And individuals entrusting their banker with money are engaging in a transaction similar to a trust game. In everyday life, we often see that contributions to the public good are encouraged by heavily fining free-riders, etc (Henrich 2006; Ostrom and Walker 2003). On the other hand, there are essential differences between the games and the real-life parallels. For instance, many passersby will look into the shopwindow, whereas the ultimatum game has only one responder (if there are several, the outcome is drastically altered).

In each of these games, reputation can play an essential role in boosting the economically advantageous strategy (just as in indirect reciprocity, see (Nowak and Sigmund 2005; Wedekind and Milinski 2000). Reputation requires an information flow in the population. This information flow extends the knowledge obtained through the games that are personally experienced by a player, and usually relies on gossip. For instance, we have seen in section 12 that as soon as it is safe to assume that a funds manager who returns less than the investment becomes publicly known, the social strategy (for the clients, to invest, and for the managers, to return more than that investment to the clients) is a global attractor. Another example concerns internet trading, such as e-Bay. It relies heavily on the possibility that clients can rate their former partners. Another argument stems from psychology. If individuals feel unfairly treated, they often vent their emotions to a large audience (see e.g., Xiao and Houser 2005). Anger is loud. The logic behind this is clear: rejecting an unfair but positive offer involves costs, which can only be recouped if they prevent others from making unfair offers. If you take the trouble of getting emotional, you should make it known.

The importance of information has been displayed in a neat experiment based on two treatments of the ultimatum game (Fehr and Fischbacher 2003). Each player engages in several such games (always against a different partner, of course). In one treatment, players are anonymous. In the other treatment, players have pseudonyms and know that their decisions, as responder, will be made known to their future proposers. Aspiration levels are significantly higher in the second treatment. Players are more likely to reject offers. They seem to expect that if they once accept a low offer, they run a high risk of encountering such offers again and again. (See also http://homepage.univie.ac.at/hannelore.brandt/ultimatum/ for online computer simulations, cf. Figs. 8a and b).

Of course, even if players know perfectly well that their action is not observed, they often act as if it were. The lingering suspicion that despite double-blind

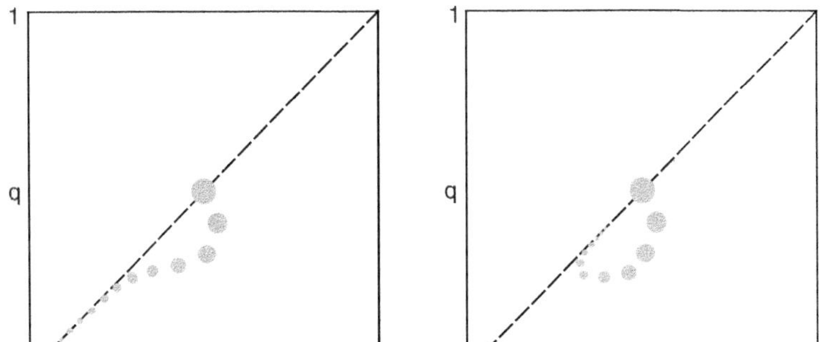

Fig. 8 Two variants of individual-based simulations on the ultimatum game. In both cases, 1,000 fictitious players with randomly chosen strategies (p, q) (where p is the size of the offer and q the aspiration level) each play 50 games against randomly chosen co-players. Then, the frequencies of the strategies are updated according to the replicator dynamics. This is repeated for many "generations." *Left*: players are anonymous. The population average of the (p, q)-values starts out close to the center (p and q close to 50%). After a drop in the p-value, the population average converges back to the diagonal and then inches along the diagonal towards $(0, 0)$. *Right*: players know the past of their co-players, and offer the minimum of their own p-value and the offers previously accepted by their co-player. The evolution begins similarly. But then, when the population average has returned to the diagonal, the p and q-values creep up, not down, and reach a value slightly below 50%

conditions etc. someone could be watching, is neatly captured in a series of experimental papers that show that the mere picture of an eye (on a poster, or on a computer screen) can activate a subconscious concern for the own reputation (Haley and Fessler 2005; Bateson et al. 2006; Burnham and Hare 2007).

But the emergence of pro-social behavior not only requires information, it also requires a certain amount of selfishness (or "self-love," to use a kinder but old-fashioned term). Without selfishness, incentives would not work. In the public good games with punishment, for instance, players must not only acquire knowledge about who is a punisher and who not, they must also be prone to defect if they know that they can get away with it. This is a finding well in the spirit of Voltaire's statement that "it is impossible that a society can emerge and subsist without self-love."

References

Bateson M, Nettle D, Roberts G (2006) Cues of being watched enhance cooperation in a real-world setting. Biol Lett 2:412–414

Berg J, Dickhaut J, McCabe K (1995) Trust, reciprocity, and social history. Games Econ Behav 10:122–142

Burnham T, Hare B (2007) Engineering cooperation: does involuntary neural activation increase public goods contributions? Hum Nat 18:88–108

Camerer C (2003) Behavioral game theory. Princeton University Press, Princeton

Camerer C, Fehr E (2006) When does "economic man" dominate social behaviour? Science 311:47–52

Fehr E, Fischbacher U (2003) The nature of human altruism. Nature 425:785–791

Güth W, Schmittberger R, Schwarze B (1982) An experimental analysis of ultimatum bargaining. J Econ Behav Organ 3:367–388

Haley K, Fessler D (2005) Nobody's watching? Subtle cues affect generosity in an anonymous economic game. Evol Hum Behav 26:245–256

Hardin G (1968) The tragedy of the commons. Science 162:1243–1248

Hauert C, Haiden N, Sigmund K (2004) The dynamics of public goods. Discrete and Continuous Dynamical Systems B 4:575–585

Henrich J (2006) Costly punishment across human societies. Science 312:176–177

Henrich J, Boyd R (2001) Why people punish defectors. J Theor Biol 208:79–89

Hofbauer J, Sigmund K (1998) Evolutionary Games and Population Dynamics. Cambridge University Press, Cambridge, UK

Kagel JH, Roth AE (eds) (1995) The handbook of experimental economics. Princeton University Press, Princeton

Nowak MA (2006) Evolutionary dynamics. Harvard University Press, Harvard

Nowak MA, Sigmund K (2005) Evolution of indirect reciprocity. Nature 437:1292–1298

Nowak MA, Page K, Sigmund K (2000) Fairness versus reason in the ultimatum game. Science 289:1773–1775

Olson M (1965) The logic of collective action. Harvard University Press, Harvard

Ostrom E, Walker J (2003) Trust and reciprocity: interdisciplinary lessons from experimental research. Russel Sage Funds, New York

Sigmund K (2007) Punish or perish? Retaliation and collaboration among humans. Trends Ecol Evol 22:593–600

Sigmund K, Hauert C, Nowak MA (2001) Reward and punishment. Proc Nat Acad Sci 98:10757–10762

Wedekind C, Milinski M (2000) Cooperation through image scoring in humans. Science 288:850–852

Xiao E, Houser D (2005) Emotion expression in human punishment behaviour. Proc Natl Acad Sci USA 102:7398–7401

Groups and Networks: Their Role in the Evolution of Cooperation

Brian Skyrms

Abstract We study networks rather than groups, consider network ties as probabilistic rather than deterministic, and treat networks as dynamic rather than static. A dynamic model of network formation introduced in Skyrms and Pemantle (2000) is used to analyze paths to cooperation in Stag Hunt games and some versions of Prisoner's Dilemma. Relative rates of evolution of structure and strategy are found to be crucial.

1 Groups and Networks

Man is a social animal, and social groups have been and are an important part of the picture of human-human interactions. There are the family, the neighborhood, the guild, the corporation, church and state. These group affiliations and connections do not, in general, partition society into mutually exclusive groups. In the first place, they are usually a matter of degree, rather than an all-or-none affair. You may interact with those in your own neighborhood more frequently than with those in a different one, or a different city, state or country, but people do travel. And the commercial interactions of a shopkeeper may differ in geographical scope from those of a traveling salesman. In the second place, ties are not necessarily transitive. A friend of a friend of a friend may be your friend or not.

So, instead of thinking about groups in the sense of units, a more general point of view regards them as social networks. And instead of thinking about social networks as consisting of all-or-nothing links between individuals, it is often more realistic to think of them as being probabilistic (Kirman 1997). The probabilities quantify the probability that one individual interacts with another. They are, formally, random graphs.

Where all probabilities are all equal to the reciprocal of the population size, we can recover the usual random interaction model of evolutionary game theory. Where

B. Skyrms
School of Social Sciences, University of California, 3151 Social Science Plaza, Irvine, CA 92697-5100, USA

S.A. Levin (ed.), *Games, Groups, and the Global Good*, Springer Series in Game Theory, DOI: 10.1007/978-3-540-85436-4_6, © Springer Physica-Verlag Berlin Heidelberg 2009

the probabilities are zero or one, we get deterministic links between individuals. One familiar deterministic model is the cellular automaton model where individuals interact with neighbors on a spatial lattice. With determinism and a network divided into cliques, we get models of hard and fast groups.

There are many different kinds of human interaction, and the network for one kind of interaction may be quite different from the network for another. These networks may themselves interact with one another. How they do so is a complicated theoretical question – it is not even clear what the appropriate general framework may be. Here I discuss a much simpler, but fundamental, question – how simple learning rules can cause small social networks to form and evolve, and how this may affect evolution of cooperation.

2 Partner Choice and Network Dynamics

A small group of individuals interact repeatedly, where the interactions have consequences for participants' well-being or utilities. Interactions could involve, for example, making friends, gift giving, participating in a group enterprise such a cooperative hunting, bargaining how to split a payoff, trade, gossip, division of labor. In an interaction, individuals actualize some behavior and the behavior of the individuals jointly determines the outcome of the interaction. At a high level of abstraction, we can model interactions as games. The relevant behaviors of individuals in the interaction are the strategies of the game, and the strategies of the players jointly determine their payoffs.

Since the interactions are repeated, players can modify their behavior in the light of experience – they can learn. Agents can learn two things: *with whom to interact* and *how to act.* That is to say that adaptive learning dynamics operates both on network structure and on strategy. Evolutionary game-theory models usually fix the network structure and concentrate on studying how strategies evolve. Most often the network structure is degenerate, simply assuming random encounters in a large population. Sometimes, but less often, it is assumed that individuals interact with neighbors on a circle, torus, or other spatial structure.

In Skyrms and Pemantle (2000) we first reverse the usual bias and investigate the evolution of network structure when strategies are held fixed. Then we move to the co-evolution of structure and strategy where the respective learning dynamics may be of different kinds and may proceed at different rates. Relative rates of evolution are seen to make all the difference for the evolution of cooperation in the Stag Hunt game, with a fast network dynamics favoring cooperation.

The basic model admits of many realizations, depending on the interactions modeled and the forms of learning used for modification of network structure and of strategy, and – as I have just said – the relative rates of the structure and strategy dynamics. Some of these variations are explored in Bonacich and Liggett (2003), Huttegger and Skyrms (forthcoming), Pemantle and Skyrms (2001, 2004a, b), Skyrms (2004, 2007, 2008); Skyrms and Pemantle (2000), but many others remain to be investigated. It is important in interpreting these models to pay attention not only to long-run limiting results, but also to medium-run transient behavior. In some cases limiting results are approached very quickly, but in other ones medium-run transient behavior can look very different from limiting behavior even after millions of interactions (Pemantle and Skyrms 2001; 2004a, b).

3 Making Friends, Stag Hunt

We start by considering network dynamics by simple reinforcement learning of the kind found in Erev and Roth (1998), Herrnstein (1970), Roth and Erev (1995). In the basic model, individuals start with initial propensities to choose among various options, choose with probability proportional to the propensities, and update the propensity for the option chosen by the payoff (reinforcement) received.

Now we assume that each day each of our individuals wakes up and decides to visit someone else according to some initial propensities for whom to visit. It will be assumed that the population is small enough so that there is plenty of time in the day for all choices to be satisfied. It is also assumed that all visits are received. If everyone else decides to visit Samantha, everyone else gets to visit Samantha. (Obviously each of these assumptions could be modified, and modification might well be appropriate in certain situations.) At the end of the day each individual updates her weights for visiting an individual by adding the payoffs gotten from that day from interactions with that individual (both as visitor and host).

As baseline models, consider two models of "Making Friends" (Skyrms and Pemantle 2000). In Friends I the visitor is treated well, for a payoff of 1, while the host gets a payoff of 0. In Friends II both visitor and host enjoy themselves equally, and each has a payoff of 1. These may be viewed degenerate games, where the visitor and host only have one available strategy, with payoff matrices:

Friends I	Host
Visitor	1,0

Friends II	Host
Visitor	1,1

(The case where only the host gets a positive payoff is put forward as a model of gift-giving in Bonacich and Liggett (2003) and analyzed using a different kind of reinforcement learning (Bush and Mosteller 1955)).

We begin investigation of the network formation for these two interactions by starting ten individuals with equal initial weights of 1 for visiting each other individual. The initial network structure is one of random encounters. In this setting, it is easy to run computer simulations of the Friends I and Friends II processes, and it is a striking feature of such simulations that in both cases non-random interaction structure rapidly emerges. Furthermore, rerunning the processes from the same starting point seems to generate different structure each time. We see the emergence of structure without an organizer, or even an explanation in terms of payoff differences. The state of uniform random encounters with which we started the system does not persist, and so must count as a very artificial state. Its use as the fixed interaction structure in many game theoretic models should be suspect.

We can understand the behavior of the Friends I process if we notice that each individual's learning process is equivalent to a Polya urn. We can think of him as having an urn with balls of different colors, one color for each other individual. Initially there is one ball of each color. A ball is chosen (and returned), the designated individual is visited. Because visitors are always reinforced, another ball of the same color is added to the urn. Because only visitors are reinforced, balls are not added to the urn in any other way. The Polya urn converges to a limit with probability one, but it is a random limit with uniform distribution over possible final probabilities. Anything can happen, and nothing is favored! *In Friends I the random limit is uniform for each player, and makes the players independent.* (Theorem 1 of Skyrms and Pemantle (2000)) All interaction structures are possible in the limit, and the probability that the group converges to random encounters is zero.

In Friends II, both visitor and host are reinforced and so the urns interact. If someone visits you, you are reinforced to visit him – or to put it graphically, someone can walk up to your door and put a ball of his color in your urn. This complicates the analysis considerably. Nevertheless, the final picture is quite similar. *In Friends II the limiting probabilities must be symmetric, that is to say X visits Y with the same probability that Y visits X, but subject to this constraint and its consequences anything can happen.* (Theorem 2 of Skyrms and Pemantle (2000))

The Making Friends games provide building blocks for analyzing learning dynamics where the interactions are games with non-trivial strategies. Consider the two-person Stag Hunt. Individuals are either Stag Hunters or Hare Hunters. If a Stag Hunter interacts with a Hare Hunter no Stag is caught and the Stag Hunter gets zero payoff. If a Stag Hunter interacts with another Stag Hunter the Stag is likely caught and the hunters each get a payoff of one. Hare Hunting requires no cooperation, and its practitioners get a payoff of 0.75 in any case. It makes no difference who is visitor or who is host.

The Stag Hunt game is of special interest for social theory. Stag Hunting is both mutually beneficial and an equilibrium, but it is risky. Deciding to hunt Stag requires a measure of *trust* that the other player will cooperate, Skyrms (2004, 2007). In game theoretic terminology, Stag hunting is payoff dominant and Hare hunting is risk dominant. In a large population composed of half Stag Hunters and half Hare Hunters with random interactions between individuals, the Hare Hunters would get an average payoff of 0.75 while the Stag Hunters would only get an average payoff

Table 1 Stag hunt

	Stag	Hare
Stag	1	0
Hare	0.75	0.75

of 0.50. The conventional wisdom is that in the long run evolution will strongly favor Hare hunting.

But suppose that the players *learn to network*. We use exactly the same model as before, except that the payoffs are now determined by the individuals' strategies: Hunt Stag or Hunt Hare. We start with an even number of Stag Hunters and Hare Hunters. *In the limit, Stag Hunters always visit Stag Hunters and Hare Hunters always visit Hare Hunters* (Theorem 6 of Skyrms and Pemantle (2000)). Simulation confirms that such a state is approached rapidly. Although on rational choice grounds Hare Hunters "should not care" whom they visit, they cease to be reinforced by visits from Stag Hunters after Stag Hunters learn not to visit them. Hare Hunters continue to be visited by other Hare Hunters, so all the differential learning for Hare Hunters takes place when they are hosts rather than visitors. Once learning has sorted out Stag Hunters and Hare Hunters so that each group only interacts with its own members, each is playing Friends II with itself and previous results characterize within-group interaction structure.

Once Stag Hunters have learned to network, they prosper. The disadvantage that they experienced in the random encounter situation has been overcome, and now their payoff is superior to that of non-cooperative Hare Hunters. Is it not obvious that Stag Hunters will seek each other out? If they are more sophisticated than simple reinforcement learners – if they think strategically and optimize – and if they can choose to associate with other Stag Hunters, then they will certainly do so. That is to say that we are not dealing with an artifact of our choice of learning rule, but rather that we find a positive result in a worst-case scenario.

4 Co-evolution of Structure and Strategy

So far strategies have been fixed. Individuals were either Stag Hunters or Hare Hunters, and could not change their type. Now we investigate the co-evolution of network structure and strategies. There are now two interacting dynamic processes involved. The strategy revision dynamics might be qualitatively the same as the partner choice dynamics, or it might be qualitatively different. They might operate at different time rates, with one being fast and the other slow. The population might be heterogeneous with respect to the operative learning dynamics. Let us consider a few examples.

To the two-person Stag Hunt we add a strategy revision process based on imitation to get a *reinforcement-imitation* model of co-evolution (Skyrms and Pemantle (2000)). With some specified probability, an individual wakes up, looks around the whole group, and if some strategy is prospering more than his own, switches to it. Individual's probabilities are independent. If imitation is fast relative to structure dynamics, it operates while individuals interact more or less at random and Hare Hunters will take over more often than not. If imitation is slow, stag hunters find each other and prosper, and then imitation slowly converts Hare Hunters to Stag Hunters (who quickly learn to interact with other Stag Hunters).

Simulations show that in intermediate cases, timing can make all the difference. We start with structure weights equal to 1 and vary the relative rates of the dynamics by varying the imitation probability. With "fast" imitation (pr = .1) 78% of the trials ended up with everyone converted to Hare Hunting and 22% ended up with everyone converted to Stag Hunting. Slower imitation (pr = .01) almost reversed the numbers, with 71% of the trials ending up All Stag and 29% ending up All Hare. Fluid network structure coupled with slow strategy revision reverses the orthodox prediction that Hare Hunting (the risk dominant equilibrium) will take over. *Free association favors cooperation.*

The foregoing model illustrates the combined action of two different dynamics, reinforcement learning for interaction structure and imitation for strategy revision. What happens if both processes are driven by reinforcement learning? In particular, we would like to know whether the relative rates of structure and strategy dynamics still make the same difference between Stag Hunting and Hare Hunting. In this *Double Reinforcement* model (Skyrms 2004; Skyrms 2008), each individual has two weight vectors, one for interaction propensities and one for propensities to either Hunt Stag or Hunt Hare. Probabilities for whom to visit, and what to do, are both gotten by normalizing the appropriate weights. Weights are updated by adding the payoff from an interaction to both the weight for the individual involved and to the weight for the action taken. Relative rates of the two learning processes can be manipulated by changing the magnitude of the initial weights.

In the previous models we started the population off with some Stag Hunters and some Hare Hunters. That point of view is no longer correct. The only way one could be deterministically a Stag Hunter would be if he started out with zero weight for Hare Hunting, and then he could never learn to hunt Stag. We have to start out individuals with varying propensities to hunt Hare and Stag. There are various interesting choices that might be made here; we will report some simulation results for one. We start with a heterogeneous group of 10: 2 confirmed Stag Hunters (weight 100 for Stag, 1 for Hare), 2 confirmed Hare Hunters (weight 100 for Hare, 1 for Stag), and 6 undecided guys (weights 1 for Stag and 1 for Hare). Initial weights for interaction structure were all equal, but their magnitude was varied from .001 to 10, in order to vary the relative rates of learning structure and strategy. The percent of 10,000 trials that ended up All Stag or All Hare (after 1,000,000 iterations) for these various settings are shown in Fig. 1 (Skyrms 2004).

As before, fluid interaction structure and slow strategy adaptation favor Stag Hunting, while the reverse combination favors Hare Hunting. *Free association favors cooperation.*

However, a third possible strategy-revision dynamics – the Cournot dynamics – gives different results. Suppose that agents change their strategies by choosing the one which would have given the greatest payoff against the opponents plays in the previous round. Suppose, as before, that network structure adapts quickly, and that players are sorted into Stag Hunters interacting with Stag Hunters and Hare Hunters interacting with Hare Hunters. Then the Cournot dynamics simply confirms everyone in their strategy choice, since Stag Hunters do better against Stag Hunters and Hare Hunters do better against Hare Hunters. We now have two non-interacting

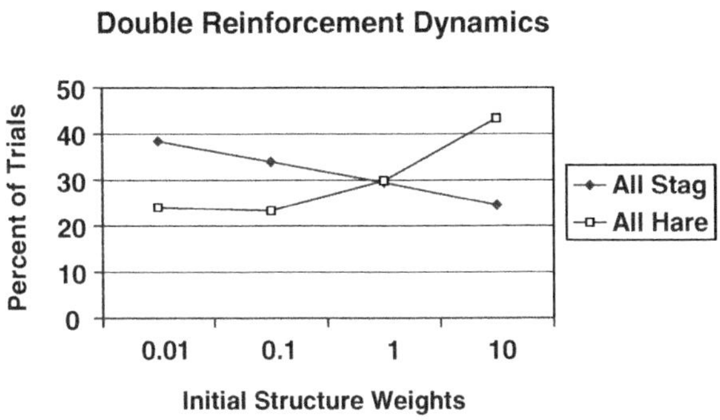

Fig. 1 Stag hunt with reinforcement dynamics for both strategy and structure

Table 2 Prisoner's dilemma

	Cooperate	Defect
Cooperate	2	0
Defect	3	1

social classes. Stag Hunters cooperate and prosper, while myopic optimization locks Hare Hunters into an inefficient equilibrium.

5 Prisoner's Dilemma

Let us see to what extent the picture changes when we vary the game. Consider the Prisoner's Dilemma.

Suppose we fix the strategies and let the network structure dynamics operate exactly as before. Then Cooperators rapidly learn to visit Cooperators, but Defectors now profit by visiting cooperators and also learn to do so (Skyrms 2004, Chap. 7). Rapid network dynamics leads everyone to visit cooperators. At the end of the day, this is not so bad for cooperation. Cooperators get a lot of visits, both from other cooperators and from defectors. The defectors' visits don't hurt or help; they carry a payoff of 0 for the cooperating partner. But cooperators always visit another cooperator, for a payoff of 2 and are visited by other cooperators for an average payoff of 2. Thus, at the end of the day cooperators have an average payoff of 4 (2 as visitor and 2 as host). Defectors are never visited and so they only have an average payoff of 3. Free association still favors cooperation.

The foregoing depends on the feature of our model that allows an agent to host as many visits as there are potential visitors. We are led to ask what would happen if we restrict our agents to hosting one visit in a day, so that an agent experiences at two interactions – one as host and one as visitor. Then, if several individuals want to visit someone, there must be some way to decide who gets to do so. It seems plausible that the potential host should choose, using the same reinforcements that he uses in choosing someone to visit.

So we now suppose that at the beginning of the day everyone chooses someone to visit according to her accumulated reinforcements as before. Then everyone who has been chosen by more that one potential visitor chooses whom to host according to the accumulated reinforcements from potential visitors. At this point, some may be left with no one to visit and some may be left with no one to host. The process is repeated until each individual gets to visit exactly one other on that day.

Now cooperators learn rapidly not to visit defectors, and when given the choice prefer to host another cooperator rather than a defector. As defectors tend usually meet other defectors (albeit as second or third choice) reinforcements from meeting other defectors build up, and the involuntary association becomes, by force of habit, voluntary.

Fast network dynamics with choice by both visitor and host now leads cooperators to associate with cooperators and defectors to associate with defectors, just as in the Stag Hunt. Cooperators prosper.

6 An Experiment

Page, Putterman and Unel (Page et al. 2005) report the results of an experiment on voluntary association in a public goods provision game. The game in question is a kind of N-person Prisoner's Dilemma. A group of four subjects are given an initial endowment of $10.00. Each decides how much to keep for herself and how much to contribute to the public good. The total contributions to the public good account are multiplied by 1.6, and then distributed to the subjects. It is to the mutual benefit of the group if everyone contributes everything, multiplying the initial endowment by 1.6. But from a selfish point of view, for every dollar I contribute I only get back $0.40. The rest of my contributions go to helping others, just as their contributions help me. As in the Prisoner's Dilemma, myopic individual self-interest leads the inferior outcome.

In a baseline treatment, individuals are assigned at random to four groups of four individuals each, and each group plays the game for 20 periods. In the regrouping treatment, every third period individuals are regrouped. They are shown other subjects average contributions and are allowed to rank others in order of preference as a partner in a new group. (Expressing preference carries a small cost.) A computer that reconstitutes groups, starting by putting the most preferred individuals in group one and working downward.

In the treatment where voluntary association was allowed, group one contributed the heavily to the public good, group two somewhat less, group three even less, and group four the least. The authors conclude that individuals had effectively sorted themselves into groups of different type. Not only did the high cooperators, group one, do better that the baseline treatment where individuals were locked into their groups. The overall average of all groups, including the lowest, in the voluntary association treatment was much higher than the baseline.

Although the setup for the experiment is a little different from the dynamic network models that I have presented, it may count as evidence for the robustness of the positive effects of free association.

7 Beyond Reinforcement

The network dynamics operating in the experiment just discussed was not driven by reinforcement. Players used information about average contributions of individuals in other groups, individuals with whom they had not yet interacted, in order to construct their preference rankings. I have concentrated on network formation by reinforcement learning so far because it is remarkable what such simple trial and error learning can achieve. (See also Argiento et al. (2009) for a proof that reinforcement leads to successful signaling in a simple signaling game.)

At some point, reinforcement learning will fall short of delivering an optimal results. Here is an example, due to Bala and Goyal (2000). Everyone in the group has different information. Information is valuable. Take someone out for a drink and he will tell you everything he knows (including what others have told him). Information is transmitted with complete fidelity. A drink costs something, but not as much as the value of one piece of information. (We abstract away timing problems; in one round everything happens instantaneously. That is to say, if A takes B out and B takes C out, A gets the information of both B and C.) The efficient equilibrium network is a circle. Everyone gets everyone else's information at the cost of only one drink.

But reinforcement learners have trouble learning the circle (Huttegger and Skyrms (2008)). What is required is some form of cognitive learning in which individuals form beliefs, perhaps probabilistic beliefs, about what others will do and choose a best response to those beliefs. Bala and Goyal (2000) show that an asynchronous version of best response to previous play, where individuals update at random times, learns the circle. Huttegger and Skyrms (2008) show that fictitious play – essentially inductive logic plus best response – learns the circle in this game. Although these are cognitive dynamics, involving beliefs rather than just reinforcement, they are based on remarkably simple cognitive rules. The basic theme of simple learning leading to efficient network structure is preserved.

References

Argiento R, Pemantle R, Skyrms B, Volkov S (2009) Learning to signal: analysis of a micro-level reinforcement model. Stochastic Processes Appl 119:373–390

Bala V, Goyal S (2000) A non-cooperative model of network formation. Econometrica 68: 1181–1229

Bonacich P, Liggett T (2003) Asymptotics of a matrix-valued markov chain arising in sociology. Stochastic Processes Appl 104:155–171

Bush R, Mosteller F (1955) Stochastic models of learning. Wiley, New York

Erev I, Roth A (1998) Predicting how people play games: reinforcement learning in experimental games with unique mixed strategy equilibria. Am Econ Rev 88:848–881

Herrnstein RJ (1970) On the law of effect. J Exp Anal Behav 14:243–266

Huttegger S, Skyrms B (2008) Emergence of information transfer by inductive learning Studia Logica 89:237–256

Jackson M, Watts A (2002) On the formation of interaction networks in social coordination games Games Econ Behav 41:265–291

Kirman A (1997) The Economy as an Evolving Network. J Evol Econ 7:339–353

Page T, Putterman L, Unel B (2005) Voluntary association in public good experiments: reciprocity, mimicry and efficiency. Econ J 115:1032–1053

Pemantle R, Skyrms B (2001) Reinforcement schemes may take a long time to exhibit limiting behavior Preprint

Pemantle R, Skyrms B (2004a) Network formation by reinforcement learning: the long and the medium run. Math Soc Sci 48:315–327

Pemantle R, Skyrms B (2004b) Time to absorption in discounted reinforcement models. Stochastic Processes Appl 109:1–12

Roth, A, Erev I (1995) Learning in extensive form games: experimental models and simple dynamic models in the intermediate term. Games Econ Behav 8:14–212

Skyrms B (2004) The stag hunt and the evolution of social structure. Cambridge, New York

Skyrms B (2007) Dynamic networks and the stag hunt: some robustness considerations. Biol Theory 2:1–3

Skyrms B (2008) Trust, risk and the social contract. Synthese 160:21–25

Skyrms B, Pemantle R (2000) A dynamic model of social network formation. Proc Natl Acad Sci USA 97:9340–9346

Skyrms B, Pemantle R (forthcoming) Learning to network In Eells E, Fetzer J (eds) Probability in Science, Open Court Publishing

Suppes P, Atkinson R (1960) Markov Learning Models for Multiperson Interactions. Stanford University Press, Palo Alto

Part II
Cooperation and Group Formation

Evolution and Construction of Moral Systems

Jessica C. Flack and David C. Krakauer

Abstract A moral system is an adaptive system for conflict management based on prescriptive, internalized social rules. We decompose moral systems into the sense of fairness, moral judgments, and rules at the aggregate level. We explore how each of these levels is constructed, including how this process is influenced by cognitive and organizational constraints and social architecture. We consider feedback across these levels as well as the implications of partial time-scale separation for reducing uncertainty about behavioral outcomes. We suggest that an appropriate theoretical framework for treating these issues is an extended theory of niche construction.

1 Introduction

Research on the evolution of morality has been dominated by three questions, which can be described as genealogical or historical, mechanistic, and functional. The first genealogical question is whether morality is a uniquely human invention or has rudimentary expression in other animals (de Waal 1996, 2006; Boehm 2000; Flack and de Waal 2000a, b). The second, mechanistic question, concerns the material basis of moral intuition (Flack and de Waal 2000; Haidt 2001; Green and Haidt 2002; de Waal 2006; Hauser 2006; Hauser et al. 2007; Koenigs et al. 2007) and goes back to Hume (Hume 1969 [1739]) and Kant (Kant 1959 [1785]). The goal of this research is to determine through behavioral (e.g., Haidt et al. 1994; Henrich 2004; Hauser et al. 2007) and neurobiological experiments (e.g., Semendeferi et al. 1997; Anderson et al. 1999; Castelli et al. 2000; Green et al. 2001; Koenigs et al. 2007), the nature of the affective-cognitive computational process generating a sense of fairness. A related line of inquiry combining both genealogical and mechanistic features, asks whether moral rules are culturally specific, or have common features across cultures (e.g., O'Neill and Petrinovich 1998; Henrich 2004), due either to

J.C. Flack (✉)
Sante Fe Institute, 1399 Hyde Park Road, Pod B-6, Santa Fe, New Mexico 87501, USA

S.A. Levin (ed.), *Games, Groups, and the Global Good*, Springer Series in Game Theory,
DOI: 10.1007/978-3-540-85436-4_7, © Springer Physica-Verlag Berlin Heidelberg 2009

cultural descent or assimilation, or to constraints imposed (and preferences generated) by underlying biological capacities shared by all humans. The third, functional question, seeks to determine the cost and benefit structure supporting the evolution of a concept of right and wrong and its relation to the evolution of cooperation (Campbell 1975; Alexander 1987; Nowak and Sigmund 1998; Boehm 2000; Nowak et al. 2000; Page and Nowak 2002; Binmore 2005; Bowles and Gintis In Press); see also (Frank 2003; Frank In Press). Missing from this literature is a "developmental" question, asking how moral systems are constructed – what dynamical processes and principles lead to the emergence of a set of interacting, prescriptive social rules at the aggregate level. The integration of the genealogical and construction questions seeks to establish how aspects of the construction processes of moral systems evolve in response to changing social dilemmas. This approach to the evolution of moral systems has a natural analogue in organic evolution: lineages of organisms evolve developmental programs and learning rules (both aspects of the construction process) that produce mechanisms allowing organisms to meet critical functional requirements. The construction component connects genealogy to mechanism and function.

So, for example, it is not yet understood how moral rules at the societal level are constructed from, and subsequently feedback to influence, individual moral judgments and the sense of fairness. The problem is difficult because of the interaction of four factors. Within-individual affective and cognitive mechanisms bias the construction process, thereby delimiting the accessible rule set. To produce aggregate level rules, individuals engaging in construction need to ensure some coordination of their perceptions of right and wrong, yet perception is noisy and competing interests and power asymmetries can undermine alignment efforts. Promotion and stabilization of the aggregate level rules requires investing in behavioral mechanisms for broadcasting and enforcement, creating the potential for a tragedy of the commons. And the rules generated must not violate stability conditions – properties that need be in evidence in order to ensure that a society persists – societal robustness requirements.

One of the consequences of neglecting an explicit constructive dynamic is that interactions between levels characterized by different rates of change are ignored. For example, many game-theoretic models of altruism, and other phenomena related to morality, tend to focus on causality operating at the individual level and not the slower change of institutions to which individuals contribute[1]. Correspondingly, many of empirical schemas for the production of moral judgments posit linear (although not necessarily single-layered) causality, almost entirely emphasizing within-individual processes (e.g., perception generates affect, which leads to a moral judgment) (see, for examples, the review in Hauser (2006)). These frameworks effectively ignore moral systems.

[1] Group selection/multi-level extensions, modeled using the price equation for example, do take into account contributions from different levels, but typically neglect the dynamics generating each level.

Consideration of aspects of the construction process should allow us to better understand how the production of moral judgments is shaped by feedback from societal level rules, and the feed-forward contribution from cognitive-emotional computational processes within individuals. In analogy with biological evolution again, the nascent field of evodevo (evolution of development) suggests that feedback from the phenotype to the genotype critically influences the pattern of gene expression, and hence the form of morphology that arises. It is not correct to treat the phenotype as a linear projection from the genotype or a simple average over genotypic contributions. In the same way, it is rather naïve to assume a moral system is a simple aggregation of individual judgments.

In the simplest moral systems observed in animal societies, such as those of great apes, the rules are implicit (in so far as they are neither written nor discussed) (de Waal 1991; Flack and de Waal 2000; Flack et al. 2004), and the sense of fairness (de Waal 1991; Brosnan and de Waal 2003) is constrained by limited (but not nonexistent) capacities for empathy (Preston and de Waal 2002; de Waal 2005), theory of mind (Heyes 1998; Lyons et al. 2006; Santos et al. 2006) and causal reasoning (Tomasello and Call 1997; Blaisdell et al. 2006). The weak instantiation of the aggregate level rules, as well as the lack of processes for generating consensus, tends to imply that the feedback from the rules to the individual sense of fairness is weak. By contrast, in human societies the feedback is often so strong between the rules and the sense of fairness within a society that accepted rules in one society can seem repugnant to other societies (Haidt et al. 1997). How the feedback between the sense of fairness and the rules intensifies, and how much coupling is possible, are unknown. That there is variation in the strength of this feedback across moral systems suggests that the concept of a moral system should by default be a *hierarchical* one, allowing for different degrees of coupling among the individual, behavioral, and aggregate levels. In addition to being empirically grounded, a hierarchical concept facilitates the development of a theory accounting for the origins and construction of moral systems by allowing for intermediate forms of coupling across the animal kingdom. It does not require the *a priori* assumption that moral systems are a human invention (de Waal 1996).

The goal of this paper is to lay the groundwork for a hierarchical theory of the construction and evolution of moral systems. To simplify this problem we decompose moral systems into components, constraints, and architectural properties. These do not correspond uniquely to questions of construction, mechanism and function but link them together. The components are level specific and include the sense of fairness, which resides at the neural-physiological level, moral judgments, which reside at the behavioral level, and moral rules, which reside at the aggregate level. Constraints and architectural properties apply at all levels.

Components are the informational building blocks of moral systems. The sense of fairness, moral judgments, and specific rules at the aggregate level interact to produce the set of interacting rules at the aggregate level we call moral systems. When the sense of fairness, moral judgments and moral rules are perfectly correlated, there are no differences in information content, and the sense of fairness perfectly predicts moral rules and vice versa. Construction is rendered a trivial mapping. In practice, the mapping between components is not one to one and not so readily compressible. Of interest are how much variation there is across human groups, and even species, in this mapping, and what the consequences of that variation are for both individuals and society.

Constraints on moral systems limit the range of values that components can assume. There are three sources of constraints in moral systems. The first includes the cognitive and affective mechanisms underlying both the sense of fairness and the capacity to generate and perceive rules. These are within-individual substrate constraints. The second source of constraint is the minimal stability requirements that must be met for the organization or society to persist in time. The point here is that some rules are more likely to jeopardize organizational robustness. The need to avoid destabilization limits which rules can arise. These are aggregate-level functional constraints. Affective-cognitive mechanisms and stability requirements appear to have interacted over evolutionary history to produce a third constraint – a set of super salient social stimuli that bias application of the sense of fairness towards a subset of social variables (Haidt and Joseph 2004). These super-salient stimuli, which we discuss in later sections of the paper, include cues for power relations, pain and suffering, purity, loyalty and group membership. These are individual, acquired, psychological constraints.

The behavioral architecture of moral systems modulates the economics and evolutionary dynamics[2] of the sense of fairness, moral judgments, and moral rules, by changing the strategy pay-off matrix. Although these architectural features are also subject to cognitive and organizational constraints, occasionally, behavioral inventions arise that overcome constraints imposed by limited cognition, robustness concerns, or reduce the high intrinsic cost of destabilizing strategies like conflict (e.g., Flack et al. 2005). Aspects of architecture include behavioral mechanisms for broadcasting rules, assaying consensus about rules, enforcing rules, resolving disputes about the priority of rules, and making rules robust. The architecture encodes principles of system dynamics.

Building a theory for the evolution and construction of moral systems requires characterizing the dynamics resulting from the interaction of components, constraints and architecture. In this paper we attempt to set the stage by decomposing the problem in the following way. We begin with a brief discussion of what a moral system is. The goal of this section is to make clear our assumptions and to minimize any confusion that might result given that most scientific work on morality has focused on the moral faculty rather than moral systems per se. We then summarize some relevant details about components, constraints and architecture. We identify fundamental issues and close by summarizing a modeling approach based on niche construction that seeks to generalize evolutionary dynamics by considering hierarchical systems with multiple overlapping time scales.

[2] We are grateful to Steve Frank for this phrase.

2 What is a Moral System?

A moral system is a set of interacting rules at the aggregate level that prescribes how individuals should act in potentially conflictual social situations. As such, the function of moral systems can be said to be conflict management (Alexander 1987; Boehm 2000; Flack and de Waal 2000). The defining feature of moral systems, setting them apart from other systems of conflict management, is that strategic social decisions are tethered to a concept of right and wrong underpinned by a sense of fairness. This concept of right and wrong can change an individual's preference function, weighting it in favor of behavior putatively increasing the social good (for discussion of how preferences are constructed, see Slovic (2000), Loewenstein (2008)). In principle moral systems reduce the frequency of conflict through the establishment of a set of rules that apply across the group, thereby bringing into greater alignment the interests of group members. When conflict does occur, due to misperception or violation of rules, or because of disagreement about how to prioritize rules, moral systems provide dispute resolution processes generally recognized as impartial. These take the form of public debates (Haidt 2001), or an appeal to agencies or individuals recognized as arbiters or at least as influential, such as the big men of small-scale societies (Godelier and Strathern 1991). And, finally, the sense of fairness encourages actors to ask the question – "right and wrong for whom and over what set of conditions". This slowing down of decision-making, and the consideration of consequences of actions for others in the present and in the future operates as an anticipatory error-buffering mechanism. It is also perhaps the most innovative and important conflict management aspect of moral systems.

Although the objective of moral systems is believed to be the establishment and maintenance of a normatively determinate code of action, *in practice moral systems often lead to normative indeterminacy.* Normative indeterminacy results because societies are almost always comprised of heterogeneous actors with different, and sometimes competing, interests, or when long and short-term societal goals are in conflict. Under these conditions, computing solutions that work for everyone in a variety of circumstances is difficult if not impossible. Furthermore, the information required to make such decisions is almost never available. One reason for this that the sense of right and wrong guiding decision-making depends on *perception*, which is typically based on incomplete, local information and biased by both the degree of uncertainty as well as its source (Ellsberg 1961; Kahneman and Tversky 2000). We discuss this at greater length in the section of the chapter on the cognitive and affective capacities underlying the sense of fairness.

There are two obvious questions that need addressing before proceeding. One is why conflict management is important. The second is why manage conflict through moral systems. This latter question is a theory question beyond the scope of this paper. We hope however that the framework we lay out here will provide the basis for constructing models comparing the efficacy, function, and accessibility of the "moral systems" conflict management strategy with other mechanisms of conflict management.

The second question, which is largely a question about the "repression of competition" (e.g., Frank In Press), is a major concern of social evolution. In order for aggregates to form and persist, whether multicellular organisms, animal aggregates, or societies, conflict among individual units competing over finite resources or time-mismatched objectives, needs to be controlled. The view we adopt (and expand on in the section of the paper "Construction Dynamics of Moral Systems") is that moral systems, by reducing uncertainty about the cost and character of social interactions, improve the quality of the social niche (Boehm & Flack, In Press). A social niche, like an ecological niche (Hutchinson 1957; Odling-Smee et al. 2003), is a resource pool. In the case of the social niche, the resources it contains are largely informational rather than energetic. A social niche can be operationalized in terms of an individual's local (degree) connections in the set of social networks in which it participates. The quality of the social niche determines access to information about food, predators, shelter, etc, and cooperation partners for coalitions, resources sharing, and other activities. We have shown elsewhere that individuals can construct, or build, their social niches, thereby controlling their quality (Flack et al. 2006). How well they build these niches depends in part on how stable or robust their social groups are to perturbations caused, for example, by conflict, and how costly interactions with other individuals in the group are. We have shown that by building conflict management mechanisms[3] that bring interests into greater alignment and reduce the intrinsic cost of interacting with unfamiliar individuals, individuals are able to build better social niches (Flack et al. 2005; Flack et al. 2006). We suggest that moral systems also play this role in social niche construction.

3 Major Components of Moral Systems

In the next three sections of the chapter, we discuss the three major components of moral systems and factors affecting their internal dynamics.

3.1 The Sense of Fairness

As we suggested earlier, a fundamental feature of moral systems separating them from other forms of conflict management is that decision-making is tethered to a concept of right and wrong underpinned by a sense of fairness. Both the concept of right and wrong and the sense of fairness are critical. The concept of right and wrong allows value to be assigned to behavior. The sense of fairness determines what the value is.

[3] Under this view, the conflict management mechanisms that individuals build become part of their social niches. This means that the social niche construction process involves both building edges in social networks that provide resources, and investment in conflict management mechanisms that allow for more efficient social edge building.

The sense of fairness can determine value by directly specifying a rule as an outcome – for example, the sense of fairness is 50–50, so the rule advocated is "split the cake in half". We call this the *content-specified sense of fairness*. Examples of content-specified fairness principles include equity in distribution, equality of outcome, needs based distribution, and effort based distribution (see Lakoff (1987), for a description of different kinds of fairness principles). The sense of fairness can also determine value *indirectly* by promoting the adoption of a process or tool for thought to produce a fair outcome, but which does not specify the outcome directly (e.g., see Flanagan for a description of this difference in alternative types of egalitarian systems (Flanagan 1989)). For example, if the sense of fairness declares that the choice of president should be by democratic means and unbiased, the rule advocated by an individual might be an anonymous voting procedure. We call the motivation for this rule the *process-based sense of fairness*[4]. Examples of process-based fairness principles include, equality of opportunity, and procedural and contractual distribution.

Different types of moral problem favor either content-specified or process-based fairness. The critical variables appear to be uncertainty about future consequences of the decision (veil of ignorance) (Rawls 1971; Kahneman and Tversky 2000; Frank In Press), familiarity with a problem (Rochat et al. 2008), and causal complexity in terms of both the ease with which factors contributing to the problem can be identified as well as potential consequences for other problems (Pizzaro and Bloom 2001). We expect that a content-based sense of fairness is invoked more frequently when the dilemma is familiar and its consequences relatively clear. In contrast, the process-based sense of fairness should be invoked when consequences are potentially very costly or when there is uncertainty about what the particular consequences of the decision will be. The content-based sense of fairness might be thought of as the limiting case of the process-based sense of fairness when the process-based sense of fairness generates a unique outcome. This would effectively eliminate the need for process and could lead to convergence onto a content-specified sense of fairness. Technically the content-based approach represents a look-up-table for the outcome, whereas the process-based approach provides a generalizing rule. When outcomes are unique rules are overly complex.

Importantly, neither the content-specified nor the process-based sense of fairness *necessarily* requires conscious causal reasoning. Once the moral dilemma is understood, both can in principle be generated by fast, automatic processing (see Pizzaro and Bloom 2001). However, a process-based sense of fairness is more likely to invoke causal reasoning and input from others in deciding outcome/producing moral judgments. In other words, an individual might "feel" that the rule "equal opportunity" (process) is better than the rule "equal outcome"

[4] Content-specified and process-based senses of fairness are related to, but do not map clearly onto, deontological and consequentialist ethics, two prominent concepts in the moral philosophy and psychology literatures. Deontological ethics focuses on the rightness of actions themselves, whereas consequentialist ethics argues that the rightness of an action is given by its consequences, regardless of the intent.

(content), but need to invoke causal reasoning to effectively implement the equal opportunity process.

Consider the procedure for electing the doge in sixteenth century Venice. That procedure, which is nicely outlined in John Julius Norwich's *A History of Venice* (Norwich 1989), involves multiple, apparently redundant rounds of voting with the set of voters in each round randomly chosen. Although the procedure seems ad hoc, and therefore poorly engineered, its obtuseness prevents any possibility of cheating and thus gives the impression of being completely fair. The Doge electoral procedure was apparently implemented in direct response to deadly and costly competition among feuding families attempting to gain power. The critical point is that whereas an individual's gut feeling might tell him the Doge election protocol is a good idea, building the procedure, with its multiple layers or redundancy, requires conscious calculation. Similarly individuals might advocate a utilitarian approach to ethics because they believe it to be fair, rather than for the utility of the outcome it produces.

The content-specified and process-based senses of fairness have different implications for the construction of moral systems. It is possible that both types of fairness will generate the same set of rules at the aggregate level. However, process-based senses of fairness are more likely to draw explicitly on reasoning processes and input from other individuals for decision-making. Consequently, when the process based sense of fairness is favored, construction of the rule set will require greater interaction between an individual's sense of fairness, her moral judgments, and aggregate level rules and norms and inter-individual signaling dynamics. The process-based sense of fairness will, as with the Doge example, require building more architecture, including behavioral mechanisms and social structures, to "find" locally optimal outcomes.

The sense of fairness varies across individuals, cultures, and even moral dilemmas, according to social context, resource distribution, and other properties of the problem to be solved (Trivers 1971; Pizzaro and Bloom 2001). In principle, there are many different content-specified and process-based fairness principles. Behavioral economics experiments using a variety of game structures suggest that the sense of fairness varies across cultures according to the degree of market integration and average payoff from cooperation in everyday life (Henrich 2004). It is unknown however to what extent the moral judgments (for a definition of moral judgment see the next section) made by individuals reflect an individual's instantaneous personal sense of fairness as opposed to an aggregate-level sense of fairness instantiated in moral rules (Sherif 1935; Asch 1956; Berger and Luckman 1967; Haidt 2001; Pizzaro and Bloom 2001). This is a hard question to answer because feedback between moral rules, individual judgments and the sense of fairness generates a complicated set of causal relations. Evaluating the mapping given this dynamic requires artificially holding the system constant for some time (patch-clamping) and assaying how perceptions of fairness at the individual level map onto established rules at the aggregate level.

Many models have been proposed to explain how an individual's sense of fairness arises (reviewed in Haidt 2001; Green and Haidt 2002; Cushman et al. 2006;

Hauser 2006; Hauser et al. 2007). Although some authors (Haidt 2001; Pizzaro and Bloom 2001) have argued that social persuasion can effect an individual's moral judgments and sense of fairness, none of the prominent models in the literature explicitly takes into account feedback from moral rules perceived to exist at the aggregate level. The focus of most models (and experiments) recently has been on describing the within-individual processes generating the sense of fairness and subsequent moral judgments. The two major axes of this argument are conscious processing versus automatic processing and calculation verses feeling.

A better understanding of the extent of variation within and across systems in the sense of fairness, the within-individual computational process generating the sense of fairness, and the factors influencing adoption of content-specified, or process-based, senses of fairness, should help parameterize suitable models of construction dynamics (see section Construction Dynamics of Moral Systems).

3.2 *Moral Judgments*

We stated earlier that the concept of right and wrong allows value to be assigned to behavior and that the sense of fairness determines what this value is. A moral judgment is a social, or broadcasted, pronouncement about whether an action (or character of an individual) is good or bad. We stipulate "social" even though individuals do not always express their moral judgments. We do so because without the commitment to a position that comes with a social pronouncement, moral judgments, and the sense of fairness, are effectively equivalent. A moral judgment can map directly onto the sense of fairness, or it can reflect a compromise between an individual's personal sense of fairness and perceived moral rules (Pizzaro and Bloom 2001). Moral judgments, as social pronouncements, can be dishonest. Moral judgments are dishonest when the pronouncement is at odds with the pronouncer's sense of fairness and is motivated by the desire to appear consistent with the status quo or to gain acceptance by some group with a particular moral code. The primary point is that a moral judgment is a signal. The sender is the individual making the pronouncement. The receiver set includes both the target of the judgment and the audience. It is the integration (in the simplest case of minimal construction, the integration is just the average) of these signals over the population that leads to the establishment of moral rules. The judgment can therefore influence both the target's behavior and act as a "vote" for a particular moral rule (for additional discussion, see Flack and Krakauer 2006, which discusses this process in the context of building power structures).

In contrast to a moral judgment, a moral justification is the post-hoc explanation given by the pronouncer for the judgment (Hauser et al. 2007). The justifications individuals provide, particularly for decisions concerning difficult moral dilemmas they encounter for the first time, are often underspecified, involve weak commitment, and do not appear to make explicit use of a cost-benefit analysis (Haidt 2001).

3.3 Rules and Codes

Moral rules are prescriptive rules perceived to exist at the aggregate level. A prescriptive rule is an internally (self) or externally (second or third party) enforced statistical regularity (de Waal 1991; Flack et al. 2004). Enforcement mechanisms are discussed in greater detail in the section of the chapter on Architecture. Briefly, the primary mechanisms of self-enforcement are "moral emotions" like shame, guilt, and embarrassment, anticipation of punishment by others, and the desire to avoid scorn (Haidt 2003). The primary mechanisms of external enforcement are second and third-party punishment (to include both physical punishment and economic sanctions), policing, and ostracism[5] (see Gruter 1986; Flack et al. 2004; Ehrlich and Levin 2005; Flack et al. 2005; Henrich et al. 2006; Frank In Press).

Explicit prescriptive rules are typically referred to as moral codes and can be arrived at either through consensus, or coercion/imposition, with the critical point being that the rules are broadly recognized and stated either verbally or posted in writing. In many cases, explicit moral codes are buttressed through legal systems (Masters and Gruter 1992). Implicit rules are rules widely recognized as having consequences when violated (e.g., Flack et al. 2004), but for which there is no verbal or written statement. The content of both implicit and explicit rules changes through cultural evolution, and in particular, learning dynamics operating on top of social networks (e.g., Nakamaru and Levin 2004; Ehrlich and Levin 2005) but is biased by constraints – including the extent to which rules are linked (Durrett and Levin 2005) or have consequences in aggregate – and architecture. Implicit rules are noisier and harder to enforce, and presumably less stable than explicit rules because they are not reinforced through broadcasting at the group level (and so are less "viscous" (Ehrlich and Levin 2005). They can be reinforced through punishment, policing and other control mechanisms, although justifying punishment will be more difficult in the absence of clearly broadcasted rules. Implicit rules have the advantage of being more closely tethered to each individual's sense of right and wrong. Explicit rules require greater coordination and so can in principle deviate substantially from any single individual's sense of fairness. The broadcasting mechanisms reinforcing explicit rules can cause them to change more slowly than implicit rules, but also allow the potential for monopolization by a few. This offers the possibility of abrupt and dramatic replacement of the rule set with an alternative rule set.

The starting point for the spread and establishment of moral rules is the signaling dynamics of the moral judgments made by individuals. We suggest that competition among individuals to promote their sense of fairness – which they signal through their moral judgments–coupled to learnablility and stability constraints, produces moral rules at the societal-level. This construction process can be democratic or biased by power or other social structures. Studies of how rules spread in structured populations have treated rules fairly generically. It is not yet known how properties specific to moral rules shape learning and epidemiological dynamics (Ehrlich and

[5] Whereas economic sanctions constitute withholding access to material resources, ostracism constitutes withholding access to social resources.

Levin 2005). Is it admissible, for example, to theorize about a moral rule in the way we theorize about a virus, or do additional constraints and properties require that we consider an expanded set of transmission mechanisms (see Odling-Smee 2007)?

Most moral systems consist of multiple rules and codes (e.g., The Ten Commandments). These rules interact and can be at odds (Ehrlich and Levin 2005). For example, concern for individuals with terminal illness can be at odds with rules that state a person should not take the life of another, and rules that state a person should not contribute to the suffering of others. Rules vary in terms of their robustness and the extent to which they are subject to interpretation due to mitigating circumstances or contextual factors. It is not yet known how adoption of particular rules constrains or shapes adoption of additional rules (Ehrlich and Levin 2005). It is also not yet known if alternative sets of rules are effectively neutral in their contribution to organizational robustness and plasticity, or if there is an optimal, accessible, rule set.

A long-standing question is whether there are universal moral rules found in all societies. Whether such rules exist, and what accounts for them, are two hotly debated issues. There are three primary hypotheses, none of which are mutually exclusive because each stresses different aspects of the problem. One is that the rules themselves are genetically encoded. This hypothesis has largely been rejected. Another (which typically ignores mechanistic/substrate considerations) is that the set of rules in any society is the outcome of an optimization procedure selecting for the rule set that will give rise to the most competitive or robust society, either through group selection (e.g., Bowles 2006) or endogenous selection at the individual level for behavior generating beneficial aggregate level traits (e.g., Ehrlich and Levin 2005 Frank 2003; Frank In Press) or for individuals that can build beneficial aggregate level traits (see Odling-Smee et al. 2003; Flack et al. 2006). The strong form of this hypothesis posits that rules with substantial implications for societal robustness or intergroup competition will be broadly shared. The weak form of this hypothesis posits that there are some rules that will not be found in any society because they are detrimental to societal persistence. A third hypothesis is that there is a universal set of moral principles shared by all humans, and that the myriad rules arising in different societies are simply the parameterized family of these principles (Haidt 2001; Hauser 2006). There are many variants on this hypothesis, but the basic idea is that the universal principal set is the outcome of constraints imposed by underlying cognitive and emotional capacities, and their tuning, over evolutionary time, to societal robustness concerns. The strong form of this hypothesis posits the existence of a "moral organ" that specifies the essential ingredients or properties of the principles (e.g., Hauser et al. 2007).

4 Constraints on Moral Systems

In this paper, the term "constraint" refers to any kind of bias imposed by substrate limitations or functional requirements that must be satisfied to maintain a minimal operational efficiency. This includes both negative biases that limit the space of accessible solutions, and positive biases, that predispose a system to evolve or learn particular solutions.

4.1 Cognitive-Emotional Capacities

There is a substantial body of work bearing on the cognitive and affective capacities underpinning and biasing the sense of fairness, the ability to perceive rules, and the ability to make moral judgments. Reviewing this body of work would constitute a paper in and of itself. In light of this, we simply enumerate the set of capacities deemed important, and why, listing useful references as we go. For detailed arguments about whether the basis of the sense of fairness lies with affect or cognition, and whether it largely the result of conscious or unconscious processing, we refer the reader to (Haidt 2001; Green and Haidt 2002; Haidt 2003; Moll et al. 2005; Cushman et al. 2006; Hauser 2006; Dupoux and Jacob 2007; Koenigs et al. 2007; Mikhail 2007).

As we noted earlier, the sense of right and wrong guiding decision-making depends on *perception*, which is typically based on incomplete, local information. A major constraint on the evolution and construction of moral systems is how well perception is correlated with some objective measure of reality. The cognitive and affective factors affecting this mapping include,

Memory for tracking interactions and effects: by this we mean the precision of the tracking system. Do individuals track by remembering only the emotional state induced by the last interaction (e.g., attitudinal reciprocity, de Waal and Brosnan 2006)? What are the implications for moral judgments of being able to track events and objects in addition to emotional states? Research on the evolution of cooperation suggests that precision tracking is not always desirable, particularly when resources fluctuate, or the goods and services exchange are only roughly equivalent or cannot be counted. The intuition for this is that stabilizing cooperation under noisy environments uncertainty, or error requires noisy solutions (e.g., Bendor 1993; Nowak and Sigmund 1993; Kahneman and Tversky 2000).

Capacity for learning: by this we mean how increasing the efficiency of the learning process through generalization, causal reasoning (Blaisdell et al. 2006), and imitation, emulation or other forms of observational learning, reduces the cognitive burden on look-up table-like tracking. Also of interest are how efficient learning influences the accuracy of perception and whether it increases the size of the salient set of physical objects, events, and internal states influencing the sense of fairness and moral judgments.

Temporal state projection: here we mean how an individual's or an organization's ability to foresee the consequences of moral judgments many time steps into the future influences both the content and the specificity of the moral rules that are adopted. Also of interest is how the ability to forecast modulates the extent to which an individual relies on intuition verses reasoning in the production of judgments.

Emotional and mental state attribution: by this we mean how the capacity for empathy (Preston and de Waal 2002), to grasp goal directed action in others (e.g., Wood et al. 2007) and a theory of mind (Siegal and Varley 2002; Young et al. 2007) facilitate causal reasoning and motivate behavior. It is not yet well understood how the size of the set of individuals to which empathy extends varies as a function of the extent of actor heterogeneity in the system or with interaction frequency and social network structure.

Integration over multiple inputs: here we mean how an individual's capacity to integrate inputs from multiple group members and over multiple events, influences the accuracy of both her understanding of the causes and consequences of moral judgments, and her perception of what others perceive to be fair. The more perceptual errors individuals make, the more difficult it will be to align perceptions of right and wrong across group members in the establishment of rules at the societal level. Although this has not yet been shown for the emergence of moral rules, the more general point that integration over multiple inputs is a way to buffer against state assessment errors has been shown to be important to the emergence of robust and accurate power structures (Flack and Krakauer 2006; Boehm and Flack In Press).

Updating: of interest is how the conscious or unconscious use of approximate Bayesian reasoning facilitates the ability to reject rules that have lost their utility. Rules can loose their utility when the social or ecological environment has changed, or when the behavioral architecture supporting them has broken down.

Automation: of interest are which neural factors and social conditions (e.g., social conditions facilitating establishment of, or obscuring, cause and effect) contribute to the capacity to shift calculation from conscious processing to unconscious processing (Bargh and Chartrand 1999).

4.2 Organizational Robustness

Cognitive-emotional constraints are generated by neural-physiological substrate limitations. Another kind of constraint is a bias imposed by minimal functional requirements. Here we discuss functional requirements at the level of social organization and how they influence the types of rules that can arise in a moral system.

An organization is said to be robust if it has the ability to persist through time despite exogenous and endogenous perturbations and the development, decay and turnover of components. The extent to which organizations can be said to be robust in the face of perturbations is given by the direct contribution of the target of the perturbations to system structure and function (causality) minus the consequences to system structure or function of disabling or removing the target (exclusion dependence). Exclusion dependence reveals the extent to which the target also can be said to make indirect contributions to system structure and function by, for example, modulating the interactions between other components. Elsewhere we have provided a formal justification for this "causality minus exclusion dependence" definition of robustness (Ay and Krakauer 2006; Ay et al. 2007) and implemented an experimental design that can tease these apart (Flack et al. 2005, 2006). The main point is that this quantity is maximized in organizations with effective robustness mechanisms.

A common problem faced by social organizations is endogenous conflict. Endogenous conflict occurs when the interests of group members are not perfectly aligned. This kind of conflict can be thought of as a chronic, low-level perturbation. If not buffered or managed, it can lead to destabilization of the system because individuals under perform or leave the group. Most social and biological organizations have conflict management mechanisms that mitigate the negative effects of conflict (Frank 2003), and can additionally promote positive social interactions (e.g., Flack et al. 2005, 2006). We have already emphasized that perhaps the most important function of moral systems is conflict management. Moral systems generate rules aligning the interests of group members and sort out disagreements resulting from individuals perceiving and implementing rules differently. However, all rules are not equally optimal. In addition to aligning interests and facilitating dispute resolution, the rules implemented should minimize the production of instabilities that require additional robustness mechanisms. In other words, *whereas some rules serve as robustness mechanisms other rules can be perturbative*. For example, the rule "no individual between the ages of twenty and fifty should work", if adopted would mean that everyone agrees which individuals should and should not work. Such a rule (if accepted) would minimize daily conflict about the distribution of effort. This would not be however the best rule from the point of view of organizational productivity, and could be deleterious under demographic perturbations (such as during war).

We can determine the contribution of a given rule to organizational robustness, and thus whether it is a robustness mechanism, neutral, or perturbative, by measuring how it contributes to some higher-level organizational structure or function in terms of causal contribution and exclusion dependence. If the causal contribution were zero and the exclusion dependence zero (this would mean that the direct effects of the rule on the system is negligible, and that removal of the rule does not seem to compromise system function), the system could not be deemed robust with respect to perturbing this rule. If however the causal contribution were high and the exclusion dependence were low, this would mean that although the direct contribution of the rule is significant, when the rule is removed, compensatory mechanisms come into play that allow the system to continue functioning.

The formal framework we have developed to give rigor to the concepts of causal contribution and exclusion dependence is not yet equipped to disambiguate causal contribution due to negative effects on system structure or function from causal contribution due to positive effects. It is also possible that exclusion dependence is positive when a rule, or, more generally, a component, is removed. This would apply when the rule is perturbative, rather than a robustness mechanism. Finally, it is important to keep in mind that rules promoting robustness at one level can undermine it at other levels (Krakauer and Plotkin 2002).

The main point of this section is that not all rules are equally effective from the point of view of robustness. The implications of the rule, or set of rules, for the robustness of different organizational features, will constrain which rules arise in moral systems.

4.3 Super-salient Social Stimuli

Over evolutionary time, the interaction of cognitive-emotional constraints and constraints imposed by social stability requirements appear to have produced a third constraint: psychological predispositions to respond to particular types of social stimuli.

Although the space of robustness concerns is in principle vast, there are a few fundamental problems characteristic of all societies. The most significant problem is how to make group living beneficial. The null model is equal-effort-equal-benefit. In practice this principal has many variants and exceptions across and within societies (Henrich 2004). Variation (e.g., equal outcome, equal opportunity, contractual fairness, etc.) in this principle is largely attributable to three factors that make the null rule unstable. Many animal and all human societies are comprised of heterogeneous actors with different degrees of relatedness and at different developmental stages. This means that individuals will not necessarily perceive costs and benefits the same way. Resources are typically limited and fluctuate in availability. This means that individuals will vary in terms of resource need and their potential to contribute to collective benefits. Physical and social environments can be unpredictable. This means that individuals will need to continually re-evaluate the utility of their strategies. These factors can be summarized as uncertainty about the quality, reliability, and honesty of information, asymmetries in access to, possession of, and need of resources, and the perceived cost of playing a strategy given the strategies others are perceived to be playing.

Data from a variety of experiments suggest that social evolution has produced individuals who, given constraints imposed by underlying cognitive-emotional capabilities, are not only able to solve social dilemmas more quickly than nonsocial dilemmas of similar difficulty (Cosmides 1989), but are particularly sensitive to those social stimuli that are informative about which of the above factors is in play (Haidt and Joseph 2004). One set of these stimuli concerns power relations, which provide information about resource possession and need. A second set of stimuli includes group membership cues, which help reduce uncertainty about the likelihood of repeated interactions and cheating. A third set of stimuli includes proxies for perfect or nearly perfect information that are articulated in terms of "purity". The final set of stimuli include signals of suffering/pain, which serve as heuristics for assessing cost.

The sense of fairness appears to be strongly biased, or parameterized, by these stimuli (Haidt and Joseph 2004). The emotional-cognitive responses to these stimuli have been called foundational moral principles by some authors (e.g., Haidt and Joseph 2004). These authors have argued that the diversity of moral rules across societies can be coarse-grained into categories of rules structured around power relations, group membership cues, minimization of suffering and pain, and purity concepts. The strong form of this argument posits that the human brain has been built by evolution to innately detect and respond to these stimuli. The weak form posits that the human brain learns faster in the context of these stimuli.

5 Architecture of Moral Systems

The "architecture of moral systems" includes those behavioral mechanisms that can stabilize and even change the economics and evolutionary dynamics of the sense of fairness, moral judgments, codes and even their underlying constraints. Aspects of architecture include behavioral mechanisms for assaying consensus about rules, broadcasting rules, enforcing rules, resolving disputes about the priority of rules, and making rules robust.

In behavioral game theory one way to modulate the economics and evolutionary dynamics of a strategy is to change the game structure (Maskin 2008). Changing the game structure can change the pay-off structure. For example, one way to ameliorate the cheating problem is to make a one-shot game a repeated game. In mechanistic terms, this means changing the interaction structure so that individuals have the opportunity to interact repeatedly. The advantage of repeated interaction is that it reduces the temptation to cheat by allowing for the possibility of dependencies forming and for the punishment of defection.

To illustrate this general point about game structure, consider the evolution of third party policing. Third party policing is the impartial monitoring and management of conflicts among group members by uninvolved third parties (Flack et al. 2005). Mathematical models of policing and of related behaviors such as punishment are able to effectively reduce their number of state variables by assuming that all individuals pay the same cost for engaging in conflict management or repressing competition. Although this is a reasonable simplifying assumption, it makes the evolution of policing appear more difficult or improbable than actually seems to be the case. There is good evidence from the study of dominance in animal societies for individual variation in resource holding potential or vigor (e.g., Clutton-Brock and Parker 1995) and, from work on third-party policing, for individual variation in power. By allowing for variation in state, Frank (Frank 1995, 2003) has shown that small differences in individual vigor can lead to large variations in individual contributions to policing when relatedness is low.

The claim that variation in individual vigor is related to variation in investment in conflict management requires the additional assumption that the cost of conflict management varies inversely with individual state. One of the findings of work on policing is that cost is not only a function of the individual intervening but also of the power values of the individuals engaged in the dispute (Flack et al. 2005). Thus, variation in individual vigor is not sufficient. A power structure must arise in which individuals also vary in the degree to which group members perceive them capable of successfully using force. This requires not only perceived differences in vigor, resource holding potential or other analogs, but a signaling system in which individuals communicate this perception. In the case of third-party policing, cost and effectiveness vary as function of state, where state is a function of a power structure that is encoded in a network of signaling interactions about the perceived capacity of any individual to successfully use force (Flack and Krakauer 2006). The main point is that the evolution of policing turns out to be an easy problem when supported by a social structure that effectively makes the cost of policing negligible for some individuals. The hard problem is evolving a social structure with the right properties. Policing in turn modulates the costs and benefits of other behaviors, modifying, for example, the edge formation rules individuals use to build their social networks and making accessible new types of social relationships and exchange networks that were too risky pre-policing (Flack et al. 2005, 2006).

Although not well studied from a quantitative perspective, many moral systems appear to have supporting architectures like the power structure and policing mechanism in the previous example. Here we note three important support-architecture inventions, none of which are necessarily specific to moral systems. One is the invention of dispute resolution systems that function to prioritize rules in conflict or to provide definitive interpretations of fuzzy or poorly articulated rules. For example, in small-scale Melanesian societies "big men" sometimes play the role of arbiter when moral disputes arise (Godelier and Strathern 1991). Another invention is the establishment of "moral police". For example, in some religious societies, religious clergy enforce moral conduct (Stark 2001). A third invention is the ability to broadcast rules through widely circulated texts, like the Bible, or myths and cultural stories transmitted orally and through symbolic objects, like the Kachina dolls of the southwestern pueblo culture in the United States.

The critical point is that evolving supporting structures can change the pay-off structure of interactions. This in turn changes the rules specifying how and under what circumstances the components of moral systems (sense of fairness, judgments, and rules) can influence each other.

6 Construction Dynamics of Moral Systems

The decomposition of moral systems into components, constraints and architecture suggests that three dynamical factors, which we discuss below, play a critical role in the construction of moral systems. With these factors in mind, we discuss new theoretical approaches in the study of evolutionary processes that we anticipate will provide insight into how the sense of fairness, moral judgments and moral rules interact to produce moral systems.

The ingredients of a moral systems theory will be components (sense of fairness, judgments, moral rules) situated at multiple levels in a hierarchy, a set of constraints (affective cognitive and robustness related) defining the limits of component behavior and their natural range of operation, and temporal stability. And finally, the theory must include the system architecture, specifying the connectivity among components and the rules transforming these connectivities. The theory should seek to explain how individual preferences and biases feed forward, generating collective moral rules that feedback to modify individual behavior, including each individual's sense of fairness and moral judgments. The theory should also explain how, and with what implications, a partial time scale separation arises between the rate of change of moral rules at the aggregate level, moral judgments at the individual level, and the within-individual sense of fairness (for further discussion of this time scale issue, see Boehm and Flack In Press). The theory seeks to explain the origin of moral rules, their stability properties and their long-term trajectories, as well as delimiting a space of degenerate rule sets – different moral system configurations all compatible with individual biases.

A natural theoretical framework for a theory of moral system dynamics is niche construction (Lewontin 1982; Laland et al. 1996; Odling-Smee et al. 2003; Odling-Smee 2007). Niche construction considers those traits of a phenotype capable of modifying the environment of the organism, thereby allowing for selection pressures (niche parameters) to be partly encoded in the organismal genome. This property has been called the extended phenotype (Dawkins 1982). Niche construction overcomes an "adiabatic" property built into Darwinian dynamics, which means that the environment and the organism have typically been treated as time-separable dynamical systems and their reciprocal interactions neglected. The adiabatic property arises because the rate of change of the environment, and the processes shaping its construction, are treated as very slow relative to the organismal lineages that it shapes. The standard Darwinian framework and the extended niche construction framework are illustrated in Fig. 1. In niche construction the niche and the organism

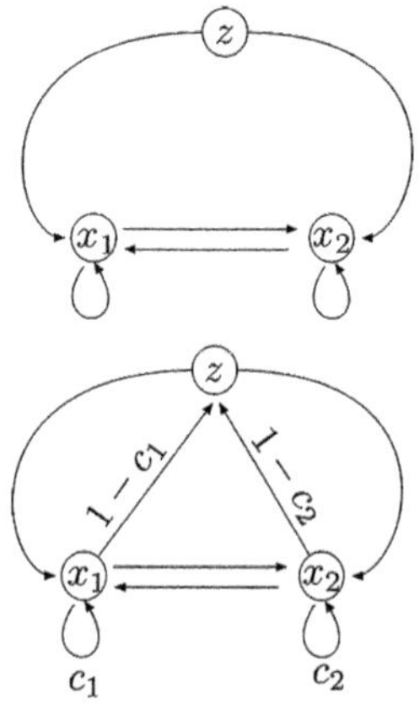

Fig. 1 The difference between a standard, "adiabatic", Darwinian model for selection (top) in contrast to a niche construction dynamic (bottom). In the top panel the social selection pressures (feedback constraints), denoted by z, are not influenced by the actions of the agents x whose future representation is determined by their interactions in the selective field provided by z. In the bottom panel, the strategies actively shape their social niche, thereby modifying the cost benefit structure of their interactions. The construction of the social niche ("reinforcing mechanisms" in the text) comes at a cost of a reduced, direct investment in the proliferation of the individual strategies

are coupled, such that the constructor can partially control the rate of change and/or the trajectory of the environmental variables (Boehm and Flack In Press). The coupling leads to the emergence of frequency-dependence, or game dynamics, from a formally density-dependent dynamic. Niche construction is, however, distinguished from standard frequency dependence by the explicit incorporation of a construction dynamic that builds the second variable, and by the explicit treatment of the emergence of a partial separation of timescales between these variables. The critical point is that by controlling the environmental variable, the constructor is able to reduce uncertainty in the environment. This in turns allows the constructor to better adapt to its environment thereby improving resource extraction.

A niche construction style theory that addresses all of the requirements discussed in this chapter is not yet available. However, we can make some progress in the direction a moral systems dynamics, by considering how low level components, say a sense of fairness (here called strategies), interact or compete with other strategies, while engaging in a construction process generating reinforcing mechanisms, policing for example. The reinforcing mechanism can either promote consensus for one particular strategy over another, or allow multiple strategies to coexist that would otherwise be outcompeted. A simple interpretation of a moral rule would then be the vector of stable strategies supported by the reinforcement mechanism. So if individuals adopting the golden rule are in competition with "turn the other cheek", advocates of both might invest in the construction of a similar mechanism for broadcasting their beliefs. The outcome of competition under this new framework, we might think of as the moral rule adopted in the society. A more sophisticated moral rule would have to involve more than a linear combination of the initial strategies and constitute an additional layer of regulation biasing the broadcast of individual judgments.

In Fig. 1 we consider the first, simple case where we have a coupled system with a single-dimensional niche, which acts as a reinforcing mechanism on a two dimensional strategy vector. In terms of differential equations we might write,

$$\dot{x}_i = f(\vec{x}, z)$$
$$\dot{z} = g(z, \vec{x}),$$

where the vector of individual rule frequencies, $\dot{x}$ contributes to the frequency of aggregate rules by constructing reinforcing mechanisms like broadcasting or policing, z. These feed back to contribute to the population densities of x_i. We might choose to consider a very simple competitive dynamic, where individuals contribute to constructing reinforcing mechanisms that bias the consensus generating process toward their particular sense of fairness. Consider two competing senses of fairness. Lets say that advocates of each of sense of fairness invest in their favored moral rule by building a broadcasting mechanism. The key here is that the broadcasting mechanism allows arbitrary, individual fairness rules to be amplified into the population level moral rule. This kind of reinforcement paradigm is effective for all fairness rules, and is not selective of rule content. Hence rule advocates can benefit from a competitor shouldering the burden of investment in building the architecture supporting this moral rule. This leads to a constructive tragedy of the commons, whereby the reinforcing mechanism and its associated benefits, fail to evolve.

We can state all of this mathematically. We consider two strategies, x_1 and x_2, and a reinforcing mechanism, z that can increase the absolute abundance of a strategy in the population (the strategic carrying capacity). Individuals can invest (c) in the direct proliferation of their own strategy, or at some cost to themselves $(1 - c)$, invest in the construction of the reinforcing mechanism. Here the parameter p seeks to capture sources of the moral rules derived extrinsic to the dynamics of $\dot{x}$.

$$\dot{x} = c_i x_i - x_i \left(\sum_j x_j \right) \Big/ kz$$

$$\dot{z} = p + \sum_j (1 - c_j) e x_j \Big/ \sum_i x_i + z - dz$$

It is straightforward to show that under these dynamics, where individual strategies differ only in the value of c, that a value $c = 1$ represents the absorbing state of evolutionary dynamics. In other words, the construction of a hierarchical system represented by a socially valuable supporting mechanism (e.g., broadcasting or policing) capable of amplifying or promoting a particular sense of fairness to the level of a slowly changing moral rule, is simply not stable under competitive interactions, and will tend towards a flat structure in which each strategy competes directly.

The solution to this dilemma is rather technical (see Krakauer, Page et al. In Press) but can be understood in terms of constraints that allow constructors to partially monopolize some proportion of the benefits derived from the reinforcement mechanism. Thus investing in reinforcing mechanisms can be supported when the benefits of the resulting structures can be differentially apportioned.

Thus we might consider the case,

$$z_i \approx z_i m + \sum_j z_j (1 - m).$$

Here the feedback from the reinforcing mechanism is partitioned by the monopoly parameter m, allowing some strategy i to gain exclusive access to m-proportion of the benefits from the construction while the remainder is shared by all. Given this constraint, it can be deduced that the evolutionary stable value of c converges on,

$$c_i = \frac{1}{1 + m}.$$

This model is a simplified introduction to a family of more inclusive models (Krakauer, Page et al. In Press) that allow for variation in the competitive strength of individual strategies, variation in the number of individual strategies, the number and strength of reinforcing mechanisms, and the diversity of strategy vectors (here simple moral rules) emerging from these mechanisms.

Hence the construction of a slow time scale variable (z) that can through feedback, act as modifier of effective payoff (total abundance due to improved broadcasting for example) changes the "rules of engagement" at the level of individual preferences.

6.1 Imperfect Information

Moral systems manage conflict under imperfect information. This has not been treated in the model outlined above. Sources of imperfect information include the fact that most moral systems are composed of heterogeneous actors with poorly aligned interests; that they are charged with dictating "good" interaction patterns when the shadow of the future is large and current solutions and societal needs are very likely to be in conflict with future needs; and that they are organized around rules arising out of an inherently noisy process for establishing consensus among group members. Consensus (here consensus means agreement whether centrally coordinated or arising through local processes; note that the critical question for the processes we have discussed is really *how much* consensus there is, rather than whether there *is* consensus) is established by integrating over the moral judgments of individuals, in the simplest case, by averaging, in more complicated cases, by weighting and competition. The moral judgments are based on an underlying sense of fairness rooted in perception based on some mixture of causal reasoning and intuition, *and* perception of what others in the system think is fair. This suggests that, from the perspective of both selfish individuals and organizational robustness, the content of rules at any given time is unlikely to be optimal. However, the focus on content as an objective is likely to be misguided. Rather it is the slowing down of decision-making and the consideration of consequences of actions for others in the present and in the future vis a vis the sense of fairness, that allows moral systems to effectively manage conflict by reducing uncertainty about behavioral outcomes. Moral systems generate temporary commitment to a partial resolution with continuous updating through a constructive dynamic.

7 Summary

We have argued for the development of a hierarchical theory of moral systems. In particular, we suggest that in addition to considering the function, evolution and mechanisms underpinning morality, it is critical to consider the construction of moral systems – a process akin to development, whereby individuals modify their social environments (social niches) by investing in an architecture that can promote greater consensus among strategies (alternative senses of fairness expressed through moral judgments), mitigating conflict, and thereby generate collective moral rules. These structures can be coercive or cooperative and need not reflect the wishes of all individuals equally. Both components and architecture are constrained by affective-cognitive limitations and social concerns. We have discussed briefly the value of a niche construction approach to moral systems as it offers a way of thinking about how the composition of a population of strategies can be modified both through direct competition, and via a social structure (social niche) that individuals build in attempt to foster their own sense of fairness. Because the social niche is built gradually, through the interaction of many individuals and often over many lifetimes, moral rules are likely to change more slowly than individual moral judgments and then the sense of fairness.

References

Alexander RD (1987) The biology of moral systems. Aldine Transaction, New Brunswick

Anderson SW, Behara A, et al. (1999) Impairment of social and moral behavior related to early damage in human prefrontal cortex. Nat Neurosci 2:1032–1037

Asch S (1956) Studies of indenpendence and conformity: a minority of one against a unanimous majority. Psychol Monog 70

Ay N, Flack JC, et al. (2007) Convergent complexity and robustness in multimodal signaling networks. Philos Trans R Soc Lond B Biol Sci 362:441–447

Ay N, Krakauer DC (2006) Geometric robustness theory and biological networks. Theory Biosci 125:93–121

Bargh J, Chartrand TL (1999) The unbearable automaticity of being. Am Psychol 54:462–479

Bendor J (1993) Uncertainty and the evolution of cooperation. J Conflict Resolut 37:709–734

Berger PL, Luckman T (1967) The social construction of reality. Doubleday, New York

Binmore KEN (2005) Natural justice. Oxford University Press, Oxford

Blaisdell AP, Sawa K, et al. (2006) Causal reasoning in rats. Science 17:1020–1022

Boehm C (2000) The origin of morality as social control. J Conscious Stud 7:149–184

Boehm C, Flack JC (In Press) Power: insights from evolutionary biology, primates, and other animals. The Social Psychology of Power A. Guinote

Bowles S (2006) Group competition, reproductive leveling, and the evolution of human altruism. Science 314:1569–1572

Bowles S, Gintis H (In Press) Cooperative homo economicus. In: Levin SA (ed) Games, groups and the global good

Brosnan SF, de Waal FBM (2003) Monkeys reject unequal pay. Nature 425:297–299

Campbell DT (1975) Conflicts between biological and social evolution and between psychology and moral tradition. Am Psychol 30:1103–1126

Castelli F, Happe F, et al. (2000) Movement and mind: a functional imaging study of perception and interpretation of complex intentional movement patterns. Neuroimage 12:314–325

Clutton-Brock TH, Parker GA (1995) Punishment in animal societies. Nature 373:209–216

Cosmides L (1989) The logic of social exchange: has natural selection shaped how humans reason? Studies with the Wason selection task. Cognition 31:(187–276)

Cushman F, Young L, et al. (2006) The role of conscious reasoning and intuition in moral judgments. Psychol Sci 17:1082–1089

Dawkins R (1982) The extended phenotype: the gene as the unit of selection. W.H. Freeman, San Francisco & Oxford

de Waal F (1991) The chimpanzee's sense of social regularity and its relation to the human sense of justice. Am Behav Sci 34:335–349 de Waal F (2005) Our inner ape. Riverhead, NY

de Waal F (2006) Primates and philosophers. Princeton University Press, Princeton

de Waal FBM (1996) Good natured: the origins of right and wrong in primates and other animals. Harvard University Press, Cambridge, MA

de Waal FBM, Brosnan SF (2006) Simple and complex reciprocity in primates. In: Kappeler PM, van schaik CP (eds) Cooperation in primates and humans: mechanisms and evolution. Springer Berlin, Berlin

Dupoux E, Jacob P (2007) Universal moral grammar: a critical appraisal. Trends Cogn Sci 11: 373–378

Durrett R, Levin SA (2005) Can stable social groups be maintained by homophilous imitation alone? J Econ Behav Organ 57:267–286

Ehrlich PR, Levin SA (2005) The evolution of norms. PLoS Biol 3:943–948

Ellsberg D (1961) Risk, ambiguity and the savage axioms. Q J Econ 75:643–669

Flack JC, de Waal F (2000) Being nice is not a building block of morality. J Conscious Stud 7:67–78

Flack JC, de Waal FBM (2000) 'Any animal whatever' – darwinian building blocks of morality in monkeys and apes. J Conscious Stud 7(1–2):1–29

Flack JC, de Waal FBM, et al. (2005) Social structure, robustness, and policing cost in a cognitively sophisticated species. Am Nat 165:E126–E139

Flack JC, Girvan M, et al. (2006) Policing stabilizes construction of social niches in primates. Nature 439:426–429

Flack JC, Jeannotte LA, et al. (2004) Play signaling and the perception of social rules by juvenile chimpanzees (Pan troglodytes). J Comp Psychol 118(2):149

Flack JC, Krakauer DC (2006) Encoding power in communication networks. Am Nat 168:97–102

Flack JC, Krakauer DC, et al. (2005) Robustness mechanisms in primate societies: a perturbation study. Proc R Soc Lond B 272:1091–1099

Flanagan J (1989) Hierarchy in simple egalitarian "societies." Annu Rev Anthropol 18:245–266

Frank S (1995) Mutual policing and repression of competition. Nature 377:520–522

Frank S (2003) Repression of competition and the evolution of cooperation. Evolution 57:693–705

Frank S (In Press) Evolutionary foundations of cooperation and group cohesion. In: Levin SA (ed) Games, groups and the global good

Godelier M, Strathern M (1991) Big men and great men: personifications of power in melanesia. Cambridge University Press, Cambridge

Green J, Haidt J (2002) How (and where) does moral judgment work? Trends Cogn Sci 6:517–523

Green J, Sommerville RB, et al. (2001) An fmri study of emotional engagement in moral judgment. Science 293:2105–2108

Gruter M (1986) Otracism as a social and biological phenomenon. Ethol Sociobiol 7:149–158

Haidt J (2001) The emotional dog and its rational tail: a social intuitionist approach to moral judgment. Psychol Rev 108:814–834

Haidt J (2003) The moral emotions In: Davidson RJ, Scherer KR, Goldsmith HH (eds) Handbook of affective sciences Oxford, Oxford University Press pp 852–870

Haidt J, Joseph C (2004) Intuitive ethics: how innately prepared intuitions generate culturally variable virtues. Daedalus 55–66 (Special Issue on Human Nature)

Haidt J, McCauley C, et al. (1994) Individual differences in sensitivity to disgust: a scale sampling seven domains of disgust elicitors. Pers Individ Dif 16:701–713

Haidt J, Rozin P, et al. (1997) Body, psyce, and culture. Psychol Dev Soc 9:107–131

Hauser MD (2006) Moral minds: the nature of right and wrong. Harper Perennial, New York

Hauser MD, Cushman F, et al. (2007) A dissociation between moral judgments and justifications. Mind Lang 22:1–21

Hauser MD, Young L, et al. (2007) Reviving rawls' linguistic analogy: operative principles and the causal structure of moral actions. In: Sinnott-Armstrong W (ed) Moral psychology. The evolution of morality: adaptations and innateness. vol 1. MIT Press, Boston

Henrich J, Boyd R, et al. (2004) Foundations of human sociality: economic experiments and ethnographic eveidence from fifteen small-scale societies. Oxford University Press, New York

Henrich J, McElreath R, et al. (2006) Costly punishment across human societies. Science 312:1767–1770
Heyes CM (1998) Theory of mind in nonhuman primates. Behav Brain Sci 21:101–134
Hume D (1969 [1739]) A treatise on human nature. Penguin, London
Hutchinson GE (1957) Concluding remarks. Quant Biol 22:415–427
Kahneman D, Tversky A (eds) (2000) Choices, values, and frames. Russell Sage Publications, Cambridge
Kahneman D, Tversky A (2000) Conflict resolution: a cognitive perspective. In: Kahneman D Tversky A (eds) Choices, values and frames. Russell Sage Foundation, Cambridge, pp 473–487
Kant I (1959 [1785]) Foundations of the metaphysics of morals. Bobbs-Merrill, Indianapolis, NY
Koenigs M, Young L, et al. (2007) Damage to the prefrontal cortex increases utilitarian moral judgments. Nature 446:908–911
Krakauer D, Page K, et al. (In Press) Diversity, dilemmas, and monopolies of niche construction. Am Nat
Krakauer D, Plotkin JB (2002) Redundancy, anti-redundancy, and the stability of genomes. Proc of the Nat Acad of Sci 99:1405–1409
Lakoff G (1987) Women, fire, and dangerous things. University of Chicago Press, Chicago
Laland K, Oldlling-Smee FJ, et al. (1996) The evolutionary consequences of niche construction: a theoretical investigation using two-locus theory. J Evol Biol 9:293–316
Lewontin RC (1982) Organism and environment. In: Plotkin HC (ed) Learning, development, and culture. Wiley, New York
Loewenstein G (2008) Exotic preferences: behavioral economics and human motivation. Oxford University Press, Oxford
Lyons DE, Santos LR, et al. (2006) Reflections of other minds: how primate social cognition can inform the function of mirror neurons. Curr Opin Neurobiol 16:203–234
Maskin ES (2008) Mechanism design: How to implement social goals. http://ideas.repec.org/s/ads/wpaper.html, Institute for Advanced Study, School of Social Science
Masters RD, Gruter M, (eds) (1992) The Sense of Justice: biological foundations of law. Sage Publications, Newbury Park, CA
Mikhail J (2007) Universal moral grammar: theory, evidence and the future. Trends Cogn Sci 11:143–152
Moll J, Zahn R, et al. (2005) The neural basis of human moral cognition. Nat Rev Neurosci 6:799–809
Nakamaru M, Levin SA (2004) Spread of two linked social norms on complex interaction networks. J Theor Biol 230:57–64
Norwich JJ (1989) A history of Venice, Vintage, NY
Nowak M, Page K, et al. (2000) Fairness verses reason in the ultimatum game. Science 289
Nowak M, Sigmund K (1998) Evolution of indirect reciprocity by image scoring. Nature 393: 573–577
Nowak MA, Sigmund K (1993) A strategy of win-stay, lose-shift that outperforms tit-for-tat in the prisoner's dilemma game. Nature 364:56–58
O'Neill P, Petrinovich L (1998) A preliminary cross-cultural study of moral intuitions. Evol Hum Behav 19:349–367
Odling-Smee FJ (2007) Niche inheritance: a possible basis for classifying multiple inheritance systems in evolution. Biol Theory 2:276–289
Odling-Smee FJ, Laland KN, et al. (2003) Niche construction: the neglected process in evolution. Princeton University Press, Princeton
Page K, Nowak M (2002) Empathy leads to fairness. Bull Math Biol 64:1101–1116
Pizzaro D, Bloom P (2001) The intelligence of the moral emotions: a comment on Haidt (2001). Psychol Rev 110:293–296
Preston SD, de Waal F (2002) Empathy: its ultimate and proximate bases. Behav Brain Sci 25:1–20
Rawls J (1971) A theory of justice. Harvard University Press, Boston
Rochat M, Serra E, et al. (2008) The evolution of social cognition: goal familiarity shapes monkeys' action understanding. Curr Biol 18:227–232

Santos LR, Nissen AG, et al. (2006) Rhesus monkeys (Macaca mulatta) know what others can and cannot hear. Anim Behav 71:1175–1181

Semendeferi K, Damasio H, et al. (1997) The evolution of the frontal lobes: a volumetric analysis based on three-dimensional reconstructions of magnetic resonance scans of human and ape brains. J Hum Evol 32(4):375

Sherif M (1935) A study of some social factors in perception. Arch Psychol 2

Siegal M, Varley R (2002) Neural systems involved in 'theory of mind'. Nat Rev Neurosci 3:463–471

Slovic P (2000) The construction of preference. In: Kahneman D, Tversky A (eds) Choices, values and frames. Russell Sage Foundation, Cambridge, 489–502

Stark R (2001) Gods, rituals, and the moral order. J Sci Study Relig 40:619–636

Tomasello M, Call J (1997) Oxford University Press, Primate Cognition

Trivers R (1971) The evolution of reciprocal altruism. Q Rev Biol 46:35–57

Wood JN, Glynn DD, et al. (2007) The perception of rational, goal-directed action in nonhuman primates. Science 371:1402–5

Young YF, Cushman F, et al. (2007) The neural basis of the interaction between theory of mind and moral judgment. Proc Natl Acad Sci 104:8235–8240

Games, Groups, Norms, and Societies

Simon Levin

Abstract The origin and evolution of social norms, social institutions (including religions), and moral systems involve an interplay among processes played out on diverse scales of space, time, and complexity. Such norms, social institutions and systems (collectively referred to here as institutions) emerge from the collective actions of individuals, and feed back to influence those behaviors, but on much faster time scales than the institutions themselves change. In evolutionary biology, this is an example of what Janzen (1980) termed "diffuse coevolution," in which an evolutionary response is not to a single agent (tight coevolution), but rather is a diffuse response to a collection of agents (or species). Dealing with such multiple dimensions requires a new kind of game theory, not only multi-player but also multi-dimensional in other ways. Group formation and the resultant collective actions may lead to diffuse benefits for group members, but are sustained by individual decisions regarding costs and benefits within a social context. To sustain individual behaviors in the collective good, groups and societies develop explicit and implicit reward and punishment schemes, including moral systems. Understanding the interplay among these various players, operating on diverse scales, will require extension of game theoretical concepts to address dynamics on multiple scales, including analysis of meta-games, in which evolved strategies are diffuse responses to collections of situations.

1 Introduction

A fundamental insight of evolutionary theory is that the ultimate and proximate explanations for observed patterns and behaviors need not coincide. The original reasons that characteristics arose may have long been obscured, simply precursors to chains of change that gave rise to current features. Indeed, those initial steps may

S. Levin
Department of Ecology and Evolutionary Behavior, Princeton University, Room 203, Eno Hall, Princeton, NJ 08544-1003, USA

S.A. Levin (ed.), *Games, Groups, and the Global Good*, Springer Series in Game Theory, 143
DOI: 10.1007/978-3-540-85436-4_8, © Springer Physica-Verlag Berlin Heidelberg 2009

have represented chance alone, and conveyed little or no selective advantage. That is the essence of evolutionary change via natural selection-undirected mutations create the variation on which selection can act, but then initial such steps provide the template for later and more substantial changes.

Thus we can understand the evolution of the remarkable shell patterns of many marine invertebrates, the striking coat patterns of the megafauna of the African savannas, or the beautiful wing patterns of myriad species of butterfly. The initial appearance of such patterns probably represented accidents in development, byproducts of other adaptations or simply chance events. But once such patterns arose, they created variation among individuals on which selection could act through mate recognition, protection from predation or other ecological factors. Much theory, though contested recently, argues that some of the most dramatic examples, such as the wondrous displays of the peacock, involve sexual selection, in which deviation from the norm becomes self-fulfilling evidence of higher fitness.

Darwin (1871), in the *Descent of Man*, wrote:

> *Sexual selection implies that the more attractive individuals are preferred by the opposite sex; and as with insects, when the sexes differ, it is the male which, with some rare exceptions, is the more ornamented, and departs more from the type to which the species belongs;—and as it is the male which searches eagerly for the female, we must suppose that the females habitually or occasionally prefer the more beautiful males, and that these have thus acquired their beauty.*

However, what constitutes beauty may be ineluctably tied up with what females associate with conveying higher fitness; so beauty becomes defined by fitness, and fitness by beauty, in a self-reinforcing cycle. The problem of how pattern originates remains unaddressed.

The most intriguing and influential model of how endogenous pattern can arise in an initially homogeneous medium was introduced by the brilliant British mathematician, Alan Turing (1952), who posited that all one needed were two interacting "species," one an *activator* and the other an *inhibitor*, provided that the inhibitor diffused at a sufficiently higher rate. Although Turing's model was presented in mathematical terms, the intuition was simple: Random fluctuations in the concentration of the activator could break symmetry, and the higher diffusion rate of the inhibitor not only removed its local inhibitory influence, but furthermore introduced further symmetry-breaking in other locations. The outcome of such events can be endogenous pattern formation due to "short-range activation" and "long-range inhibition."

More generally (Levin and Segel 1985), from pattern formation in chemical systems to pattern formation in cultural systems, the interplay between mechanisms that operate locally to break symmetry, and those that operate at longer scales to suppress deviations, can lead endogenously to structure, without a blueprint. What holds for pattern formation in the developing organism holds equally for pattern formation within societies: The ecological and evolutionary reasons groups formed initially may provide only the first glimpses of the reasons that they persist. It is impossible for a species to maintain a perfectly uniform spatial distribution, especially in a turbulent environment, so patterns of distribution of almost all species are patchy. In the plankton, this is well known, and much research has been directed to understanding both the patterns of distribution and their causes. For the phytoplankton, the small plants that occupy the upper reaches of the ocean, the patterns mirror to a large extent those seen in inanimate particles, reflecting also the physics of the oceans. But once such aggregations form, natural selection has a substrate on which to operate.

For a variety of reasons including resource acquisition, access to mates, herbivory and predation, plants and animals that live in aggregations have fitness expectations different from those in lower-density circumstances – maybe higher, maybe lower, but different. Natural selection then favors behaviors, from the buoyancy characteristics of the phytoplankton to the active swimming of zooplankton, that increase the likelihood that individuals will find themselves in higher fitness conditions. Attention to such traits has been increasing even for the study of bacteria, which form associations called biofilms based on quorum sensing of local densities (Miller and Bassler 2001; see also Holloway 2004). Such bacterial associations can involve hundreds of species, like the plaques that are the favorite targets of dental technicians. Another and classic example is provided by the cellular slime mold (Bonner 1988), a stunning example of cooperation among soil amoebae; this has provided an ideal study system both for pattern formation and for the evolution of cooperation, in large part because of the conflicts between individual and group payoffs. The slime mold is one of the most basic templates for the application of game-theoretical notions.

2 Group Formation and Dynamics

In animal societies, and especially in human groups, things are of course more complicated. Among humans, in particular, language and other forms of communication change the nature of the interactions in fundamental ways, and have dramatically accelerated the pace of social evolution.

Still, some of the same principles apply. Whether one is exploring the evolutionary history of cooperation, or simply how groups form within societies, the initial steps of group formation may involve the same sorts of random instabilities and associations that characterize simpler organisms. For species like bacteria and slime molds, exuded chemicals or exogenous environmental cues may provide the only signals needed to produce aggregation. For higher organisms like fish, birds, and grazing ungulates, simple models show that attraction and imitation also can produce remarkable consensus in movements, even when only a small fraction of individuals exercise leadership (Couzin et al. 2005). Such models are extremely suggestive for the understanding of consensus formation and cooperation in human societies, and indeed models that incorporate little more than homophilous imitation can lead to the separation of societies into distinct groups with similarly distinct attitudes and opinions (Durrett and Levin 2005).

Human societies are much more complicated than aggregates of other animals, which are in turn much more complicated than groups of plants or amoebae or bacteria. In human societies, computation, calculation, and communication all play crucial roles. Still, as has been well recognized even in the popular literature (Gladwell 2000), imitation and herd behavior are also major factors, and societies can be directed or misdirected by a few strong-minded individuals who would rather lead than follow. Thus there is much to be learned by working our way up the evolutionary ladder, from relatively simple unicellular organisms to highly complex metazoans, understanding the mechanisms that lead to grouping and cooperation in other organisms, and asking both to what extent such simple mechanisms continue to hold among more complex organisms, and what new mechanisms must be invoked to explain group behavior.

3 Cooperation, from Bacteria to Bees

Charles Darwin (1809–1882), more than anyone else, revolutionized biology, by explicating how the patterns we see in Nature could be explained, to large extent, as the result of the process of natural selection. But even for Darwin, challenges remained, especially regarding altruism. Again, in the *Descent of Man*, he wrote:

> *Selfish and contentious people will not cohere, and without coherence nothing can be effected. A tribe possessing the above qualities in a high degree would spread and be victorious over other tribes; but in the course of time it would, judging from all past history, be in its turn overcome by some other and still more highly endowed tribe. Thus the social and moral qualities would tend slowly to advance and be diffused throughout the world.*
>
> *But it may be asked, how within the limits of the same tribe did a large number of members first become endowed with these social and moral qualities, and how was the standard of excellence raised? It is extremely doubtful whether the offspring of the more sympathetic and benevolent parents, or of those which were the most faithful to their comrades, would be reared in greater number than the children of selfish and treacherous parents of the same tribe. He who was ready to sacrifice his life, as many a savage has been, rather than betray his comrades, would often leave no offspring to inherit his noble nature. The bravest men, who were always willing to come to the front in war, and who freely risked their lives for others, would on an average perish in larger number than other men. Therefore it seems scarcely possible (bearing in mind that we are not here speaking of one tribe being victorious over another) that the number of men gifted with such virtues, or that the standard of their excellence, could be increased through natural selection, that is, by the survival of the fittest.*

Darwin's writings are remarkable for their clear presentation of problems that remain central to evolutionary biology today, and for raising possible explanations in the form of reciprocal altruism and group selection, both of which remain hot topics today.

Cooperation and apparently altruistic behavior are seen at all levels of biological organization; indeed, any organism is to some extent the outcome of cooperation among its genes. Most organisms that we can see, at least without the help of a microscope, are multicellular, the ultimate in cooperation, and understanding how that multicellularity arose is a core problem in biology (Levin 2006; Bonner 1998).

The bacterial members of biofilms produce many extracellular compounds, at cost, that benefit their neighbors. The stalk cells in the cellular slime mold have foregone reproduction, while the spore cells benefit from this "altruism." Bacteria, fungi, and plants produce siderophores, which chelate iron so that it can be more efficiently imported, benefiting not only the producer, but also other organisms. Nitrogen fixation involves similar externalities. Any such situation immediately raises game-theoretical issues, since cheaters can reap some benefits at least without paying the price of cooperation. Any committed Darwinian seeks to understand the evolution of such characters within the framework of natural selection, and microbial systems provide perhaps the most basic candidates for study (West et al. 2007; Wingreen and Levin 2006; Nadell et al. 2008).

Many explanations exist for the evolution of such cooperative behavior. The great biologist, Haldane, caught the essence of one argument when he reportedly said that he would lay down his life for two brothers (but not one), or for eight cousins. His point was that, if he had had two brothers, they would have had as may of his genes as did he; and so would eight cousins. Thus, from the viewpoint of natural selection at its most basic, evolution would treat Haldane and his two brothers, or Haldane and his eight cousins, equivalently. Hamilton (1964a, b), in a pair of fundamental papers, formalized this notion, later termed kin selection, by showing that individuals will engage in behaviors beneficial to others if the cost of doing so is less than the benefits to others, multiplied by their genetic relatedness to the actor. This particularly helped illuminate the most dramatic examples of cooperation, eusociality in the haplodiploid insects (the bees, the ants, and the wasps). Because males of these species are haploid, arising from unfertilized eggs, sisters share $^3/_4$ of their genes, and hence the payoff for cooperation with siblings is 50% higher than in diploid species like humans. Though the importance of haplodiploidy is debated as insufficient to cause or maintain eusociality in those species (Gadagkar 1991), it seems obvious that it must have an influence.

More generally, kin selection is obviously neither necessary nor sufficient to explain the high level of cooperation in biological communities, even at the level of bacteria. We know from our own experience that reciprocal altruism ("You scratch my back, and I'll scratch yours") can play a role, and that reciprocity can be implicit in a structured environment, in which individuals interact more with a subset of neighbors, even if they are not aware of with whom they are interacting. This can explain much of the cooperation seen among bacteria, among plants, and among animals. Indeed, in spatially structured environments, the potential benefits to the individual actor itself are enhanced and the externalities are reduced, increasing the likelihood of prudent behavior such as reduced resource use. In an age of globalization, the breakdown of such structure, and of traditional cooperative groups, is to some extent at the root of our global environmental problems, since broader spatial mixing leads to higher individual discount rates (Levin 1999). Solving our global problems must involve understanding how cooperative groups have arisen in animal societies, including humans; why they break down as groups get larger; and what is needed to extend cooperation and moral systems to higher levels of organization. Much research has been directed to understanding how the social context guides and constrains individual behaviors; much less has been directed to understanding how the collective consequences of individual behaviors leads to the creation, evolution and maintenance of the social context. Rousseau (1762), influenced deeply by John Locke, put the evolution of such social contexts and contracts at the very core of human civil societies, given legitimacy by the collective will of the people.

4 Animal Schooling and Swarming, and the Role of Leadership

Non-human animal groups provide ideal systems for study, both from an ecological and from an evolutionary perspective. Groupings may result as a consequence of common responses to environmental signals, such as wind drafts or chemical gradients; but they also arise endogenously, as the emergent consequence of inter-individual interactions. Flocks of birds, schools of fish and swarms of insects exhibit fluid-like motions that suggest immediate analogies with statistical mechanical approaches to fluids (Flierl et al. 1999); but the most basic of such approaches ignore variation among individuals in terms of their rules of behavior, and hence make impossible examination of the evolution of strategies that lead to grouping.

The increased ability to carry out large-scale simulations has made possible the development of agent-based models, in which every individual is assigned specific characteristics. Using such an approach, Couzin et al. (2005) are able to explore the tradeoffs between behaviors (such as chemotaxis) based on using information about the external environment, and behaviors (such as imitation, attraction, or repulsion) gained from watching other individuals. Perhaps the most striking insight from these models is that a very small group of individuals ("leaders") with strong predilections to move in a specific direction can organize large groups of others ("followers") to do the same. It matters not at all in such models whether the direction is a good one or not; strongly held opinions are all that matter. Consider the implications for opinion formation in human societies, and the power that rests in self-appointed leaders. These models also form powerful templates for the study of consensus formation when there are differing opinions, and how compromise can give way to sudden and dramatic shifts in direction or opinion.

Individuals aggregate for a number of reasons, from facilitation of sexual reproduction to protection from predation to the benefits of collective foraging for food or information. From the perspective of the group, to optimize foraging success, too few leaders is not a good thing, but neither is too many. However, except for eusocial species, this is not the right way to look at the problem: Collective behavior must reflect the emergent consequences of the resolution of individual costs and benefits calculations. This is again a game-theoretic problem, in which evolutionarily stable strategies or more complicated outcomes define the leader-follower conflict, and in which game-theoretical analyses can shed light on the emergence of leadership and group organization.

The approach of this paper is to assume that the genetic determinants of human behavior have been determined over long time scales, and in response to a wide range of situations, and that these evolutionary responses shape the rules that guide individual decisions. Social behavior is thus seen as the consequence of these rules, and the issue of how and why those rules evolved is set aside. The fascinating question of gene-culture coevolution has been dealt with elegantly by others (Cavalli-Sforza and Feldman 1981; Boyd and Richerson 2005; Wilson 1975).

5 Groups and Norms in Human Societies

Whether in inanimate or animate systems, pattern formation involves two basic mechanisms: Autocatalytic influences that allow symmetry breaking and the initiation of aggregations, and on larger and longer scales, inhibitory influences that prevent those aggregations from engulfing the whole system. Both of those dimensions are in evidence in the dynamics of groups. The mechanisms already discussed – simply imitation and attraction – can provide the autocatalysis that initiates groups. Once groups form, the differential fitnesses realized inside vs. outside groups can lead to selection, cultural or genetic, for behaviors that increase or decrease group adherence, or that limit admission to the groups. The benefits of belonging to a group do not in general increase indefinitely with group size: Larger groups control more resources, but their members have less influence on group direction, and may share less in the utilization of the resources. Belonging to a political party whose membership is 45% of the population is probably not as good as belonging to one with 55% in our society, but belonging to one that held nearly 100% of the population, as in nations where true democracy is suppressed, would be like belonging to none at all. Group foraging success may increase indefinitely with group size, but individual rewards for members will not. Thus, typically, from the viewpoint of members, there will be an optimal group size that maximizes individual fitness. The corollary of this is that as the group reaches or exceeds that value, group members have an incentive for restricting access to new members, even though solitary individuals still would be better off joining the group than going it alone. This conflict is a nearly universal property of groups in our societies, and explains the exclusive policies of private clubs as well as anti-immigration attitudes currently in evidence in our nation.

What are the implications of these conclusions? First of all, when groups are small, they need to attract and hold members. This leads to proselytism and inducements to join, and possibly penalties for leaving. Group members may pay small prices for belonging, as signals of their commitment, in expectation that group membership will bring benefits that far exceed those costs. For example, in a variety of social situations, I willingly don a tie, although it makes me much less comfortable, as a signal that I accept the social norms of the group and am not going to cause trouble. Were I to ignore this dress code, it might be taken as a signal that I was on the road to revolution, which might lead to my exclusion from the group and its benefits.

Secondly, when groups get too large, their desire to attract new members may be replaced by a desire to become exclusive. Fences make good neighbors, or at least they make neighbors, so the motivation both to retain current members and to exclude new ones leads to the building of fences: Individuals within groups are discouraged from excessive fraternization with members of other groups, and mixing taboos emerge. Of course, it is not far from this step to the encouragement of conflict among groups, and that conflict is facilitated by and facilitates the maintenance of group integrity and sharp boundaries between groups. I believe that this is what Anthony Appiah means when he talks of "the other," and the fact that hostile relations are more likely between groups that have been in historical contiguity rather than those who have had little interaction with one another. No fences are needed if realms do not abut.

The simplest model of imitation is the "voter model," the staple of percolation theory. In this model, individuals have one of two attitudes, but feel no attachment to their particular view. At random, one individual interacts with another, generally drawn from those in some local neighborhood (say on a grid), and adopts that individual's attitude (in other words, changes attitude if the neighbor's is different). In one or two dimensions, this always leads to consensus for one attitude or the other. To deal with the erection of fences, Richard Durrett and I considered a modification of this model: Individuals sample all of their neighbors, and change their attitudes only provided the fraction of their neighbors of a different opinion exceeds some predetermined threshold. If thresholds are low, individuals are always changing their attitudes, and no groups form. If thresholds are high, individuals never change their attitudes, and no groups form. At intermediate thresholds, however, the system can be observed (on the fast time scale) to break into groups, which then interact with each other across their boundaries on much slower time scales (see Ehrlich and Levin 2005). This of course is only a crude beginning; with Adrian de Froment and others, I have been investigating the evolution of these thresholds themselves, which interacts with the changing benefits individuals receive as a consequence of changing group membership.

The evolution of languages shares many features with the above dynamic, and in fact is a special case of it. According to the biblical story of the Tower of Babel, the Lord caused peoples to speak different languages in order to confuse their interactions with one another because of their arrogance in building a tower to heaven. However one interprets this allegory, there are elements of truth. By the same sorts of hydrodynamic instabilities that lead to pattern formation in any system, the viscous and distributed nature of human societies makes small dialect differences a virtual certainty. These dialects, for geographical reasons, become associated with particular groups, and help to define those groups. It is natural then for new verbal expressions to arise that are only to be shared within the group, and that create barriers to communication with other groups.

6 Formalizing Rules and Codes of Conduct: The Evolution of Moral Systems

As groups are established and solidified, the mechanisms for sustaining group structure can become more and more formalized. In general, the ways to do that often are to increase harmony within groups, and conflict among groups. The latter is not essential, but the correlation is hard to avoid: Inter-group competition is a strong driver of within-group cooperation, and indeed even temporary coalitions among historical enemies may emerge in the face of a common threat. This raises the hope that humanity can recognize that the threats of environmental degradation, nuclear conflagration and other aspects of global decay should provide the incentives sufficient for humanity to bond together to oppose these common enemies. The game-theoretic considerations should similarly encourage cooperation, whether the opponents are other humans or the emergent fallout from anthropogenic causes.

The formalization of mechanisms typically goes through stages, though steps can be skipped. Informal norms and customs can arise, like codes of dress, in which the consequences of deviation are unwritten and softly enforced, but still costly to the deviant. It is only a short step from there to the regularization of these into laws with legally mandated penalties. As I write this, today's local paper in Trenton, New Jersey, reported that some in our local legislative body would outlaw baggy pants, and impose penalties on young boys who wear them. Of course, such laws and statutes are only feasible if formal institutions exist to enact them and enforce them, so there is an obvious mutualism in the emergence of institutions and of the laws that help define them.

The existence of laws stabilizes behaviors that society deems appropriate because of the threat of punishment; punishment is not an abstract threat, as long as it is used, because potential violators can observe its application. Of course, there are old laws on the books that societies consider archaic, and that are not enforced; as a consequence, they tend to be ignored. Religious systems, in contrast, carry this evolution a step further, inferring rewards and penalties whose application is not testable, in some cases because they are only applied in an afterlife. Since the reality of the payoffs is not testable, they are less likely to be viewed as archaic and hence tend to be robust to change. Many behaviors mandated by religious practice have clear basis in moral, ethical, and social behavior; but others do not, and adherence to them is more a test of one's faith and unquestioned fidelity than of support for any moral principle. Indeed, in my own tradition, Judaism, orthodox teaching would argue against seeking a rational basis for many such practices and prohibitions, relying instead on unquestioning acceptance or rabbinical teachings. It would reject, for example, efforts at health-based explanation of prohibitions against eating certain foods like pork and shellfish, regarding these prohibitions simply as necessary to follow because they were divinely given. If nothing else, this makes them inseparable from orthodox practice, and immutable to change within that part of Judaism. I am not arguing that religious laws and customs arose only to maintain the integrity of groups, but the effects are clear in the same way that prohibitions against intermarriage and even interfaith dating help maintain group integrity through the construction of fences.

Understanding how social norms, social institutions (including religions), and moral systems arise and evolve requires consideration of the interplay among processes played out on diverse scales of space, time, and complexity. Such norms, social institutions and systems (collectively referred to here as institutions) emerge from the collective actions of individuals, and feed back to influence those behaviors, but on much faster time scales than the institutions themselves change. In evolutionary biology, this is an example of what Janzen (1980) termed "diffuse coevolution," in which an evolutionary response is not to a single agent (tight coevolution), but rather is a diffuse response to a collection of agents (or species). Dealing with such multiple dimensions requires a new kind of game theory, not only multi-player but also multi-dimensional in other ways.

One benefit of group membership is collective intelligence: Animals forage in groups because they profit from the information learned by and from others. When difficult decisions are to be made, group wisdom can be helpful (or occasionally misleading). Underlying moral systems are principles, like fairness, that allow individuals to develop heuristics that provide the basis for decision-making when time or complexity makes *de novo* computation difficult or impossible. This is bounded rationality, in the sense of Herbert Simon (1957). The rules that govern individual behaviors in these complex systems often produce apparently anomalous behaviors, analogues of optical illusions that result in the misapplication of general principles that work well on the average (or that may have evolved in response to circumstances no longer relevant), but that can misfire in particular situations. Efforts to understand these apparent anomalies, and how individuals respond in contests like the ultimatum game, cannot be understood within the unique circumstances that define these anomalies. They require consideration of meta-games, in which evolved strategies are diffuse responses to collections of situations. This is new territory for game theorists, but one that is essential for understanding the fascinating emergence of community norms and moral systems.

Acknowledgments Simon Levin is in the Department of Ecology and Evolutionary Biology at Princeton University, and is a Fellow of the Beijer Institute of Environmental Economics and of Resources for the Future. He gratefully acknowledges the support of the National Science Foundation and the helpful comments of Adrian de Froment, Herbert Gintis, Carey Nadell and Elinor Ostrom.

References

Bonner JT (1988) The evolution of complexity by means of natural selection. Princeton University Press, Princeton

Bonner JT (1998) The origins of multicellularity. Integr Biol 1:27–36

Boyd R, Richerson PJ (2005) The origin and evolution of cultures. Oxford University Press, Oxford

Cavalli-Sforza LL, Feldman MW (1981) Cultural transmission and evolution: a quantitative approach. Monographs in population biology 16. Princeton University Press, Princeton
Couzin ID, Krause J, Franks NR, Levin SA (2005) Effective leadership and decision making in animal groups on the move. Nature 433:513–516
Darwin C (1871) The descent of man, and selection in relation to sex. John Murray, London (See the Complete Works of Charles Darwin Online)
Durrett RT, Levin SA (2005) Can stable social groups be maintained by homophilous imitation alone? J Econ Behav Organ 57:267–286
Ehrlich PR, Levin SA (2005) The evolution of norms. PLoS Biol 3(6):e194
Flierl G, Grünbaum D, Levin S, Olson D (1999) From individuals to aggregations: the interplay between behavior and physics. J Theor Biol 196:397–454
Gadagkar R (1991) On testing the role of genetic asymmetries created by haplodiploidy in the evolution of eusociality in the hymenoptera. J Genet 70(1):1–31
Gladwell M (2000) The tipping point. How little things can make a big difference. Back Bay, MA
Hamilton WD (1964a) The genetical evolution of social behavior I. J Theor Biol 7:1–16
Hamilton WD (1964b) The genetical evolution of social behavior II. J Theor Biol 7:17–52
Holloway M (2004) Microbes seem to talk, listen and collaborate with one another–fodder for the truly paranoid. Bonnie L. Bassler has been eavesdropping and translating. http://www.sciam.com/article.cfm?chanID = sa006&colID = 30&articleID = 0001F2 DF-27D8–1FFB-A7D883414B7F0000
Janzen, DH (1980) when is it coevolution? Evolution 34(3):611–612
Levin SA (1999) Fragile dominion: complexity and the commons. Perseus, MA
Levin SA (2006) Unity from division: In search of a collective kokoro. Paper given at 2006 Kyoto International Culture Forum. http://www.forumkokoro.jp/2006/index_e.html#
Levin SA, Segel LA (1985) Pattern generation in space and aspect. SIAM Rev 27:45–67
Miller M, Bassler B (2001) Quorum sensing in bacteria. Ann Rev Microbiol 55:165–199
Nadell C, Bassler B, Levin SA (2008) Observing bacteria through the lens of social evolution. J Biol 7:27
Rousseau, JJ (1762) Du Contrat Social; ou Principed du Droit Politique
Simon H (1957) A behavioral model of rational choice. In models of man, social and rational: mathematical essays on rational human behavior in a social setting. Wiley, New York
Turing A (1952) The chemical basis of morphogenesis. Phil Trans Royal Soc London, Series B 237:37–72
West S, Diggle SP, Buckling A, Gardner A, Griffin A (2007) The social lives of microbes. Ann Rev Ecol Evol Syst 38:53–77
Wilson EO (1975) Sociobiology: the new synthesis. Harvard University Press, Cambridge
Wingreen N, Levin SA (2006) Cooperation among microorganisms. PLoS Biol 4(9):e299. doi:10.1371/journal.pbio.0040299

Evolutionary Theory and Cooperation in Everyday Life

David Sloan Wilson and Daniel Tumminelli O'Brien

Abstract A rapid process of integration is taking place for theories of cooperation in both evolutionary biology and the human social sciences. It includes a return to the concept of social groups as like single organisms, which was once commonplace but was eclipsed by various forms of individualism that became dominant during the second half of the twentieth century. So far, the integration has taken place mostly within academia, but it is highly relevant to everyday life, as we show with our research on cooperation and its consequences at a city-wide scale in Binghamton, New York.

Theories of cooperation in both biology and the human social sciences have had a turbulent history. During the nineteenth and early twentieth centuries, it was common to regard societies as like organisms in their own right. This holistic worldview was largely replaced during the middle of the twentieth century by a more reductionistic view that sought to explain as much as possible in terms of individual self-interest. In evolutionary biology, this trend was represented by the rejection of group selection in favor of "the theory of individual selection" and ultimately "selfish genes." In the human social sciences, the trend was represented by "methodological individualism" (Sober and Wilson 1998; Wegner 1986) and especially rational choice theory in economics, which assumes that all human preferences can be understood in terms of individual utility maximization, with the utility usually conceptualized as material gain (e.g., income).

These various forms of individualism did not develop in a coordinated fashion and differ from each other in important details. Scientists like to think that they are immune from cultural influences. We know that this is not the case for Darwin and his contemporaries, who were influenced by the cultural assumptions of the Victorian Age (e.g., Browne 1995, 2002). Historians will probably look back upon the twentieth century and marvel at the degree to which formal scientific theories reflected widespread cultural assumptions about individualism.

D.S. Wilson
Departments of Biology and Anthropology, Binghamton University (State University of New York), Binghamton, NY 13902-6000, USA
e-mail: Dwilson@binghamton.edu

S.A. Levin (ed.), *Games, Groups, and the Global Good*, Springer Series in Game Theory, DOI: 10.1007/978-3-540-85436-4_9, © Springer Physica-Verlag Berlin Heidelberg 2009

Thankfully, science has a way of correcting itself, even if decades are sometimes required. In evolutionary theory, the concept of major transitions has turned individualism on its head (Maynard Smith and Szathmary 1995, 1999; Hammerstein 2003). We now know that evolution takes place not only by small mutational change – individuals from individuals – but by groups becoming so well integrated that they become higher-level organisms in their own right – individuals created from groups. The seemingly antiquated concept of society as an organism has become so well established that the organisms of today are literally the societies of past ages. Moreover, it is quite possible that human evolution represents nature's newest major transition. Mechanisms evolved that suppressed selection within groups, causing selection among groups to become a strong evolutionary force (Bingham 1999; Boehm 1999; Richerson and Boyd 2005; Wilson 2006; Wilson and Wilson 2007). Our ancestors became the primate equivalent of bodies and beehives. As with previous major transitions, such as multicellular organisms and eusocial insect colonies, our capacity for cooperation within groups enabled us to become ecologically dominant, occupying the entire planet and displacing many other species along the way.

Since human cooperation includes the social sharing of information, the process of adaptation takes place primarily through cultural evolution rather than directly by genetic evolution (Richerson and Boyd 2005). An analogy with the mammalian immune system is apt; it is an elaborate set of adaptations that evolved by genetic evolution, including the process of antibody formation and selection that counts as an evolutionary process in its own right. The human capacity for behavioral change similarly consists of an elaborate architecture that evolved by genetic evolution, including an open-ended process that counts as evolutionary in its own right. Plotkin's (1994) term "Darwin Machine" elegantly captures the concept of a fast-paced process of evolution, built by the slow-paced process of genetic evolution.

The human social sciences have also become disenchanted with individualism (e.g., Fehr and Fischbacher 2003; Gintis 2005). Many economists concluded on their own that the self-regarding actor model wasn't working and that humans are guided by a more complex set of social preferences. The fields of experimental and behavioral economics represent an empirical effort to move beyond the self-regarding actor model (although the more general concept of optimization still plays an important role). Empirical research must be guided by theory, however. What is the theoretical framework for experimental and behavioral economics, if the self-regarding actor model is false? One answer is "psychology," but that merely identifies proximate psychological mechanisms. Where did they come from? As soon as we embark upon this kind of reasoning, all roads lead to evolutionary theory.

Formal theoretical models of cooperation are diverse in their specific approaches, but all of them center upon the fact that cooperation tends to be locally disadvantageous (Wilson and Wilson 2007, 2008). It is simply a fact of life that for a group to function well as a unit, members must perform services for each other, which requires time, cost and energy. These "for the good of the group" behaviors are vulnerable to passive free-riding and active exploitation. Individual-level adaptations such as better eyesight or sharper teeth provide an advantage to individuals, compared to other individuals in their immediate vicinity. They can evolve straightforwardly in a single group. If group-level adaptations are locally disadvantageous, how can they evolve? Only by providing a positive fitness advantage at a larger scale. Groups of cooperative individuals robustly outcompete other groups, even if individual cooperators are vulnerable to free-riding and exploitation within single groups. This has always been the central problem of multilevel selection theory, which is explicitly formulated in terms of selection differentials within and among groups. Other theoretical frameworks appear to explain cooperation in individualistic terms, but a closer look reveals the same logic. All of them assume that evolution takes place in multiple groups, that cooperation is selectively disadvantageous within each group, and evolves only by virtue of cooperative groups differentially contributing to the total population. In n-person game theory, for example, n refers to the size of the groups within which social interactions occur. Individuals employing cooperative strategies such as tit-for tat lose or at best match the fitness of the social partners with whom they actually interact.

It is important to stress that cooperation and other behaviors that are "for the good of the group" need not involve a high-degree of self-sacrifice. In some cases, public goods can be provided at minimal cost to the public good provider. The lower the cost, the less variation among groups is required for the behavior to evolve. For very low-cost public goods, random variation among groups is sufficient. Natural selection at the individual level proceeds primarily on the strength of random variation among individuals, which can also be a potent form of variation at the level of groups, especially when initial random variation becomes amplified by complex social dynamics within each group (Wilson 2004).

In addition to low-cost cooperative behaviors that evolve easily by between-group selection, complex social interactions can also result in multiple stable equilibria. These varied groups become separate environments, each featuring its own set of selective pressures on social behavior, but the equilibria can differ dramatically in their adaptedness at the group level (e.g., the selfish and cooperative equilibria of an iterated prisoner's dilemma model). Group-level selection is still required to select among equilibria (Boyd and Richerson 1992; Gintis 2000; Bowles 2003).

All of these cases – extreme altruism that cries out for an explanation, low-cost forms of cooperation, and multiple stable equilibria – share a common feature. The traits that cause whole groups to function well as units seldom maximize relative fitness within the groups. This robust conclusion follows from the basic concept of trade-offs. Adaptation at level X usually requires a process of natural selection at level X and is undermined by selection at lower levels (except in the case of multiple local equilibria), however weakly. The mechanisms that cause human groups to function as adaptive units might not look or be altruistic, but at some stage of their origin and maintenance, they probably required a process of group-level selection.

Theories of cooperation often appear individualistic because individual fitness is conceptualized as fitness averaged across groups, rather than relative fitness within groups. From this perspective, traits that are locally disadvantageous (a relative fitness comparison) can appear individually advantageous because "individual fitness" now includes the individual's share of the public goods that it provides its own group, plus the contribution of other cooperators to its own fitness. Regardless of the perspective, all models agree that for cooperation to evolve, *cooperators must interact with other cooperators and avoid interactions with non-cooperators*. This is the fundamental requirement, regardless of whether the segregation is accomplished by genealogical relatedness, behavioral sorting, or conditional strategies that govern the choice of behaviors.

Against this background, human social preferences emerge as a set of genetically evolved psychological adaptations that are impressively designed to facilitate cooperation and suppress free-riding and exploitation within groups, especially at the scale of small face-to-face groups characteristic of the ancestral human social environment (Boehm 1999; Hammerstein 2003; Gintis 2005). The experimental economics literature elegantly shows how cooperation falls apart or robustly asserts itself, when the mechanisms of social control that come naturally in small-scale human society are removed or provided. (e.g., Bowles 2008; Brosig 2002).

Moreover, human history since the invention of agriculture can be interpreted as a process of multilevel cultural evolution that has increased the scale of cooperative human society by many orders of magnitude, albeit with many reversals along the way (e.g., Boehm 1999; Wilson 2002; Turchin 2005). It is bracing to contemplate that so many human-related subject areas, such as psychology, cultural anthropology, history, economics, political science, religious studies, and sociology, are becoming integrated with each other and the biological sciences under the broad umbrella of evolutionary theory.

1 Human Cooperation in Everyday Life

These foundational developments in basic science are only beginning to be reflected in studies of cooperation in everyday life. Theoretical models and laboratory experiments need to be supplemented with the equivalent of field studies of non-human species in their natural environments. People from all walks of life need to be studied as they go about their daily lives. Not only can this naturalistic approach contribute to basic scientific research, but it is maximally relevant to improving the quality of life in a practical sense, resulting in a positive rather than a negative trade-off between basic and applied research.

In 2006, we began to study the city of Binghamton, New York from an evolutionary perspective with a focus on prosociality. We define "prosociality" as any attitude or behavior that is oriented toward others or society as a whole. It therefore includes but also goes beyond narrow definitions of cooperation. In particular, it is agnostic about whether helping others requires self-sacrifice on the part of individuals. In this chapter, we will provide an overview of our results in relation to the basic scientific literature on cooperation outlined in the previous section. Detailed results for each study are provided elsewhere (e.g., O'Brien et al. 2008 unpublished data).

The Binghamton neighborhood project: Binghamton is a small city (population approximately 50,000) in a region of New York that has been economically depressed over the last few decades. It is ethnically diverse, both from past immigrations from various parts of Europe and current immigrations from all over the world. Over 18 different primary languages are spoken by students in Binghamton's single high school. Binghamton's proximity to New York City introduces elements of the drug trade and other criminal activity, in addition to more positive influences such as an art scene. In short, Binghamton shares the same problems and potentials as many other cities, but its relatively small size makes it manageable as a "field site" for basic and applied research. The Binghamton Neighborhood Project (http://evolution.binghamaton.edu/bnp/) was initiated in 2006 to create a general infrastructure for community-based research from an evolutionary perspective, in coordination with EvoS, Binghamton University's campus-wide evolutionary studies program (http://evolution.binghamton.edu/evos/).

Measuring prosociality and its correlates: Our initial measure of prosociality was based on a survey given to nearly 2000 middle and high school students in collaboration with the Binghamton City School District (Wilson and O'Brien 2009). The survey included a 58-item "Developmental Assets Profile (DAP)" developed by Search Institute, a non-profit organization dedicated to the scientific study and improvement of communities (http://www.search-institute.org/). The DAP is widely used and highly regarded nationwide as an instrument that can help school districts and other community organizations measure and improve the quality of life for youth. Items on the DAP include a number of questions about prosocial attitudes and behavior that were used to create a prosociality subscale (e.g., "I am sensitive to the needs and feelings of others," "I am serving others in my community"). Other items on the DAP and background variables included in the survey were used to create subscales measuring various forms of social support (family, school, religion, neighborhood, extracurricular activities), personal psychological wellbeing, and performance on state-mandated math and English tests. Finally, residential location enabled us to create spatial maps of the variables using geographical information systems (GIS) technology and to link the data with other spatially based information such as US census statistics. Student identity was protected according to guidelines approved by both the school district and Binghamton University's Human Subject Review Board. The results can be summarized as follows:

1. *The prosociality of the individual correlates highly with the prosociality of the individual's social environment.* The overall correlation between individual prosociality and total social support is $r = 0.723$ (Pearson correlation coefficient, $p < 0.001$). Very simply, individuals who give to others are also very likely to get from others. This correlation is the fundamental requirement for prosociality to succeed as a behavioral strategy in a Darwinian contest, as outlined in the previous section of this chapter. The size of the correlation is surprising. At a very crude level, it is comparable to the coefficient of relatedness (r) in a theoretical model such as Hamilton's rule or the Price equation (Hamilton 1975). Remarkably, the chance of a highly prosocial student interacting with other highly prosocial individuals in the city of Binghamton is considerably higher

than the chance of an altruist having an altruist for a full sibling in a simple genetic model!

2. *Prosociality comes from multiple sources*. The most highly prosocial individuals receive social support from all their social environments – family, neighborhood, school, religion, and extracurricular activities (see Table 1). The total correlation coefficient reflects the separate contribution of each of these sources; evidently it really does take a village to raise a child. It is worth stressing that the contribution of religion appears quite modest compared to other sources of social support. Religion often attracts a disproportionate amount of attention in discussions of cooperation, as if cooperation would be impossible without religion. It is therefore important to put religion in perspective as one of several possible forms of social support.
3. *Space matters*. Given the fluid nature of modern social interactions, it might seem that one's physical neighborhood doesn't matter. On the contrary, a GIS map of the prosociality subscale shows a rugged landscape of hills and valleys, as shown in Fig. 1. The top figure was generated by a technique called kriging, which creates a continuous surface by calculating a value for each location based on the values of the neighboring data points. The peaks (dark areas) and valleys (light areas) are neighborhoods in which students score high and low on the prosociality subscale, respectively. The krig map approximates the actual spatial variation but is difficult to analyze statistically. For statistical analysis, the city is divided into 63 discrete census block groups as shown in the bottom map. This impressive spatial heterogeneity does not necessarily reflect the effect of neighborhoods per se; for example, family social support can itself be spatially structured. However, neighborhood quality is a statistically significant predictor of prosociality at both the individual and group levels. In other words, a student's prosociality score correlates not only with how that student rates his or her neighborhood, but also

Table 1 Two stepwise regression models using social support subscales to predict prosociality. In the first model, a subscale measuring general social support appears to make family social support insignificant (second column). When this subscale is removed, family social support becomes highly significant. See Wilson and O'Brien (2009) for more detailed discussion

Subscales in model	Standardized beta (semi-partial)[a]	Standardized beta (semi-partial)[b]
General	0.522 (0.376)***	–
Extra-curricular activities	0.146 (0.133)***	0.218 (0.204)***
School	0.119 (0.094)***	0.248 (0.208)***
Neighborhood	0.109 (0.091)***	0.174 (0.145)***
Religion	0.070 (0.066)***	0.115 (0.109)***
Family	0.015[c]	0.258 (0.211)***

[a] Total model was significant at $p < 0.001$ with $R = 0.758$
[b] Total model was significant at $p < 0.001$ with $R = 0.691$
[c] Not entered into model
***$p < 0.001$

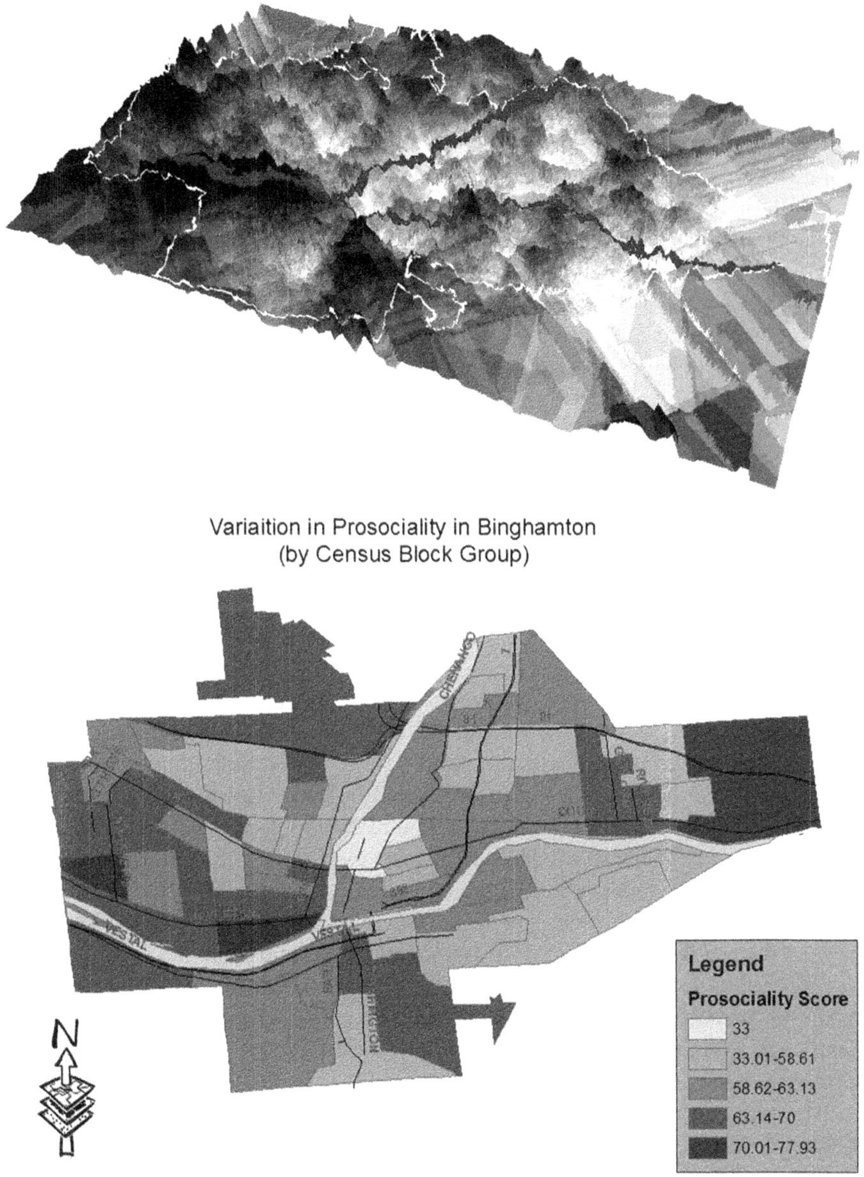

Fig. 1 Mapping Binghamton's prosociality with two different methods. On *top* is a continuous map using kriging, *underneath*, the city is split into discrete census block groups with scores. Both use the responses from the DAP

on how the neighborhood is rated by other resident students. See Sampson (2008) for a more general discussion of how space continues to matter in modern human social interactions.

4. *Median income does not matter, except insofar as it contributes to social support.* GIS technology enables all spatially based information to be interrelated. The median income of a neighborhood (obtained from US census statistics) positively correlates with self-report estimates of neighborhood quality (obtained from our survey). However, median income does not correlate with individual prosociality, once neighborhood quality is entered into the regression analysis, suggesting that median income does not have a direct effect on individual prosociality. In the following section we will provide evidence that median income has a direct negative effect.
5. *Validating the survey results.* We have validated our results based on self-reported prosociality with a number of additional studies. For example, the lost letter method involves dropping stamped addressed envelopes on sidewalks and using return rate as a measure of prosociality (Milgram et al. 1965). There is a 20-point difference in the return rate of envelopes in the best and worst neighborhoods, as measured by the self-report survey (Wilson and O'Brien 2009). Self-reported neighborhood quality also correlates strongly with crime rates, school disciplinary cases, and so on (unpublished data). There can be little doubt that small-scale spatial heterogeneity in the prosociality of the social environment actually exists, which correlates strongly with the prosociality of the individual – the basic requirement for prosociality to survive as a behavioral strategy in a Darwinian contest.

Using experimental economics games to study prosociality: Economists have developed a number of elegant experimental "games" for measuring social preferences such as cooperation, trust, and risk-taking. These games are an important supplement to surveys because they measure actual behavior rather than self-reported attitudes. They can also be given across cultures and contexts in a standardized fashion. In a landmark study, Henrich (2004) used the ultimatum game to measure tendencies to cooperate in 15 small-scale traditional societies. Variation among the societies was greater than among modern market economies, which appeared to be explained by two factors. First, members of societies that had a greater need to cooperate (e.g., whale hunting vs. slash-and-burn horticulture) cooperated more in the ultimatum game. Second, members of societies accustomed to trade evidently had adopted norms of reciprocity that caused them to cooperate more in the ultimatum game.

We are using the same methodology to study variation among neighborhoods within a single city (Unpublished data; see also Carpenter et al. 2005; Carpenter and Cardenas 2008; Falk and Zehnder 2007). We use the sequential prisoner's dilemma game rather than the ultimatum game because it more clearly distinguishes trust, trustworthiness, and self-sacrificial altruism as separate traits. In this game, one player chooses whether to cooperate or defect first, enabling the second player to choose on the basis of the first player's decision. Deciding to cooperate as the first player is an "offer of cooperation," and may be seen as the manner in which one perceives the balance between the advantage of cooperation and the risk of exploitation. Deciding to cooperate as the second player, given that the first player cooperates, can be described as reciprocation and demonstrates a degree of trustworthiness.

Deciding to cooperate as the second player, given that the first player defects, demonstrates a degree of self-sacrificial altruism.

In one of our studies (Unpublished data), the game was played with 182 public school students in grades 9–12 in their health and global studies classes. In each class, the students were introduced to the game and shown a payoff matrix with dollar values of $30 for mutual cooperation, $45 for the temptation to defect, $10 for the sucker's payoff, and $15 for mutual defection. The students indicated on paper how they would play as first mover, second mover in response to cooperation, and second mover in response to defection. The papers were collected, a pair of responses was chosen at random, one of the responses was randomly selected to be first mover, the game was played, and the two players were paid real money.

Figure 2 relates the proportion of students who cooperate as first movers to neighborhood quality (as measured by the DAP) and median income (as measured by US census statistics). The most cooperative students come from neighborhoods that are high in quality and low in income. The results for reciprocating cooperation as a second mover were similar but did not reach statistical significance. We interpret these results as similar to the study of worldwide variation by Henrich (2004). Low-income students have a greater need to cooperate in their daily lives than high-income students, who can pay for what they need. Yet, the need to cooperate can only be satisfied if there are also norms of cooperation that presumably exist in high quality neighborhoods.

We have also used experimental games in a study that measures the psychological response to photographs of the neighborhoods (unpublished data). College students unfamiliar with the city of Binghamton were shown photographs of a sample of

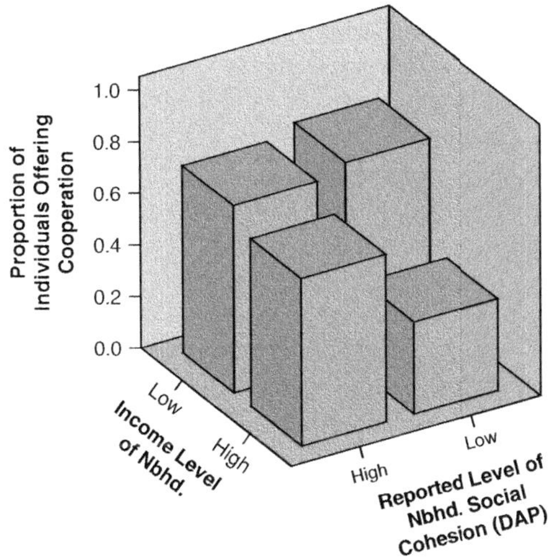

Fig. 2 Proportion of individuals offering cooperation in the sequential prisoner's dilemma game, grouped by income and neighborhood quality of census block group of residence. High and low are measurements of being above or below the mean of all census block groups. See O'Brien, Wilson, Eldakar, and Carpenter (ms) for additional details

neighborhoods. In one version of the study, they were asked to rate the quality of the neighborhoods based on the photographs, using the same survey items that were completed by students who actually live in the neighborhoods. In a second version, the college students were asked to play a sequential prisoner's dilemma game with a member of each neighborhood depicted in the photographs. This was not an imaginary exercise but could take place in reality, based on our previous study of the public school students. First the college students indicated how they would play the game for each neighborhood depicted in the photographs. Then a single college student was chosen at random, one of the neighborhoods was chosen at random, and the response of the college student was paired with the response of a public school student from that neighborhood. The order of play was randomly determined and the college student was paid real money.

The results of this study are shown in Fig. 3. Estimates of neighborhood quality based on the photographs correlate strongly with estimates of neighborhood quality by people who actually live in the neighborhoods. In addition, the photographs had a strong influence on the tendency to cooperate in the sequential prisoner's dilemma. Remarkably, variation in prosociality among neighborhoods might be based in part on the instantaneous psychological response to the neighborhoods, in addition to longer term effects of living in the neighborhoods.

Individual well-being, efficacy and the need for directed prosociality: Studies of religion frequently show that religious believers surpass non-believers in psychological well-being, health, and efficacy at accomplishing goals (e.g., Post 2007; Wilson and Csikszentmihalyi 2007). These are often regarded as individual-level advantages of religion because they directly benefit the individual, in contrast to benefits that require cooperating with others. This interpretation is potentially misleading, however, because the so-called "individual-level" advantages might be the result of past social support. If Jane gives Harry a million dollars, Harry thrives as an individual but only thanks to Jane's prosociality. The mere fact of individual thriving cannot be used as evidence for individual-level advantages; we need to know the causes of individual thriving.

Our database includes measures of individual psychological wellbeing (based on DAP survey items) and personal efficacy in the real world (e.g., academic performance) in addition to measures of individual prosociality and social support described in the previous sections of this chapter. Our results indicate that social support – the prosociality of the social environment – is like the hub of a wheel with spokes representing individual psychological wellbeing, academic performance, and individual prosociality. Social support correlates strongly with each of these variables, which have little direct influence on each other. In other words, once social support is entered into the multiple regression analysis, the other variables explain little, if any, of the residual variance (unpublished data).

In a more detailed study of academic performance (Unpublished data), we have found the strongest correlates to be parent's education and extracurricular activities. Other forms of social support, including family, neighborhood, religion, and even school do not explain any of the residual variation after parent's education and extracurricular activities are entered into the regression model. Moreover, parent's

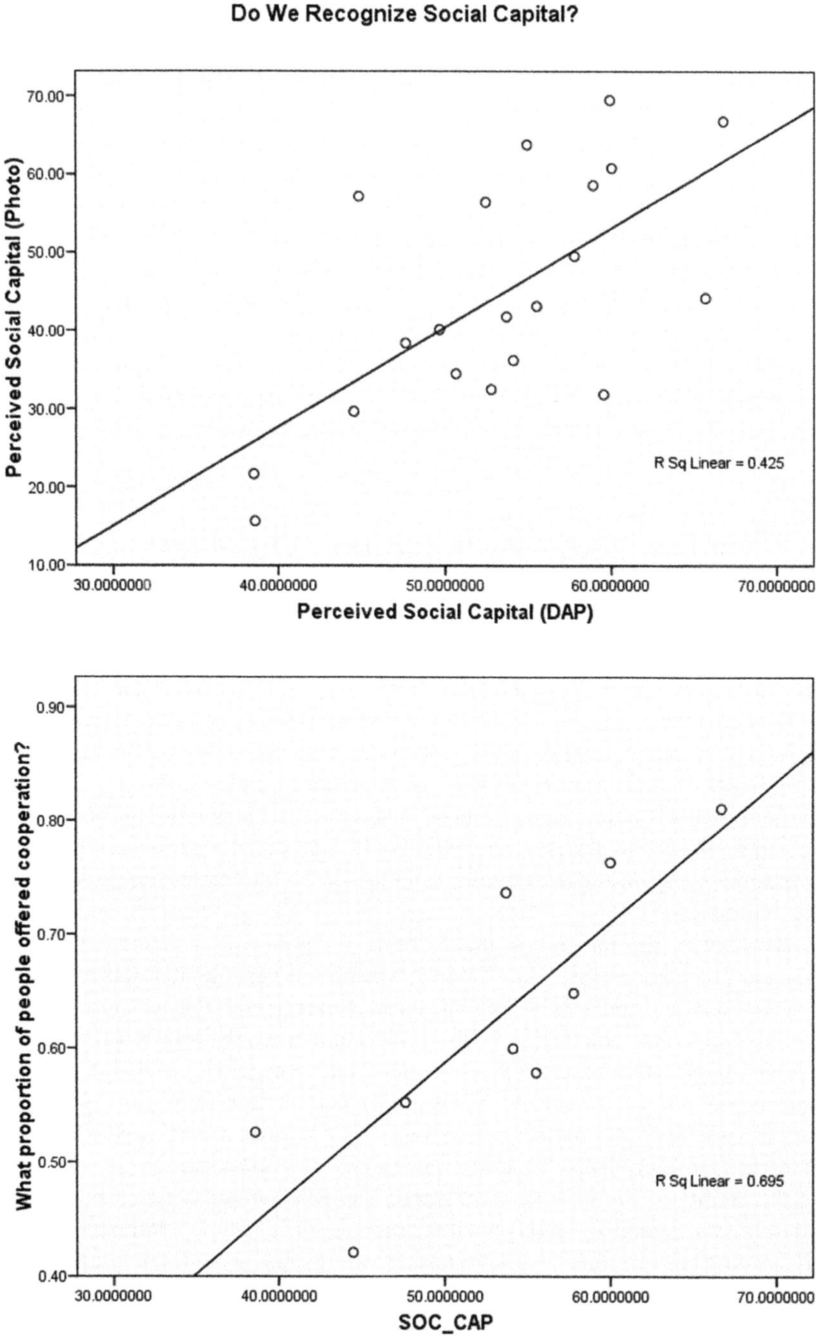

Fig. 3 *Top* figure shows ratings by college students of neighborhood quality based on photographs (*y-axis*) compared to ratings by public school students who actually live in the neighborhoods. *Bottom* figure shows the proportion of college students who offer cooperation as first mover in a sequential prisoner's dilemma with a resident of a neighborhood, after viewing a photograph of the neighborhood

education is significant as both an individual- and group-level variable in a hierarchical regression analysis. In other words, academic performance correlates not only with the educational level of one's own parents, but also the educational level of other parents in the neighborhood.

This result suggests that social support is not a generic substance but must be oriented toward specific goals. Academic performance requires certain normative values that motivate particular activities. Families that value academic performance and are familiar with how to achieve it will be successful with their own children and will provide more of a resource for other children, compared to families who are supportive in other respects or who are generally unsupportive. Similarly, neighbors who like each other and frequently get together for potlucks will not reduce crime in their neighborhood unless they specifically orient their activities toward the reduction of crime. Even neighbors who don't interact in other respects can become collectively efficacious with respect to specific goals (e.g., Sampson 2004).

2 Toward the Integration of Academic Disciplines and a Positive Tradeoff Between Basic and Applied Research

These are exciting times for the academic study of humans. Previously isolated disciplines are becoming integrated with each other and with evolutionary theory. The study of proximate mechanisms, represented by fields such as cognitive psychology and neurobiology, are becoming integrated with the study of ultimate causation based on theoretical frameworks such as game theory and multilevel selection theory. Minimalistic assumptions about human nature, such as rational choice theory, are being replaced by a richer conception of human social preferences that evolved by genetic evolution and provide the building blocks for ongoing cultural evolution (e.g., Gintis 2005).

Even better, the integration makes basic scientific research more relevant to practical applications than ever before. Typically, a negative tradeoff is imagined between basic and applied research; the more "fundamental" the question, the longer the time lag before practical benefits result. Human-related research from an evolutionary perspective creates a positive rather than a negative tradeoff. Evolution is fundamentally about organisms in relation to their environments, making it crucial from a basic scientific perspective to study people from all walks of life, as they go about their daily lives. This kind of research is also most relevant for improving the quality of life in a practical sense. The Binghamton neighborhood project is designed to capitalize on this positive tradeoff. A human population the size of a city is being used as a laboratory for basic scientific research, which in turn is being used to foster prosociality in the real world.

References

Bingham PM (1999) Human uniqueness: A general theory. Q Rev Biol 74:133–169
Boehm C (1999) Hierarchy in the forest: egalitarianism and the evolution of human altruism. Harvard University Press, Cambridge, MA
Bowles S (2003) Microeconomics: behavior, institutions, and evolution. Princeton University Press, Princeton
Bowles S (2008) Policies designed for self-interested citizens may undermine the moral sentiments: evidence from economic experiments. Science 302:1605–1609
Boyd R, Richerson PJ (1992) Punishment allows the evolution of cooperation (or anything else) in sizable groups. Ethol Sociobiol 13:171–195
Brosig J (2002) Identifying cooperative behavior: some experimental results in a prisoner's dilemma game. J Econ Behav Organ 47:275–290
Browne J (1995) Charles Darwin: voyaging. Knopf, New York
Browne J (2002). Charles Darwin: the power of place. Knopf, New York
Carpenter JP, Cardenas J-C (2008) Behavioral development economics: lessons from field labs in the developing world. J Dev Stud 44:337–364
Carpenter JP, Harrison GW, List JA (eds) (2005) Field experiments in economics. JAI, Greenwich, London
Falk A, Zehnder C (2007) Discrimination and in-group favoritism in a citywide trust experiment, IZA Discussion Paper Series, no. 2765
Fehr E, Fischbacher U (2003) The nature of human altruism. Nature 425:785–791
Gintis H (2000) Game theory evolving. Princeton University Press, Princeton, NJ
Gintis H, Bowles S, Boyd R, Fehr E (eds) (2005) Moral sentiments and material interests. MIT, Cambridge, MA
Hamilton WD (1975) Innate social aptitudes in man, an approach from evolutionary genetics. In: Fox R (ed) Biosocial anthropology. Malaby Press, London
Hammerstein P (ed) (2003) Genetic and cultural evolution of cooperation. MIT, Cambridge, MA
Henrich J, Boyd R, Bowles S, Camerer C, Fehr E, Gintis H (2004) Foundations of human sociality: economic experiments and ethnographic evidence from fifteen small-scale societies. Oxford University Press, Oxford, UK
Maynard Smith J, Szathmary E (1995) The major transitions of life. W.H. Freeman, New York
Maynard Smith J, Szathmary E (1999) The origins of life: from the birth of life to the origin of language. Oxford University Press, Oxford
Milgram S, Mann L, Harter S (1965) The lost-letter technique: a tool of social research. Public Opin Q 29:437–438
Plotkin H (1994) Darwin machines and the nature of knowledge. Harvard University Press, Cambridge, MA
Post SG (ed) (2007) Altruism and health: perspectives from empirical research. Oxford University Press, Oxford
Richerson PJ, Boyd R (2005) Not by genes alone: how culture transformed human evolution. University of Chicago Press, Chicago
Sampson RJ (2004) Neighborhood and community: collective efficacy and community safety. New Economy 11:106–113
Sampson RJ (2008) After-school Chicago: space and the city. Urban Geogr 29:127–137
Sober E, Wilson DS (1998) Unto others: the evolution and psychology of unselfish behavior. Harvard University Press, Cambridge, MA
Turchin P (2005) War and peace and war. Pi Press, Upper Saddle River, NJ
Wegner DM (1986) Transactive memory: a contemporary analysis of the group mind. In: Mullen B, Goethals GR (eds) Theories of group behavior. Springer, New York
Wilson DS (2002) Darwin's cathedral: evolution, religion, and the nature of society. University of Chicago Press, Chicago
Wilson DS (2004) What is wrong with absolute individual fitness? Trends Ecol Evol 19:245–248

Wilson DS (2005) Natural selection and complex systems: a complex interaction. In: Hemelrijk C (ed) Self-organization and evolution of biological and social systems. Cambridge University Press, Cambridge, UK, pp 151–165

Wilson DS (2006) Human groups as adaptive units: toward a permanent concensus. In: Carruthers P, Laurence S, Stich S (eds) The innate mind: culture and cognition. Oxford University Press, Oxford, pp 78–90

Wilson DS, Csikszentmihalyi M (2007) Health and the ecology of altruism. In: Post SG (ed) The science of altruism and health. Oxford University Press, Oxford, pp 314–331

Wilson DS, O'Brien DT (2009) Human prosociality from an evolutionary perspective: variation and correlations at a city-wide scale. Hum Behav Evol in press

Wilson DS, Wilson EO (2007) Rethinking the theoretical foundation of sociobiology. Q Rev Biol 82:327–348

Wilson DS, Wilson EO (2008) Evolution for the good of the group. Am Sci 96:380–389

The Error of God: Error Management Theory, Religion, and the Evolution of Cooperation

Dominic D.P. Johnson

> *Let us weigh the gain and the loss in wagering that God is... If you gain, you gain all; if you lose, you lose nothing. Wager, then, without hesitation that He is.*
>
> –Blaise Pascal (Pascal's Wager)

Abstract The punishment of free-riders is widely regarded as central to the evolution of cooperation, but the problem of who pays the costs of punishment remains controversial. I have previously proposed that: (1) human cooperation was promoted by a fear of *supernatural* punishment for selfish actions (Johnson and Kruger 2004); and (2) such beliefs increased *Darwinian fitness* because they reduced the probability of *real-world* detection and punishment for selfish actions or violations of social norms (Johnson and Bering 2006). Here, I explore the role of "Error Management Theory" (Haselton and Buss 2000; Haselton and Nettle 2006; Nettle 2004) in the evolution of beliefs in supernatural punishment, which offers a complementary perspective. Error Management Theory (hereafter EMT), which is derived from signaling theory, suggests that if the costs of *false positive* and *false negative* decision-making errors have been asymmetric over human evolutionary history, then natural selection would favor a bias towards the least costly error over time (in order to avoid whichever was the worse error). So, for example, we have a bias to sometimes think that sticks are snakes (which is harmless), but never that snakes are sticks (which may be deadly). Applied to religious beliefs and behaviors, I derive the hypothesis from EMT that humans may gain a fitness advantage from a bias in which they tend to assume that their every move (and thought) is being watched, judged, and potentially punished by *supernatural* agents. Although such a belief would be costly because it constrains freedom of action and self-interested behaviors, it may nevertheless be favored by natural selection if it helps to avoid an error that is even *worse*: committing selfish actions or violations of social norms when there is a high probability of *real-world* detection and punishment by victims or other group members. Simply put, supernatural beliefs may have been an effective

---

D.D.P. Johnson
Politics and International Relations, School of Social and Political Science, University of Edinburgh, Chrystal Macmillan Building, 15a George Square, Edinburgh EH8 9LD, UK

S.A. Levin (ed.), *Games, Groups, and the Global Good*, Springer Series in Game Theory,
DOI: 10.1007/978-3-540-85436-4_10,

mindguard against excessively selfish behaviour – behavior that became especially risky and costly as our social world became increasingly transparent due to the evolution of language and theory of mind. If belief in God is an error, it may at least be an adaptive one. I present theoretical and empirical support for the hypothesis.

1 Error Management Theory (EMT)

When you set a smoke alarm in your house, you deliberately set it to go off slightly too often. The cost of this is occasional false alarms when you burn your toast, but this is a small price to pay to make sure the alarm will not fail to go off in a real house fire. In order to avoid the latter error, you tune the sensitivity of the alarm to make sure the only mistakes are ones that err on the side of caution. This "smoke detector problem" has a number of analogues in decision-making problems ranging from engineering to public policy (Nesse 2005; Pacala et al. 2003).

As it turns out, evolution may have encountered exactly the same problem (and devised similar solutions) with a number of human *physiological* and *psychological* dispositions. In a given domain, if the costs of *false positive* and *false negative* errors have been asymmetric over human evolutionary history, then natural selection would favor a bias towards whichever was the least costly error over time (Haselton and Nettle 2006; Nesse 2001, 2005).

There are two conditions for this effect to operate: (1) there must be uncertainty about the true signal (otherwise evolution or engineers would simply build a perfect device that guessed right 100% of the time); and (2) false positive and false negative errors must entail different costs over time (otherwise they would cancel each other out). Wherever these conditions are fulfilled, then we would expect the natural selection of biases in decision-making.

Haselton and Nettle's (2006) review found that the EMT model accounts for a large number of psychological biases in three broad domains: (1) protective effects in perception, attention, and learning (the sound of approaching objects; bodily protection from harm; dangerous animals; dangerous people; food aversions; avoidance of the ill); (2) biases in interpersonal perception (the illusion of animacy; the sinister attribution error, overweighting of social gaffes, and negative forgiveness bias; the fundamental attribution error; the social exchange heuristic; sex-differences in interpreting courtship signals); and (3) self-related biases (positive illusions; the illusion of control).

As one example of many, experimental evidence demonstrates that men tend to assume that women are sexually interested in them, whereas women tend to assume that men are not. From an EMT perspective, this makes evolutionary sense because men's investment in reproduction is negligible whereas a woman's investment in reproduction is enormous (Trivers 2002). In Darwinian fitness terms, for men the false negative (missing a real sexual opportunity) is much *more* costly that the false positive (assuming sexual interest when there is none). For women, the false negative (missing signs of a genuine desire to commit) is much *less* costly that the false positive (assuming a willingness to commit when there is none). Men have little to lose in striving for all possible sexual relationships, whereas women have everything to lose in having sex too readily with a man who is unlikely to provide long-term care for her and her offspring (Haselton and Buss 2000) – an asymmetry that may help to understand the persistence and prevalence of sexual harassment (Browne 2008). This example is instructive because it shows how EMT may lead to biases that either *under*estimate or *over*estimate, depending on the cost/benefit ratios of false positives and false negatives in a given case.

2 Application to Religious Beliefs

I propose that EMT may help to explain the evolution of belief in God (or any other supernatural agents). The belief that God exists may simply be a less costly "error" than the belief that he does not exist. This is not in the C. S. Lewis sense – that if God turns out to be real, then the costs of disbelief (eternity in Hell instead of eternity in Heaven) are infinite – rather, my argument is that there may be tangible Darwinian fitness costs in the real world associated with *not* believing in God. This is, therefore, an *adaptive* theory of religion focusing on selective advantages at the individual-level (no group selection is or need be invoked).[1]

Obviously, I am not proposing a theory that can explain all aspects of religion, nor one that can explain the enormous diversity of the world's religious beliefs and behaviors. However, it can explain why a general belief in supernatural agency may have adaptive advantages, and thus why such beliefs may have been favored by natural selection over human evolutionary history.

According to the conditions for EMT outlined above, it remains for me to explain: (1) why the signal of whether God exists or not is one of uncertainty; and (2) why the costs of not believing in God are higher than those of believing in God.

2.1 *God's Existence as Uncertain*

Smoke alarms face uncertainty because different things burning in different locations around the house will give off highly variable signatures. The content and concentration of aerial particulates to which the smoke alarm is sensitive will vary from case to case, independent of the *actual* size or severity of the fire.

[1] Haselton and Nettle (2006) suggest that EMT may explain the "illusion of animacy," which is the propensity for people to perceive agency even where there is none. Guthrie (1993) and Atran and Norenzayan (2004) suggest that this may result from an adaptive bias to be alert for agency in the environment – a source of danger from predators or other humans. However, they suggest that inferring *supernatural* agency is an *accidental byproduct* of the bias. My paper, in contrast, suggests that inferring supernatural agency is not an accidental byproduct at all, but rather a highly adaptive belief in itself.

A conflagration at the far end of the house might emit a similar fire signature to a single piece of toast burning in the same room, for example. This means that sensitivity settings have to be set to approximate some unknown true probability that a given input means there is a real fire.

Whether God exists or not is also a judgment made under uncertainty, because there are variable signals that He may or may not exist. On the one hand, natural and social phenomena may be caused by physical, chemical or biological forces in the environment. But on the other hand, many phenomena appear to the human mind as inexplicable or impossible chance. One therefore has to make a judgment about the extent and scope of natural vs. supernatural causation.

This may strike the scientific reader as odd, because many natural phenomena have well-known scientific explanations in today's world. Even if not, they may simply be awaiting scientific explanation. However, this is, in fact, beside the point. If we are interested in the evolutionary origins of religious beliefs, then our analysis is firmly rooted in the realm of our *pre-scientific* past. Before scientific explanations emerged for natural phenomena, things we now take for granted – the sun, stars, seasons, lightning, thunder, eclipses, rain, fire, droughts, births, deaths, disease – were more or less unfathomable miracles. As is evident from indigenous religions around the world, supernatural agents were routinely *assumed* to be responsible for such events (indeed they are commonly appealed to in order to alter them). For example, George Murdock's (1980) analysis of 186 pre-industrial societies around the globe found that *all* of them attributed the causation of illness to supernatural sources of one form or another. Until fairly recently in human history, supernatural agents were automatically thought to be responsible for much of what life threw at us. Even today, many people – including atheists – consciously or subconsciously continue to believe this (think of superstition, folklore, karma, Just World beliefs, "comeuppance" and so on).

Of course, interpretations of evidence for supernatural activity are also hugely bolstered by *cultural* narratives, norms, and beliefs. Why would you not believe that ancestral spirits were real if everyone else did, and if your forebears had always done so before you? Indigenous peoples do not discuss theology the way we do. There is no question of *whether* supernatural agents exist or not. Instead, what we describe as "religion" is part and parcel of their everyday thinking and living (see Appiah, this volume). There is little division between what is "religious" and what is "non-religious." It is not really a matter of a search for evidence of supernatural agents, but rather a search for ways to live peaceably alongside them. As psychologist Jesse Bering (2002, 2006) notes, the logic is actually reversed: negative life events afflicting oneself or afflicting others are often interpreted as *evidence* of wrongdoing, betrayed by inevitable supernatural punishment itself.

To summarize, in the pre-scientific era supernatural agency was as good an explanation as any other of many natural and social phenomena. Among indigenous cultures, however, not all gods are completely omniscient or omnipresent. Indeed, the degree to which gods are moralizing and thought to be concerned with human behaviors varies across the globe (Johnson 2005; Roes and Raymond 2003). People therefore have to make judgments about whether a supernatural agent may or may not be observing and judging their actions at a given moment. These are judgments made under uncertainty.

2.2 *Costs of Belief and Non-belief*

2.2.1 Costs of False Positives (Belief in God)

The costs of a false positive error – believing in God when he does not exist – are not insignificant in terms of biological fitness. Considerable time, energy, and opportunity costs are incurred by religious beliefs and behaviors (e.g., taboos, special clothing, prayer, rituals, sacrifices) which diverts precious resources from being invested into survival and reproduction (Sosis and Alcorta 2003). From a natural selection point of view, any such costs, however small, should be selected against. Unless, that is, the costs of not holding the belief are worse. All I intend to do for this chapter is to make the case that while belief in God may incur some fitness costs, not-believing in God may incur fitness costs as well. If so, then it becomes a goal for future research to work out which of those two costs is greater.

2.2.2 Costs of False Negatives (Non-Belief in God)

In most theorizing and applications of EMT, the focus is explicitly on false positives and false negatives. In the current case, however, we are not strictly talking about a "false negative". Here's why. Applied to belief in God, the errors would be represented as follows (see Table 1): false positive = a belief that God exists when he does not (bottom-left cell); false negative = a belief that God does not exist when in fact he does (top-right cell). For the purposes of this paper, I am making the assumption that God does not exist, so false negatives are not possible (in other words, if you do not believe in God, it is not an "error"). This is in order to test the hypothesis on purely scientific grounds – can a belief in God be favored by natural selection even if the belief is false? Thus, we are only interested in the bottom row of Table 1.

The important point here is that the logic of EMT holds regardless: we are simply interested in the differential costs of believing in God or not believing in God. Which belief has the higher costs? As I argue below, a belief in God may be adaptive even if it is false.

Table 1 The four possible combinations of actual status of God and belief in God

		Belief	
		God exists	God does not exist
Actual Status	God exists	*Correct*	*False negative*
	God does not exist	*False positive*	*Correct*

Why would a belief in God be advantageous? The basic argument presented below is that selfishness accrued significant costs in the human lineage as the evolution of language and theory of mind made social life increasingly transparent and reputations increasingly vulnerable.

Standard game theoretical models show that selfish individuals will outcompete altruistic individuals because they can exploit others' cooperation while not contributing anything themselves (Axelrod 1984; Olson 1965). However, the great caveat to this is that there are reputational costs to selfishness. Being seen as selfish can reduce the probability and value of future interactions, and thus selfish individuals can do less well than others over time. As a result, reciprocity and indirect reciprocity (via third parties) between known cooperators tend to be advantageous strategies (Ehrlich and Levin 2005; Nowak and Sigmund 2005; Nowak and Sigmund 1998; Trivers 1971). This is particularly true for humans as compared to other species, because language and theory of mind (that is, the ability to infer intentionality and knowledge in others people's minds) makes the reputational stakes significantly higher – interaction partners and third parties can infer, remember, report, gossip, scheme, and retaliate on the basis of your behavior, even long after the event (Bering and Shackelford 2004). This means that selfish actions are: (1) more likely to be detected (because people can work out, inform on, and discuss what you have done); and (2) more likely to be punished (because people can spread the word and form punitive coalitions that punish at low cost to themselves). With the evolution of language and theory of mind, punishment became cheaper just as selfishness became more evident.

Cooperation is, of course, well known to predate the evolution of theory of mind and language. For a start, cooperation is widespread among numerous other species of mammals, birds, insects, cells, microscopic organisms, even different organs of the body (Dugatkin 1997; Wilson 2000). We also know that cooperation is common among monkeys and apes and is thus likely to have been a well developed feature of our common ancestor (de Waal 1996; Wrangham and Peterson 1996). Therefore, there is no suggestion here that human cooperation *emerged with* the evolution of theory of mind and language – in fact, cooperation and social life were probably a necessary precursor for both. Rather, the claim is simply that selection for cooperation (and selection against selfishness) *increased* to a higher level that whatever it was before because of the social consequences of theory of mind and language. As Martin Nowak recently suggested, the role of language (and by implication theory of mind) was decisive in attaining the levels of cooperation – via the power of indirect reciprocity – that is unique to humans (Nowak 2006).

Faced with such a sophisticated social environment, people needed to be careful about what they did – *much more careful than their evolutionary forebears* who lacked the language and theory of mind to carry out complex detective work. One needed to be especially careful to avoid selfish actions if and when there was a high probability that others would see, discover, or infer what one was doing.

Any trait which made you more sensitive to being observed and punished would increase your alertness to the possibility of detection, and/or make you less likely to be selfish in the first place. A belief in supernatural agents is one way of achieving these precautionary behaviors. Although you would often be committing a false positive – believing that you are being observed when in fact you are free to commit rampant selfishness – this cautionary mind guard may pay off over time. The cost is missing out on some opportunities for selfish gain. But the costs of atheism would be an increased probability of detection and punishment by angry victims, their kin, or other group members whom you may have exploited or offended. Obviously there is a trade-off here. I am not able to prove that the costs of missed opportunities are, in reality, less than the costs of occasional retribution – that is a question for future empirical work to examine. The point is that, where that is the case, god-fearing individuals will outcompete non-believers (Johnson and Bering 2006).

3 Towards a more Formal Model

To put the above argument into more precise terms, Table 2 compares the performance of three alternative strategies: "Ancestrals," "Atheists," and "God-fearers." Ancestrals represent a baseline strategy prior to the evolution of language and theory of mind. These individuals *cannot* fear God, but nor can they utilize language and theory of mind to their own ends. Atheists have the same level of selfishness as Ancestrals (both are unconstrained by any fear of God), but they are able to exploit language and theory of mind to their own ends.[2] God-fearers are also able to exploit language and theory of mind to their own ends, but they reduce their selfish behavior because they believe that supernatural agents may observe and punish them.

Table 2 Three strategies come into competition with the advent of theory of mind and complex language. Atheists outcompete Ancestral individuals, and God-fearers outcompete Atheists as long as $pc > m$. See text for details

Strategy	Theory of mind and language present?	Can exploit theory of mind and language for personal gain?	Probability of detection (p)	Cost of punishment (c)	Cost of missed opportunities (m)	Payoff
Ancestrals	No	No	High	Same	None	Lowest
Atheists	Yes	Yes	High	Same	None	Highest (if $pc < m$)
God fearers	Yes	Yes	Low	Same	Some	Highest (if $pc > m$)

[2] Note that Ancestrals and Atheists do not *lack* cooperation. They only lack theory of mind and language, the consequence of which leads them to be *somewhat less cooperative* and *somewhat more selfish* than God-fearers. The model works just the same if, say, Ancestrals and Atheists are fantastically cooperative already – God-fearers are just even more cooperative than that. For the same reason, Atheists are not automatically "cheats."

Atheists would clearly outcompete Ancestrals because, while everything else is identical between them, ancestrals cannot exploit the new cognitive features (language and theory of mind) for personal gain. More importantly however, Table 1 indicates that God-fearers can, in well defined circumstances, outcompete Atheists. These strategies differ in just two respects: God-fearers have a lower probability of detection, but miss out on some opportunities for selfish rewards. Therefore, God-fearers will outcompete Atheists *as long as* the total expected costs of punishment (i.e., the probability of detection (p) multiplied by the cost of punishment (c)) is greater than the cost of missed opportunities for selfish rewards (m). In other words, when the inequality $pc > m$ is true. This would occur wherever the rewards of selfishness were relatively small compared with the costs of public exposure (which may include social sanctions, seizure of property, physical harm, ostracism, imprisonment, punishment of kin, or death). Even a small p can mean selfishness does not pay on average.

Weighing up the relative costs and benefits suggests that there are at least some conditions where belief in God would outcompete atheism – in purely Darwinian fitness terms. However, EMT makes the story even more interesting, because it predicts that, where $pc > m$, we should expect *exaggerated* estimates of p (such as a hyperactive belief that supernatural agents are watching) to outperform *accurate* estimates of p, given that the latter will engender more mistakes under conditions of uncertainty (Haselton and Buss 2000; Haselton and Nettle 2006; Nettle 2004).[3] That is, EMT does not just predict a balance of costs and benefits. Rather, it predicts a *bias* to systematically overestimate the probability that God is watching. This would account for the human brain's so-called "Hyperactive Agency Detection Device" (Atran and Norenzayan 2004; Barrett 2004; Boyer 2001).

My argument so far begs the question of why natural selection would favor such a complex solution to the problem of avoiding the social costs of selfishness. Why evolve a belief in supernatural punishment rather than simply tuning down the overall level of selfish behavior in the first place? Perhaps EMT led to a reduction in selfish behavior through secular mechanisms, such as emotional or cognitive heuristics for cooperation (Fessler and Haley 2003; Yamagishi et al. 2007). However, there are several observations that point to a special role of supernatural agency in the evolution of human cooperation.

First, it was language and theory of mind that caused the increased costs of selfishness in the first place, casting aside the veils of social life and increasing the probability of detection and the severity of punishment. Second, it is theory of mind that makes belief in God possible at all, and language that underlies the *sharing* of beliefs and organized *religion*. Only with theory of mind would people be concerned about the existence of God and the possibility that he knows what they are doing; only with language would these beliefs become part of culture. These two points strongly suggest to me that, since theory of mind and language created the problem (increasing the probability of detection and punishment for selfish behavior), it may also be part of natural selection's solution in redressing the balance.

[3] For a mathematical derivation of EMT, see Haselton and Nettle (2006) and references therein.

Theory of mind made people worry about the contents of unseen minds, whether these were human, animal or any other type of agent (Bering 2002). We would certainly expect such a significant social shift to be accompanied by many psychological traits adapting and counter-balancing to fit the new social "ecology" (Bering and Shackelford 2004). Finally, there is good evidence that religious beliefs are *more* effective than secular beliefs in promoting cooperation (Rappaport 1999; Sosis 2005; Sosis and Bressler 2003). If so, then natural selection may simply have discovered the persuasive power of religion well before it ever came to puzzle scholars.

4 Predictions and Evidence

If a belief in supernatural agency was favored by natural selection following the emergence of theory of mind and complex language, then we can derive some predictions for what we should see today as evidence.

First, fear of supernatural punishment should be a feature common across diverse and widely dispersed religions around the world. Indeed, although the concepts of *omnipresence* and *omniscience* are not evident in all religions, there is a universal predominance of supernatural agency, religious taboos (oughts and ought nots), and supernatural sanctions across the world's modern, ancient, and pre-industrial cultures suggesting that these beliefs have deep and common origins (Atran 2004; Bering and Johnson 2005; Boehm 2008; Boyer 2001; Johnson 2005; Whitehouse 2008).

Second, human brains should be susceptible to *underestimating* the probability of detection and punishment for selfish actions. A recent study of criminal evidence indicates that a major factor in offenders' decision to commit a crime is that they underestimated the probability of being caught and the costs of punishment (Robinson and Darley 2004). Most of us are not criminals, of course, but it suggests that there is at least scope for a corrective mindguard to avoid unchecked selfishness (perhaps necessary to overcome evolutionary older causes of behavior such as emotional or heat-of-the moment reactions).

Third, cues of supernatural observation or presence should decrease selfishness and promote cooperation. Bering and colleagues (2005) found that people were significantly less likely to cheat in a computer task when subtly primed with the idea that a ghost was present in the laboratory. Shariff and Norenzayan (2007) also found that people primed with religious concepts gave significantly more money to strangers in an anonymous economic game.

Finally, from a cross-cultural perspective, societies with gods that play a more active role in their moral lives should be more cooperative. A study of 186 pre-industrial cultures around the world found positive correlations between the extent to which a society's gods were "moralizing" or not, and the degree of societal cooperation (Johnson 2005).

5 Conclusion

The argument of this chapter is very simple. A belief in God may be adaptive because it helps people avoid the social costs of selfish behavior (and I think the *costs* of selfish behavior have been underestimated in the evolution of cooperation literature). Selfish behavior became especially (and uniquely) costly for humans with the evolution of theory of mind and complex language, which made human social life significantly more transparent than before (Bering and Shackelford 2004; Dunbar 1996). Cheating, free-riding, and selfishness became much more obvious and socially reprehensible – sentiments that remain powerfully rooted in our cognition and behavior today (Nowak and Sigmund 2005; Price et al. 2002). It was harder to be selfish and get away with it once other people could infer, remember, report, gossip, scheme, and retaliate on the basis of your behavior, even long after the event. What is more, selfishness was less likely to go unpunished as well as less likely to go undetected. Language and theory of mind meant that punishment could be requested, planned, and synchronized, allowing for powerful real-world punishment that was cheap to carry out (further elevating the costs of selfish actions).

With such significant Darwinian fitness costs associated with selfish behavior, a belief in *supernatural* detection and punishment may have been a very effective psychological device for curtailing selfishness, or at least for reflecting on potential costs and benefits before committing selfish acts. What better way than to equip the human mind with a sense that their every move – even thought – is being observed, judged, and potentially punished? Error Management Theory cannot explain *why* we believe in supernatural agents, or why we fear their punishment – that requires other aspects of human cognition (Bering 2006; Johnson and Bering 2006). The value of EMT, however, is that it neatly shows how a belief in God, even if false, may be favored by natural selection if its costs are less than those of assuming one is alone and free to do as one wishes. A fear of the fires of hell may be a very effective smoke alarm against getting burnt for real in this world.

References

Atran S (2004) In Gods we trust: the evolutionary landscape of religion. Oxford University Press, Oxford

Atran S, Norenzayan A (2004) Religion's evolutionary landscape: counterintuition, commitment, compassion, communion. Behav Brain Sci 27:713–730

Axelrod R (1984) The evolution of cooperation. Penguin, London

Barrett JL (2004) Why would anyone believe in God?. Altamira Press, MD

Bering JM (2002) The existential theory of mind. Rev Gen Psychol 6:3–24

Bering JM (2006) The folk psychology of souls. Behav Brain Sci 29:453–462

Bering JM, Johnson DDP (2005) Oh Lord, you hear my thoughts from afar: recursiveness in the cognitive evolution of supernatural agency. J Cogn Cult 5:118–142

Bering JM, Shackelford T (2004) The causal role of consciousness: a conceptual addendum to human evolutionary psychology. Rev Gen Psychol 8:227–248

Bering JM, McLeod KA, Shackelford TK (2005) Reasoning about dead agents reveals possible adaptive trends. Hum Nat 16:360–381
Boehm C (2008) A biocultural evolutionary exploration of supernatural sanctioning. In: Bulbulia J, Sosis R, Genet C, Genet R, Harris E, Wyman K (eds) The evolution of religion: studies, theories, and critiques. Collins Foundation Press, Santa Margarita, CA, pp 143–150
Boyer P (2001) Religion explained: the evolutionary origins of religious thought. Basic Books, New York
Browne KR (2008) The evolutionary psychology of sexual harassment. In: Duntley JD, Shackelford TK (eds) Evolutionary forensic psychology: darwinian foundations of crime and law. Oxford University Press, New York pp 81–100
de Waal FB (1996) Good natured: the origins of right and wrong in humans and other animals. Harvard University Press, Cambridge
Dugatkin LA (1997) Cooperation in animals. Oxford University Press, Oxford
Dunbar RIM (1996) Grooming, gossip and the evolution of language. Faber & Faber, London
Ehrlich P, Levin S (2005) The evolution of norms. PLoS Biol 3:e194
Fessler DMT, Haley KJ (2003) The strategy of affect: emotions in human cooperation. In: Hammerstein P (ed) The genetic and cultural evolution of cooperation. MIT, Cambridge, MA pp 7–36
Guthrie SE (1993) Faces in the clouds: a new theory of religion. Oxford University Press, New York
Haselton MG, Buss DM (2000) Error management theory: a new perspecitive on biases in cross-sex mind reading. J Pers Soc Psychol 78:81–91
Haselton MG, Nettle D (2006) The paranoid optimist: an integrative evolutionary model of cognitive biases. Pers Soc Psychol Rev 10:47–66
Johnson DDP (2005) God's punishment and public goods: a test of the supernatural punishment hypothesis in 186 world cultures. Hum Nat 16:410–446
Johnson DDP, Bering JM (2006) Hand of God, mind of man: punishment and cognition in the evolution of cooperation. Evol Psychol 4:219–233
Johnson DDP, Kruger O (2004) The good of wrath: supernatural punishment and the evolution of cooperation. Polit Theol 5.2:159–176
Murdock GP (1980) Theories of illness: a World survey. HRAF, University of Pittsburgh Press, Pittsburg
Nesse RM (2001) Natural selection and the regulation of defensive responses. Ann NY Acad Sci 935:75–85
Nesse RM (2005) Natural selection and the regulation of defenses: a signal detection analysis of the smoke detector problem. Evol Hum Behav 26:88–105
Nettle D (2004) Adaptive illusions: optimism, control and human rationality. In: Evans D, Cruse P (eds) Emotion, evolution and rationality. Oxford University Press, Oxford pp 193–208
Nowak M, Sigmund K (2005) Evolution of indirect reciprocity. Nature 437:1291–1298
Nowak MA (2006) Five rules for the evolution of cooperation. Science 314:1560–1563
Nowak MA, Sigmund K (1998) Evolution of indirect reciprocity by image scoring. Nature 393:573–577
Olson M (1965) The logic of collective action: public goods and the theory of groups. Harvard University Press, Cambridge
Pacala SW, Bulte E, List JA, Levin SA (2003) False alarm over environmental false alarms. Science 301:1187–1188
Price ME, Cosmides L, Tooby J (2002) Punitive sentiment as an anti-free rider psychological device. Evol Hum Behav 23:203–231
Rappaport RA (1999) Ritual and religion in the making of humanity. Cambridge University Press, Cambridge, UK
Robinson PH, Darley JM (2004) Does criminal law deter? A behavioural science investigation. Oxf J Leg Stud 24:173–205
Roes FL, Raymond M (2003) Belief in moralizing gods. Evol Hum Behav 24:126–135

Shariff AF, Norenzayan A (2007) God is watching you: supernatural agent concepts increase prosocial behavior in an anonymous economic game. Psychol Sci 18:803–809
Sosis R (2005) Does religion promote trust? Interdiscipl J Res Relig 1:1–30 (article 7)
Sosis R, Alcorta C (2003) Signaling, solidarity, and the sacred: the evolution of religious behavior. Evol Anthropol 12:264–274
Sosis R, Bressler ER (2003) Cooperation and commune longevity: a test of the costly signaling theory of religion. Cross Cult Res 37:211–239
Trivers R (2002) Natural selection and social theory: selected papers of robert trivers. Evolution and cognition. Oxford University Press, Oxford; New York
Trivers RL (1971) The evolution of reciprocal altruism. Q Rev Biol 46:35–57
Whitehouse H (2008) Cognitive evolution and religion; cognition and religious evolution. In: Bulbulia J, Sosis R, Genet C, Genet R, Harris E, Wyman K (eds) The evolution of religion: studies, theories, and critiques. Collins Foundation Press, Santa Margarita, CA
Wilson EO (2000) Sociobiology: the new synthesis. Belknap Press, Harvard
Wrangham R, Peterson D (1996) Demonic males: apes and the origins of human violence. Bloomsbury, London
Yamagishi T, Terai S, Kiyonari T, Mifune N, Kanazawa S (2007) The social exchange heuristic: managing errors in social exchange. Rationality and Society 19:259–292

Moral Motivation

John E. Hare

Abstract This paper is about moral motivation from the perspective of Kantian moral philosophy. It looks at recent literature on the development of human sociality from within game theory, and argues that a retrieval of Kant's views on moral theology would be helpful in understanding an aspect of this sociality. In particular, Kant's view that we need to postulate the existence of God as sovereign of the kingdom of ends helps us understand the role of religion in making self-indexed and non-self-indexed motivation consistent (i.e., motivation towards a good which is specified with essential reference to the self and motivation towards a good which is specified without such reference). Kant's complex views on divine punishment are also helpful, and his views on divine assistance in the production of moral motivation can help us understand the invocation of such assistance in signaling difficult commitment.

I write this paper as a moral philosopher, not a game theorist or a biologist or an economist, and I am aware of the risks of writing about matters outside my discipline. But practitioners of those other disciplines sometimes take the risk of writing about matters in moral philosophy, and the exchange can be fruitful. I want to start with some claims made by Ken Binmore. I want to suggest that within one tradition of moral theory, the tradition of Kantian moral philosophy that he attacks, it makes sense to relate moral philosophy with theology, and I will attempt to do that. I am not going to try to defend Kantian morality here, for that would be too large a project. But Kant's account is one of the three accounts (together with consequentialism and virtue theory) usually appealed to in the Western ethical literature of the last half-century, and it has been, in my view, the most influential of the three in the modern period. Any account of morality that is inconsistent with it needs some way to explain away its power. I am going to point to an unfamiliar feature of Kant's account, namely its relation to theology. I want to suggest that bringing in some theological premises will have explanatory power, in the sense that it will help us understand better some otherwise puzzling aspects of our morality.

J.E. Hare
Yale University Divinity School, 409 Prospect Street, New Haven, CT 06511, USA

S.A. Levin (ed.), *Games, Groups, and the Global Good*, Springer Series in Game Theory,
DOI: 10.1007/978-3-540-85436-4_11,

Ken Binmore, in *Game Theory and the Social Contract* argues against Kant on the grounds that "Nature cannot achieve the first-best outcomes to which those like Kant aspire because the latter are not incentive-compatible. That is to say, they are achievable only if the human beings who live in the society act in a manner that is incompatible with their nature."[1] What is this nature? It is *Homo economicus*. "The assumptions about human nature to be made in this book are those of neoclassical economics. People are assumed to act in their own enlightened self-interest."[2] This last phrase is multiply ambiguous. But Binmore is opposed to Kant, and endorses Hobbes's view, "of the voluntary acts of every man, the object is some Goode to himselfe." This shows that Binmore is not merely claiming what Kant would also claim, that morality *is* in an agent's interest, and not merely claiming what Kant would also claim, that the moral agent is made happy by the happiness of others. The anthropology behind the theory of *Homo economicus* requires that the object as purposed be self-indexed (i.e., it is specified with essential reference to self-interest), and this is something Kant cannot agree to. The heart of the *categorical* imperative is that it is binding on an agent *whatever else* she wants, and so whatever she wants for herself.

This is not the proper place for a detailed account of Kantian moral theory, but some elements of the theory will be necessary in order to understand what follows. Kant claims to have discovered, in what he calls "the categorical imperative," the supreme principle of morality. An imperative is categorical if it is binding on us without reference to any other end or purpose that we might have. "For instance, when it is said that you should not make a false promise, the assumption is that the necessity of this avoidance is no mere advice for escaping some other evil, so that it might be said that you should not make a false promise lest you ruin your credit when the falsity comes to light. But when it is asserted that an action of this kind must be regarded as bad in itself, then the imperative of prohibition is therefore categorical."[3] His first statement (what he calls his first "formula") of the categorical imperative is, "Act only according to that maxim whereby you can at the same time will that it should become a universal law."[4] The notion of the "maxim" of an action is key to Kant's idea here. The maxim of an action is the prescription of the action together with the actually intended reason for it.[5] In the case of the false promise, "the maxim of his action would then be expressed as follows: when I believe myself

[1] Binmore (1994, vol. 1, p. 152). The title of the chapter is a quotation from Samuel Johnson, "My Dear Sir, Clear Your Mind of Cant."

[2] Binmore (1994, p. 18).

[3] *Groundwork of the Metaphysics of Morals*, 4: 419. I will cite Kant's works in the Berlin Academy edition, by volume and page number, and I will abbreviate the present work as *Groundwork*.

[4] *Groundwork of the Metaphysics of Morals*, 4: 421.

[5] The description of a maxim as giving the "actually intended reason" raises important questions about "proximate" and "ultimate" causes of behavior, and about self-deception. See Trivers (2000). In this paper I am distinguishing reasons (not necessarily articulated to herself) that an agent gives to herself for acting in a certain way, from external causes of behavior. There are large philosophical questions about whether an agent's reasons for action are ultimately caused by forces (for example the evolutionary imperative to reproduce) outside the agent's rational will. If they are, then Kant's

to be in need of money, I will borrow money and promise to pay it back, although I know that I can never do so."[6] To will the maxim of an action as a universal law is to will that *anyone* in such circumstances may will such an action for such a reason. The universal law thus does not allow any special reference to the agent herself or her interests. In what follows I will say that moral motivation is, in this sense, "non-self-indexed." Kant has an argument, which I will not now describe in detail, to show that willing the maxim of a false promise as a universal law is self-contradictory, because it requires willing the undermining of the institution of making promises, which is itself a means required for even a false promise to be successful. For the purposes of the present paper, it is important to see that the maxim is what determines the moral status of the action that follows from it. This means that no account that abstracts from what is going on "inside" a person (in her will), and looks only at her behavior, is going to be able to give an account of moral motivation that captures what Kant was after.

The third formula of the categorical imperative is, "Act in such a way that you treat humanity, whether in your own person or in the person of another, always at the same time as an end and never simply as a means."[7] Kant goes on to give the same examples as he did for the first formula. "The man who intends to make a false promise will immediately see that he intends to make use of another man merely as a means to an end which the latter does not likewise hold." To treat another person as an end in herself requires sharing her ends (her purposes) as far as possible, which is as far as the moral law itself allows (the categorical imperative does not require sharing her immoral ends). Again, it is central to moral motivation that the agent *wills* to respect what Kant calls the "dignity" of another person, which derives from the other person's rational agency, namely her capacity to set her own ends.

Another Kantian principle that will be required in understanding what follows is the principle that "ought" implies "can"; if it is the case that a person ought to do something, it must be the case that she can do it.[8] There are different ways to understand this principle, but the way I shall treat it is that if it is not the case that she can do something, the question of whether she ought to do it does not properly arise. It is, for example, a cardinal principle of child-rearing that we should not hold children accountable to standards they are unable to meet.

Because of Kant's emphasis on the will, moral action is more than mere cooperation, though it is one form of cooperation. Suppose we define "cooperation" as Martin Nowak and Sarah Coakley do, as "a form of working together in which one individual pays a cost (in terms of fitness, whether genetic or cultural) and another gains a benefit."[9] This definition does not say anything about motivation. Note that this definition is different in this respect from their definition of "altruism," as

---

moral philosophy fails. But we do not yet know that they are, and in the meantime we need a theory of reasons for action.

[6] *Groundwork of the Metaphysics of Morals*, 4: 422.

[7] *Groundwork of the Metaphysics of Morals*, 4: 429.

[8] *Critique of Practical Reason* 5: 114.

[9] Nowak and Coakley (2009), introduction.

"a form of costly cooperation in which an individual is *motivated* by good will or love for another" (emphasis added). The terms in the literature are confused. Cooperation as defined by Nowak and Coakley is what Elliott Sober and David Sloan Wilson call "evolutionary altruism" and "altruism" as defined by Nowak and Coakley is what Sober and Wilson call "psychological altruism."[10] In this paper I am not using the term "altruism," but "moral motivation." Kant sometimes talks about moral motivation as "practical love," so his account will fit under Nowak's and Coakley's definition of altruism, except that he insists that the moral agent is motivated by respect for humanity both in her *own* person and in the person of others. If 'altruism' is defined in a way to exclude *this* kind of self-concern, then Kant's account is inconsistent with it.

Sometimes writers use terms that suggest that they want to talk about more than cooperation, as defined above, but it turns out that their data is actually about strategies for cooperation and not about internal states of motivation. Thus Samuel Bowles and Herbert Gintis end their paper for the present volume, "For students of human cooperation, the challenge thus shifts from that favored by biologists and economists over the last half century – showing why self-interested individuals would nonetheless cooperate – to explaining how the other-regarding preferences and group-level institutions that sustain cooperation could have emerged and proliferated in an empirically plausible evolutionary setting."[11] One of their most conspicuous targets is Ken Binmore, and his insistence that society developed its moral rules to deal with repeated interactions by self-regarding individuals.[12] I am in sympathy with Bowles and Gintis in the negative part of their project. I think that Kant is right about moral motivation, and therefore Binmore cannot show that morality is self-interested. But a Kantian is also going to be skeptical about whether we can show *empirically* that some practice is *not* self-interested, though we can certainly show empirically that it involves cooperation. The reason for this is that Kant holds that the will is in principle beyond the range of sensory observation.[13] Bowles and Gintis do not claim to be defending Kant against Binmore, and they may well reject this part of Kant's account. But if they do intend to reach an account of the development of a Kant-style morality, their *positive* claims will have to be more modest.

I want to talk briefly about how Kant sees the relation between morality and religion, and then I want to suggest that recent work in game theory, including that by Gintis, makes Kant's position about this interesting, and potentially useful for explanatory purposes. Kant says, throughout his published work, that "we have to

[10] Sober and Wilson (1998, pp. 4–6).

[11] Bowles and Gintis (2009) final paragraph.

[12] Binmore (2005).

[13] There are complex questions in the philosophy of mind about whether brain imaging would help determine empirically what is going on "inside a person" in the relevant sense. This is not a debate I can get into in this paper.

recognize our duties as God's commands."[14] He defines "religion" as such a recognition.[15] In terms of the five formulas of the categorical imperative that I mentioned earlier, the fifth formula is that "a rational being must always regard himself as legislator in a kingdom of ends rendered possible by freedom of the will, whether as member or as sovereign."[16] Moral agents are to regard themselves as making law for a kingdom, the kingdom that consists of all rational beings. Kant says that this formula sums up the other four, taking the notion of law from the first formula and the notion of humans as ends in themselves from the third formula. This kingdom in which we regard ourselves as making laws has both members and a sovereign, and Kant continues that the position of the sovereign "can be maintained not merely through the maxims of his will but only if he is a completely independent being without needs and with unlimited power adequate to his will." Kant thinks that moral agents need to believe in the existence of such a sovereign, because morality commits them to an end which Kant calls "the highest good," which we can rationally believe is possible only if we postulate the sovereign's existence. This is where the principle that "ought" implies "can" comes in. If it is not the case that the highest good can be achieved, then the question whether we ought to achieve it does not properly arise.

The highest good has two components, our happiness and our virtue. Kant is here rejecting two ancient theories, as he interprets them. The Stoics, he says, thought that our highest good consisted in our virtue and our knowledge of our virtue; if we meet these two conditions, we will be happy. Kant rejects this view as inhumane. The Stoic sage, tortured on the rack, can be virtuous and can know that he is virtuous, but surely he is not *happy*. On the other hand, the Epicurean, Kant says, thinks that happiness is pleasure, and virtue is the means to that. For Kant, as we have seen, virtue or morality has to be pursued for its own sake. The Stoic reduces happiness to virtue and the Epicurean reduces virtue to (the means to) happiness. Kant insists that these two components of our highest good are different from each other and both are necessary, even though only virtue is good without qualification. Moral agents are rationally required to believe that it is really possible for these two components to go together, or, to put the same point the other way round, that we do not have to do what is morally wrong in order to be happy. Unlike the sovereign of the kingdom of ends, we mere members are creatures of need, and "to be happy is necessarily the demand of every rational but finite being and therefore an unavoidable determining ground of its faculty of desire."[17] So we desire to be happy in everything else that we desire, but it is not this desire that gives our actions *moral* value. Rather, what gives moral value is our obedience to the categorical imperative. But our experience of the world does not, Kant thinks, license the conclusion that happiness and virtue

---

[14] For example, *Critique of Practical Reason*, 5: 129, but there are similar statements in the *Critique of Judgment*, the *Metaphysics of Morals*, *Religion*, *Conflict of the Faculties*, and elsewhere in the *corpus*.

[15] *Religion within the Boundaries of Mere Reason*, 6: 154. I will abbreviate this work as *Religion*.

[16] *Groundwork*, 4: 434.

[17] *Critique of Practical Reason*, 5: 25.

go together, and he thinks we do not know how to produce a world in which they do. "Real" possibility, in his usage, is not merely logical possibility, but a possibility that is grounded in something that we can rationally believe actual. Kant suggests that we have to postulate the agency of the sovereign of the kingdom of ends, who can bring about the correspondence of happiness and virtue because such a being has a holy will and also "has power adequate to his will," as we do not.

Some readers of Kant think that his bringing in happiness to the highest good pollutes his account of morality. But his position can be defended as balancing in a subtle and convincing way the two incentives of happiness and duty. He is saying that we are in fact pulled by both incentives, and there is nothing wrong about this. Morality, he says, "on its own behalf in no way needs religion."[18] But because we are creatures of need, and not purely rational beings with a holy will (like the sovereign of the kingdom of ends) Kant continues two paragraphs later, "morality inevitably leads to religion, and through religion it extends itself to the idea of a mighty moral lawgiver outside the human being, in whose will the ultimate end (of the creation of the world) is what can and at the same time ought to be the ultimate human end."[19]

A brief historical excursus may make Kant's theological roots clearer. Very probably Kant took his formulation of the categorical imperative from a version in Christian August Crusius, a Prussian pietist of the previous generation.[20] Crusius in turn took the root idea from the Lutheran appropriation of the medieval Franciscan distinction, found for example in Duns Scotus, between "the affection for justice" and "the affection for advantage."[21] The affection for advantage is an inclination towards our own happiness and perfection, and the affection for justice is an inclination towards what is good in itself, regardless of its relation to us. The affection for justice is therefore an example of non-self-indexed motivation. Humans have both affections, on this view, and there is nothing wrong about that, but the question is how the affections are *ranked*. The right ranking makes the affection for justice unconditional, and Scotus tells us that this ranking is most clearly seen counterfactually; if God were to require us, which God does not, to sacrifice even our salvation for the sake of God's glory, we should be willing to do so. Jonathan Edwards (echoing words of Moses and Paul in the Hebrew and Greek Scriptures) says, "I would be willing to be damned for the sake of the glory of God."[22] Any Kantian account of human moral motivation has to accept that our motivation is irreducibly double in this way (corresponding to Scotus's two affections).

It is reasonable to think that religion is connected to the origins of morality, when we consider the conjunction of the following two theses. These theses take me outside the competence of the professional philosopher, and I readily concede that I may have misunderstood the non-philosophical literature I have tried to read. The

[18] *Religion* 6: 3.

[19] *Religion* 6: 6. This does not mean, in Kant's view that only theists can be morally good, but he holds that the position of the morally good atheist is "rationally unstable." See Hare (2005).

[20] Schneewind (1998) 447, 520–521.

[21] I have laid out the history more fully in Hare (2007) especially Chap. 2.

[22] *Religious Affections*, x. The words of Moses are at *Exodus* 32: 32, and of Paul at *Romans* 9: 3.

first thesis is that humans have a capacity for non-self-indexed motivation that has not yet been observed in non-human animals. For reasons stated earlier, we are not going to get an empirical demonstration of the capacity for Kantian moral motivation. We may find, however, that humans use strategies that are consistent with a preference for non-self-indexed values. We may find more than mere consistency, that there are observable preferences that have game-theoretic explanations as part of a strategy directed at a non-self-indexed goal. Gintis, for example, claims to find a capacity that he calls "strong reciprocity," "a predisposition to cooperate with others, and to punish (at personal cost, if necessary) those who violate the norms of cooperation, even when it is implausible to expect that these costs will be recovered at a later date."[23] "Strong reciprocity" is an unhappy label, since the preference in question does not need to involve anticipated reciprocal benefit to the agent. But the "Ultimatum" game and various public goods games do seem to show that humans are capable of adopting preference strategies that do not have any obvious link to self-advantage, and this has been documented (if we can trust the testing conditions) cross-culturally on a large scale, including hunter-gatherer societies that are assumed to be the closest in form to the societies of our original human ancestors.[24] The term "preference" here has to be distinguished from what Kant calls a "maxim." The preference is demonstrated in choice-behavior, or the adoption of a strategy of a certain kind, and the strategy is defined in terms of the achievement of something that is not itself self-indexed, even though the people who adopt the strategy may or may not have self-indexed motivation (and so a self-indexed maxim) for adopting the strategy. Those who adopt the strategy may or may not, for example, have what Kant calls "the good will," which is motivated by love for the moral law for its own sake.

Non-human primates do not seem to display this preference-predisposition (though I know this is controversial).[25] They certainly display cooperation, as indeed do much simpler organisms. Kin-selection necessarily involves benefits to kin (and so is self-indexed), "reciprocal altruism" involves expected benefit to the organism that cooperates, as do "indirect reciprocity" (mediated through reputation), and "network reciprocity" mediated through interactions with those within the network. But these forms of cooperation leave the question of motivation untouched. We do not know how to reach "inside" the minds of non-human animals, assuming that there is an "inside" there to be reached. We can probably say, however, that these forms of cooperation are consistent with the absence of non-self-indexed motivation, and have game-theoretic explanations in terms of strategies directed entirely at

[23] The quotation is from Gintis (2005, p. 8). Non-self-indexed motivation also needs to be distinguished from the benefactor's and recipient's so-called "reciprocal altruism" (where there is expectation of benefit from the recipient) and "indirect reciprocity" (in which one's reputation for cooperation is rewarded indirectly through the favor of third-party observers).

[24] One study worked in 12 countries on four continents, and recruited subjects from 15 small-scale societies, consisting of three foraging groups, six slash-and-burn horticulturalists and agropastoralists, four nomadic herding groups, and two sedentary, small-scale agricultural societies, Henrich (2004).

[25] The claim about primates is defended by Silk (2005, pp. 63–64).

self-indexed goods. Silk concludes, after looking at the evidence for "strong reciprocity" among primates, "Strong reciprocity in humans seems rooted in a deep sense of fairness and concern for justice that is extended even toward strangers, but we have no systematic evidence that other animals have similar sensibilities."[26] With humans, by contrast, cooperation seems to continue when kin-selection, "reciprocal altruism," indirect reciprocity and network reciprocity are not at stake. Fehr and Gächter say, "People frequently cooperate with genetically unrelated strangers, often in large groups, with people they will never meet again, and when reputation gains are small or absent."[27]

The second thesis, that can usefully be added to the first, is that the human groups that now seem closest to the form of life of the earliest groups of *Homo sapiens*, and which do display the preference-predisposition I have been talking about, also all seem to display some form of religion.[28] The evidence for this is strong, given the inevitable limitations produced by the distance in time. "Religion" is here broadly defined to include a belief in "high gods" but also lower grades of divinity, ancestors who are still active, and witches and sorcerers. A "high god" is defined as "a spiritual being who is believed to have created all reality and/or to be its ultimate governor, even though his sole act was to create other spirits who, in turn, created or control the natural world." High gods are also supposed to be "moralizing" and to have an interest in people's conduct. The question then arises whether religion has anything to do with the development of specifically human forms of cooperation, and it is initially plausible to think that it does.[29] I am going to suggest that Kant's view of the relation between morality and religion has a contribution to make to this discussion. I will focus on some ways in which the connection of morality to religion might help us understand moral motivation. I will connect these with Kant, but my thought is that they can be generalized to cover forms of religion and morality that Kant knew nothing about. The importance of finding these features *also* in Kant is that it becomes plausible to think they are still part of the deep structure of the morality system "we" are familiar with. This is true even though ancient morality systems were no doubt radically different in other ways from the Kantian system.

As I explained earlier, Kant thinks that we are both under the law and creatures of need, aiming at our happiness whatever else we aim at, and we therefore have to be able to believe that the members of the kingdom of ends can be virtuous *and happy*, if we are rationally to persist in the life of virtue. This is why he says that "morality inevitably leads to religion."[30] Our morality itself requires us to postulate the existence of a sovereign of the kingdom of ends, who has the role of bringing about

[26] Silk (2005, p. 63).

[27] Fehr and Gächter (2002, pp. 137–140).

[28] The claim about the pervasiveness of religion is in Brown (1991) and Boyer (2001). See also Bering and Johnson (2005). In the standard cross-cultural sample of 186 societies (from the Ethnographic Atlas of 1,267 entries) all have gods and 168 have "high gods."

[29] A number of writers have suggested such a connection. See Johnson and Kruger (2004), Roes and Raymond (2003), Sosis and Alcorta (2003), and Wilson (2002).

[30] *Religion* 6: 6.

the consistency of virtue and happiness.[31] This being also has the roles I have not yet mentioned of enabling in us a revolution of the will (which reverses the innate ranking of happiness above duty, and so of Scotus's two affections), and of coordinating the ends of all rational beings so that the happiness of every virtuous person is consistent with his or her virtue, even if not all rational beings are virtuous.[32] If we return now to Binmore's claim, mentioned at the beginning of this paper, we can see that his claim that Kant's requirements are not incentive-compatible is invalid once Kant's moral theology is taken into account, since this theology has the function of making self-indexed and non-self-indexed motivation compatible. I am suggesting a functional relation between religion and the development of what I have called "non-self-indexed motivation" in humans. The term "functional" is tricky here. The claim that religion is *necessary* for the development of this kind of motivation is too strong, because we do not know what all the alternative are.[33] Perhaps non-self-indexed motivation is consistent with self-indexed motivation in evolutionarily stable games even without belief in high gods and their role in maintaining cosmic order. But there will still be a functional relation if belief in these gods in fact plays the role I have described. One practice can be functionally related to another if they co-occur and the working of one is helpfully understood through the working of the other, even if the first could exist in some possible world without the second.

One part of Kant's account of the relation of morality to religion is his account of the role played by a belief in divine punishment. Kant approaches this by analogy with punishment by the State for those who break the laws of the State. Kant's moral agent needs the State to punish, but not because her moral motivation is fear of such punishment. Rather, she values freedom (which in Kant's view is the capacity of her will to choose in obedience to the moral law) and so she values external freedom (which is the outward expression of her inner freedom in her actions). But general lawlessness prevents or hinders her external freedom, because, for example, she is not externally free to walk in the street if the streets are full of criminals who will rob her or assault her. Punishment is valued morally as "the hindrance to the hindrances to freedom."[34] The lawlessness is a hindrance to her external freedom, and punishment is a hindrance to this hindrance.

Analogously the moral agent is not given moral motivation by fear of God's punishment. Here we can return to the formulas of the categorical imperative in Kant's *Groundwork*. I have mentioned the first, third, and fifth. The fourth formula centers on the notion of autonomy, according to which we are both subject to the moral law and authors of it.[35] If our motivation is fear of punishment, for example fear of hell, the source of our action is not a free willing of the moral law, but mere

---

[31] It is possible that there is empirical evidence of some kind of human predisposition to entertain the idea of divine agency. See Bering (2006).

[32] I have explained this in more detail in *God and Morality*, *op. cit.* Chap. 3.

[33] Here I depart from Kant, who does make the necessity claim, e.g., *Critique of Practical Reason* 5: 138. See also footnote 38 of the present paper.

[34] *Metaphysics of Morals* 6: 396.

[35] *Groundwork* 4: 431.

inclination, for example the interest we have in avoiding eternal torment. But the moral agent aims at the highest good, and this requires believing that the system by which virtue is consistent with happiness is in place and the apparent disproportion of virtue and happiness that we experience is not final. She believes that the good in the universe is more fundamental than the evil, and will in the end prevail. Religion in a very broad sense is the combination of a belief about how things ought to be with a belief that this is supported by how they fundamentally are. However, as with life in the State, the moral agent has to live in a world in which not all her fellow-humans are law-abiding. The threat of punishment can motivate those who are *not* motivated by respect for the moral law, and so can be a hindrance to the hindrances to freedom. This Kantian account allows us to distinguish two different motivations related to divine punishment. One is fear, because "punishment can force the costs of free-riding above the costs of cooperation."[36] But the other (more satisfactory to the Kantian) is hope, because a belief in divine punishment is part of a belief in "a world morally governed."[37] Even if hunter-gatherer groups do not have developed theories of human autonomy, as seems most likely, we can still make a similar distinction on their behalf.[38] There is a difference between being motivated by fear of divine punishment, and being motivated by love of justice or fairness, which is a system that divine punishment maintains.[39]

I want to make three other points about the relation Kant establishes between morality and religion (and there is more to be said, but not space to attempt an exhaustive list). The first involves atonement, which can be interpreted as a kind of self-punishment if performed by the offender, or as vicarious punishment if someone else performs it. Suppose an agent has violated the moral law, and thus put herself at odds with "the high gods." She can repent, apologize, and sometimes make reparation to her human victims, but there will be many cases where she cannot (e.g., her victim is dead).[40] We could construct a "forgiveness game" that will specify the tasks of the offender and victim and the payoffs from the mutual completion of these tasks. Where she cannot, the problem is that the offender is not in a position to forgive herself unilaterally without her tasks being performed. But in Kant's terms,

[36] Johnson (2005, p. 411). The point, made by Johnson, that punishment is more effective motivationally than reward (sticks than carrots) does not affect the distinction I have just made. Being motivated by a reward in heaven is no more autonomous than being motivated by punishment in hell.

[37] Sidgwick (1981, p. 505).

[38] Kant, like other contemporaries in the Enlightenment, is too optimistic that he can identify what reason tells every human at every time and place. See *Religion* 6: 9. He did think that the categorical imperative has been known to all humans at all times and places.

[39] We might expect that if God is believed to be the ultimate guarantor of justice, human punishers, though they will punish failures to cooperate, will not punish so readily failures to punish those who fail to cooperate. The reason is that the second-order punishing (and perhaps third-order punishing of those who fail to punish those who fail to punish those who fail to cooperate) is closer to the systems maintenance which is the role given to the gods. In such a set of beliefs, the rules of cooperation might survive without large-scale human second-order and higher-order punishing.

[40] For a more detailed treatment of these tasks see Swinburne (1989, 25f), and Hare (1996) Chap. 9.

an agent needs to be able to sustain a belief in her "worthiness to be happy" even in the face of moral failure, if she is to rationally sustain her belief that she will be happy, given her commitment to the highest good as I described it earlier.[41] Religions can provide a method of atonement, a process by which the agent can be reconciled with the divinities that have also been offended along with the human victims when the offence is committed. There is a complex relation here between self-forgiveness and forgiveness by the gods. The agent may also be able to signal genuine remorse to her human victims by engaging in costly acts of atonement.[42]

A second point is that the commitment to cooperation is hard to form and to sustain. Although Kant thinks we are born with both the predisposition to good (which is part of the concept of "human") and the propensity to the wrong ranking of happiness and duty (which is universal, but not part of the concept of "human"), he thinks the latter is dominant in our initial condition at birth. Kant says that we need divine assistance to overcome this ranking (either "a positive increase of power" or the removal of obstacles), and we may receive this assistance even though we do not recognize it.[43] This is the role of the sovereign of the kingdom of ends that I mentioned previously in accomplishing (together with us) a revolution of the will, by which our initial ranking of happiness over duty is reversed. But given the difficulty of commitment, signals of commitment are hard to make credible. A signal of commitment is more likely to be credible if the signaler and the recipient both believe in the availability of this divine assistance. Thus in our own culture God's help is often invoked at marriage services, where difficult commitments are made. The same is true at times of large-scale communal undertakings such as new presidencies, or going to war.

Finally, religion can give us a model of the kind of thinking that morality requires. "Impartial spectator" theories in ethics are just one example, holding that moral thinking is an approximation to the thinking of a being with complete information and no bias (especially bias towards the self).[44] It is typical of divine command theories of moral obligation to hold that what is right is what a benevolent being would prescribe who knew the desires (the ends) of the relevant parties and what action(s) or type(s) of action would satisfy those desires, and who loved those parties equally. If such a being could also coordinate the realization of morally permissible ends, the concept of obligation (as Kant acknowledges) would fit best our being under the authority of, or being commanded by, such a being.[45]

I have not been trying to prove, in any of this, that God(s) exist(s), but that there is one kind of moral theory, I would claim a dominant theory in Western culture, which is intimately tied with religion and which reflects an original tie between the

[41] *Religion* 6: 72.

[42] There has been an interest, in the literature, on the benefits of religion in providing costly signaling. See Sosis and Alcorta (2003).

[43] *Religion* 6: 44.

[44] Carson (2000) Chap. 8.

[45] Anscombe (reprinted 1981). Adams (1999) 253f argues that Durkheim parodied this concept in order to get a concept of society as the source of moral obligation.

two. It is always possible to agree with this (as Nietzsche and more recently Bernard Williams did) and reject both the morality system and religion along with it.[46]

References

Adams RM (1999) Finite and Infinite Goods. Oxford University Press, Oxford

Anscombe E (1981) Modern moral philosophy, In: Ethics, religion, and politics. University of Minnesota Press, Minneapolis

Bering JM (2006) The cognitive psychology of belief in the supernatural. Am Sci 94:142–149

Bering JM, Johnson DDP (2005) O Lord . . . You perceive my thoughts from afar: recursiveness and the evolution of supernatural agency. J Cogn Cult 5:118–142

Binmore K (1994) Game theory and the social contract. MIT, Cambridge

Binmore K (2005) Natural justice. Oxford University Press, Oxford

Bowles S, Gintis H (2009) Cooperative Homo economicus, present volume

Boyer P (2001) Religion explained: the evolutionary origins of religious thought. Basic Books, New York

Brown DE (1991) Human universals, McGraw-Hill, New York

Carson T (2000) Value and the good life. University of Notre Dame Press, Notre Dame

Edwards J (1746) Religious affections

Fehr E, Gächter S (2002) Altruistic punishment in humans. Nature 415:137–140

Gintis H et al (2005) Moral sentiments and material interests. MIT, Cambridge

Hare JE (1996) The moral gap. Clarendon Press, Oxford

Hare JE (2005) Kant on the rational instability of atheism. In: Dole A, Chignell A (eds) God and the ethics of belief. Cambridge University Press, Cambridge, pp 202–218

Hare JE (2007) God and morality: a philosophical history. Blackwell, Oxford

Henrich J et al (2004) Foundations of human sociality: ethnography and experiments in fifteen small-scale societies. Oxford University Press, Oxford

Johnson DDP (2005) God's punishment and public goods. Human Nature 16

Johnson DDP, Kruger O (2004) The good of wrath: supernatural punishment and the evolution of cooperation. Polit Theol 5:159–176

Kant I (1785) Groundwork of the metaphysics of morals

Kant I (1788) Critique of practical reason

Kant I (1790) Critique of judgment

Kant I (1793/4) Religion within the boundaries of mere reason

Kant I (1797) Metaphysics of morals

Nietzsche F (1967) On the genealogy of morals. Vintage Books, New York

Nowak MA, Coakley S (2009) Evolution, games and God: the principle of cooperation. Harvard University Press, Cambridge

Roes FL, Raymond M (2003) Belief in moralizing Gods. Evolution and human behavior 24:126–135

Schneewind JB (1998) The invention of autonomy. Cambridge University Press, Cambridge

Sidgwick H (1981) The methods of ethics, 7th edn. Hackett, Indianapolis

Silk JB (2005) The evolution of cooperation in primate groups, In: Gintis (2005)

Sober E, Wilson DS (1998) Unto others. Harvard University Press, Cambridge

[46] Nietzsche (1967, pp 90–91), "There is no small probability that with the irresistible decline of faith in the Christian God, there is now also a considerable decline in mankind's feeling of guilt." Williams (1985, pp. 191–196). I am grateful to Dominic Johnson for comments on the present paper.

Sosis R, Alcorta C (2003) Signaling, solidarity, and the sacred: the evolution of religious behavior. Evol Anthropol 12:264–274
Swinburne R (1989) Responsibility and atonement. Clarendon Press, Oxford
Trivers R (2000) The elements of a scientific theory of self-deception. Ann NY Acad Sci 907:114–131
Williams B (1985) Ethics and the limits of philosophy. Harvard University Press, Cambridge
Wilson DS (2002) Darwin's cathedral: evolution, religion, and the nature of society. Chicago University Press, Chicago

Explaining Religion: Notes Toward a Research Agenda

Kwame Anthony Appiah

Abstract I begin by arguing that our model of religion is often based on Christianity. A Christian model of religion is going to look for gods and creeds, churches, priests, prayer, collective worship, moral codes, each of which is absent in some of the things we might want to call religions. And it may well ignore dietary and sumptuary rules or cult for ancestors, which are important in some of them. Religion is a paradigm of what Wittgenstein taught us to call a "family-resemblance" concept: each religion, like each member of a family, is like every other, in some respect, but there are few, if any, characteristics they all share. So the first thing we need to do in trying to decide what it is we're explaining is disaggregate the elements that come together in Christianity; if we find that they usually come together that will be one of the things that we need to explain. What then are the questions worth focusing on? I think that, from an evolutionary point of view, it will be two families of issues. First will be the social and the cognitive features of religions that make their explanation challenging. A second family of issues worth exploring, once we have identified these components, is how they fit together. Why, for example does belief in invisible beings go with rituals dealing with disaster? Why does agreement in creeds go with creating powerful social groups that last across the generations? When one finds broad patterns across many societies there are usually two natural types of explanation that spring to mind. One is that the pattern reflects shared solutions to common problems, independently discovered: evolutionary homology, as it were. The other is diffusion from common sources: in a word, copying. I suspect that much of what is share in the organization of religions globally today is the result of diffusion. But, of course, why some patterns diffuse successfully and others don't is itself something that needs explaining.

There's been a great deal of discussion recently both about the causes and about the effects of religions. On the causing side of things, ambitious evolutionary psychologists ask: What accounts for the evolution of religion? (Or, more precisely, for the

K.A. Appiah
Department of Philosophy and the University Center for Human Values, Princeton University, 219 1879 Hall, Princeton, NJ 08544, USA

S.A. Levin (ed.), *Games, Groups, and the Global Good*, Springer Series in Game Theory, DOI: 10.1007/978-3-540-85436-4_12, © Springer Physica-Verlag Berlin Heidelberg 2009

human propensity to create and sustain religions?)[1] They wonder whether religion evolved independently of morality or whether the evolution of morality was a component of the evolution of religion... or vice versa. They ask what accounts for the astonishing success of the religious meme (or memes). On the effect side, both defenders and enemies of religion ask: Is religion responsible for more suffering and evil or more good?[2]

These approaches tend to suppose that what needs explaining is something that is going to turn out to be more or less the same for all the great diversity of cultural phenomena we call "religion." They presuppose we are interested in the causes and effects of Buddhism, Christianity, Judaism, Hinduism, Islam and Taoism, as well, no doubt, as the less-often-mentioned range of "traditional religions" from Patagonia to Hudson's Bay, from the Cape of Cairo, from Sri Lanka to Mongolia, from New Guinea to New Zealand.

That there are things properly called "religions" in all these places is itself a fairly substantial claim. In my home-town of Kumasi in Asante, in the mid-nineteenth century, for example, when my great-grand-mother was young, people had all sorts of beliefs about Nyame (the sky God), Asaase Yaa (the earth goddess) and other divinities and spirits of divers kinds; and there were ritual practices and prayers addressed to them and professional priests and shrines of varying degrees of sophistication, ranging from the domestic to the national: but it would never have occurred to people to define themselves by an identity based in these shared traditions of belief. They might have identified *others* in this way – "Kramo" is an old word for Muslim in our language – but precisely because the traditions were so widely shared, participation in these various religious forms didn't mark anyone out as special ... except for those few, mostly strangers and the crazy, who diverged from them. Being Asante was a serious identity and Asantes shared these "religious" beliefs and engaged in these "religious" practices. Those things we might now pick out and call "Asante religion," however, were a constituent of Asante-ness not an independent source of identity; and, indeed, you would be hard put to find a word in nineteenth-century Asante-Twi (as in most of the world's languages until recently) to translate our English word "religion."[3]

Much of what is called religion in the world today was first identified *as* religion by Europeans in their global expansion since 1492: and, by and large, they called anything a religion that was in competition with the Christian ideas, practices and traditions they had left behind. The result is that, many people in our intellectual tradition unwittingly take as the paradigm of religion one particular set of religious sects – those that developed out of medieval Christendom in the coincidence of the age of exploration and the reformation. As a result, for example, belief in God is taken to be one mark of religion, even though there are atheistic forms of Buddhism and most of the world's non-Abrahamic religions are basically polytheistic. Now

[1] Atran (2002).

[2] Dawkins (2008) or Harris (2005).

[3] Twi is the name of the language spoken in Asante. It is one of a family of dialects spoken in the wider world of Akan cultures in Ghana, and belongs to the wider group of Akan languages.

I say *basically* polytheistic, because the very idea that all those beings in the sky, those forces in the earth, those spirits of lakes and forests and rocks, should be called "gods" is not much help until we say a little more about what we want this term to mean. Missionaries, for example, tended to identify them as spirits, or, more hostilely as idols and devils, unless they showed strong affinities with Jehovah. Some scholars have argued, for example, that the spirits of much of African religion are not supernatural beings, but part of nature, continuous with the rest of nature: and, if that is so, perhaps they aren't poly*theistic* at all.[4]

The focus on gods is only one mark of the Christian origins of the modern Western concept of religion. We often look for creeds that define not just the existence of at lease one divinity but also claim a great deal more specifically about him (or her or it). The history of Christianity makes this natural enough, since councils – Nicea, Trent, Chalcedon – and creeds – Athanasian, Nicene again – are central ... as are the multifarious heresies; and getting on the wrong side of one of these debates was often fatal. But, once again, the obsession with getting other people to accept your creed is much less central to others of the things we call religions: especially, as you might expect, in traditions that have no central religious authorities or no written documents. Chinua Achebe, the Nigerian novelist, once observed that he couldn't imagine his Igbo ancestors "traveling 4,000 miles to tell anybody their worship was wrong!"[5]

A Christian model of religion is going to look for more than creeds: it will expect churches, priests, prayer, collective worship, moral codes. And it may well ignore things that people in other traditions might think obviously important: dietary and sumptuary rules or ancestral cults. One could go on. Religion is a paradigm of what Wittgenstein taught us to call a "family-resemblance" concept: each religion, like each member of a family, is, no doubt, like every other, in some respects, but there are few, if any, characteristics they all share.[6]

So it seems to me that the first thing we need to do in trying to decide what it is we're explaining is disaggregate the elements that come together in Christianity; if we find that they usually come together that will be one of the things that we need to explain.[7]

---

[4] Wiredu (1980, p. 2).

[5] Achebe (1982, p. 209).

[6] People (including Dominic Johnson in his comments on an earlier draft) often express skepticism when I say this. But until we have agreement on the range of things we're going to call "religions," I think that – at the very least – we shouldn't assume that there is a list of features that they all share. The Bambuti of Zaire don't seem to believe in a high God – see Turnbull (1968) – nor do many Buddhists. Ritual is of very little importance for many Quakers. Many Unitarians are agnostic at best. Most Lutherans don't believe in spirit possession. Early Judaism doesn't seem to have involved belief in an after-life. We could decide that, for this reason, these aren't religions, I suppose; or that the concept of religion is incoherent. But the view that it's a family resemblance concept still strikes me as the best option.

[7] Dominic Johnson has drawn my attention, rightly, to recent work that takes this challenge seriously. For example, Whitehouse (2008), tries to identify a "cross-culturally recurrent religious repertoire" of regularly recurring elements whose disaggregation makes more deeper and more fine-grained analysis. I think that disaggregation of this sort will prove a useful strategy; though

I want to underline, in closing this section, that my claim is that it was the Christian framework of many of Europe's first modern encounters with the non-West that explains how certain combinations of belief, emotion and behavior got to be called "religion." So even those evolutionary psychologists who have focused on religions from the ethnographic record and have no inclination to focus on Christianity are encumbered, I claim, with a concept that faces these problems. Let me repeat: there's no word for "religion" in many of the so-called traditional religions of the world, for reasons that I explained in the case of Asante. So I'm not just objecting to the framing of religions from elsewhere by Christian theologians or by others with a Christian world-view. It's our concept not *us* that has the problem.

Dominic Johnson suggests another problem for those who start with Christianity or any of the other "world religions" as their model. It's that these all developed in historical times. And so, as he puts it nicely,

> Since these religions were founded and developed *after* humans had moved out of the physical and social environment in which we evolved, they may actually tell us very little about the adaptive role of religious beliefs and behaviors as evolution originally "designed" them. If so, our only window onto the relevant selection pressures in human evolution is to look at the "religions" of hunter-gatherer societies.[8]

This worry makes it clear that, in taking current world religions as the lens through which we look at the evolution of the tendencies of our natures that produce religious belief, emotion and behavior, we are assuming that what is central to these world religions exemplifies what was adaptively important in the longer pre-historical period in which our natures were mostly shaped. But current world religions are the product, among other things, of substantial exchange – the diffusion of ideas and practices since the Axial Age – as well as of substantial shaping by political forces that have taken up religions to organize states. Christianity spread within the Roman Empire and then as the religion of the Empire; Islam, likewise, spread as the religion of an expanding Arab empire; Buddhism is associated with state formation in China, Central Asia and Japan. If you want, in particular, to focus on biological or psychological evolution – rather than the cultural evolution of religions – these are crucial issues to bear in mind.

What then are the questions worth focusing on? I think that, from an evolutionary point of view, it will be two families of issues. First will be the social and the cognitive features of religions that make their explanation challenging. For example, whether we call them gods or not, religious traditions tend to invoke appeal to forces that are personal (unlike magnetism or gravity, which are impersonal). Furthermore most, if not all, of these personal forces do not actually exist (though, in the case of ancestors and the unborn, they may have existed once).[9] So the obvious

---

if what we are trying to explain is religion, it will be important to see how the disaggregated components fit back together; and to explain in particular, as I say in the text, why these components come together so often in a package.

[8] Personal communication, Sept 8 2009.

[9] John Hare wondered why I simply assumed this. The main reason is because I believe it and would be willing to defend the claim if asked. But I think the claim is a good deal less controversial than it

explanation of belief in them – they're there and people have noticed them – can be ruled out. Unfortunately, while this obvious explanation would fit these beliefs into the general story of the evolution of normal cognition, once we've ruled it out we need to have some special story to explain these beliefs in invisible beings.

Similarly many religious traditions involve ritual practices aimed at handling moments of significance in human life: birth, puberty, marriage, harvest, death; responding to natural disasters: locusts, earthquakes, floods, droughts, plagues, disease; and preparing for man-made crises, especially war. Here, too, if these ritual practices actually worked as they were supposed to – if religious rituals could increase the probability of survival of a child or a marriage or a food crop or a herd, or of success in warfare – their existence would be less puzzling. And we think that even where such rituals do in fact raise the probability of success of some enterprise, it isn't for the reason that the practitioners expect it to work, and so there is an interesting question as to what the mechanism is.[10]

And – to give a final sort of example – religions in complex societies often play a role in the organization of social groups whose identities can then be mobilized

---

may appear. While the vast majority of human beings believe in some sort of non-human agencies of the sort we would normally call "gods" or "spirits," they tend to believe in different ones. So it's true of most of the spirits people invoke that there are more (usually many more) people who don't believe in them than people who do. More people, then, should agree with me that "most, if not all" the spirits others invoke don't exist than should disagree. (Of course, it depends a little on how you individuate spirits: is Jehovah really the same spirit as the Christian God or as the Moslem Allah and the Akan Nyame? Opinions diverge on this topic.) At all events, I believe myself that most of the invisible agents that people believe in do not exist; and, about most of them, taken one by one, I believe most people will agree with me. The result is that belief in each of them will not be taken by most people, including me, to be best explained by supposing that belief to be true. I grant that there is a possible intellectual approach to the question of the ontological components of religious belief that takes gods and other spirits as given and asks how human awareness of them came to be embodied in social institutions. My favorite expression of this idea is in Sir Richard

Burton's couplet:

All Faith is false, all Faith is true: Truth is the shattered mirror strown
In myriad bits; while each believes his little bit the whole to own.

Burton (1880, p. 10). This has been a standard position in Christian missiology, which takes religious views outside the Christian world to have grasped the central truth of God's existence "through a glass darkly." (It has also been common to advert to Satan as one of the explanations of why the non-Christian view is distorted). I'm not aware of work based in game theory or evolutionary psychology that takes this approach, which is why I don't discuss it further. But it is, as I say, a possibility.

There is, by the way, an interesting parallel here to debates in the sociology of science between those who believe in the so-called "Strong Program" – associated, inter alia, with Bloor (1991) – which holds that the development of theories that are (currently taken to be) true should be explained in the same way as the development of theories that are (currently taken to be) false, on the one hand, and those who believe that the history of science can appeal to scientific truth, on the other. The strong program is a kind of methodological agnosticism about the truth-claims of science. Many scientists worry that the strong program is anti-science; I suspect that many religious people will worry that methodological agnosticism, as a research method in the study of religion is similarly anti-religious.

[10] One classic discussion that aims to make belief in invisible agencies less puzzling is Evans-Pritchard (1951).

in competition with other groups, and are often central to ethnic and other political identities. Here, I think, there is much to be learned by thinking about the social psychology of identity in general and applying it to religious identities.[11]

A second family of issues worth exploring, once we have identified the components that we take to be constitutive of religion, is how they fit together. Why does belief in invisible beings go with rituals dealing with disaster? Why does agreement in creeds go with creating powerful social groups that last across the generations? When one finds broad patterns across many societies there are usually two natural types of explanation that spring to mind. One is that the pattern reflects shared solutions to common problems, independently discovered: evolutionary homology, as it were. The other is diffusion from common sources: in a word, copying. I suspect that much of what is share in the organization of religions globally today is the result of diffusion: everybody knows that Judaism, Christianity and Islam have a common root, as do Hinduism and Buddhism. I believe that the modern forms of many of these religions are shaped by their awareness of each other, as well as by direct interaction through institutions like the Parliament of the World's Religions. But, of course, why some patterns diffuse successfully and others don't is itself something that needs explaining.

So here is an agenda for research on the historical processes that have generated the wide diversity of what we call religions today. First, we should explore the persistence of belief in non-existent personal entities, often supposed to be either invisible or usually manifest only rather indirectly. This is the major cognitive dimension of religion, I believe. It was identified as such by the great nineteenth century anthropologist Sir Edward B. Tylor, who wrote in his *Primitive Culture* that although "animism," (by which he meant a "general belief in spiritual beings") "... may seem to afford a bare and meagre definition of a minimum or religion, it will be found practically sufficient; for where the root is, the branches will generally be produced."[12]

I believe the best theory here is one essentially due to Robin Horton: it is that these invisible agents are theoretical postulates originally invoked to explain phenomena, their personal character being a reflection of our having a relatively strong predisposition to adopt the intentional stance towards things.[13]

A second important topic of research, from an evolutionary psychological point of view, is the role of religion in creating social solidarity. If – and the evidence here does not seem to me very decisive – religion sustains pro-social behavior, that is also worthy of investigation.[14] Here the question is whether there is anything distinctive about the shared practices and beliefs (and emotions) of co-religionists that makes religious – as opposed, to, say, merely ethnic – identity especially inclined to produce solidarity.

---

[11] See Appiah (2005), passim. And see also my essay, Appiah (2008, pp. 41–64).

[12] Tylor (1899, pp. 424–426).

[13] Horton (1967, 1993) which gives a general account of the development of such beliefs in invisible agency, based on case studies in West Africa. Dennett (1989).

[14] Phillips (2007).

Finally – a question for social psychology, anthropology and history – is there reason to believe that inter-group conflicts based on religion are especially troublesome in some important way? Genocide in the last century or so has been undertaken largely without religious bases. (Hitler, Stalin, Pol Pot, Mao and the proponents of Hutu Power: none depended, so far as I know, on religion especially.) But there are many especially recalcitrant religious divides in our world: and one might think that the combination of a powerful mechanism for solidarity with a cognitive tendency to form beliefs of a sort that are extremely resistant to inter-subjective evaluation might be part of the explanation.[15]

To proceed in this way, it is worth insisting, is to rule out one perfectly possible response to the question why there are religions, which is a religious answer. Many people believe in God, they say, for the same sort of reason as you believe in your earthly father. He speaks to them. They believe that the Bible is God's revealed truth and that there is a good deal of evidence in the history of the world for God's presence. The main reason for religious belief, in their view, is that God is there, he has made these truths known, and they have simply responded sensibly to the evidence. If there is a God who did these things, and who makes himself known to his people, the story of why people believe in him isn't all that interesting. In fact, the interesting question is why there are people who don't believe in him. And, as we know, there are religious explanations for that, too: for example, the work of Satan.

These are religious explanations for the causes of religion. There are religious explanations for the effects of religion, too. If you think religion makes people behave better, you may think that it's because most religions share some core understanding of the true morality vouchsafed in its full correctness only to those who have the correct religion.

Sometimes when one mentions these views in an academic setting it's assumed that one must think they're simply unreasonable. That's not my view. My view is that, given the extreme difficulty of the task of forming an approximately correct view of the universe, it's amazing that we do as good a job as we do. I am drawing these views to your attention as explanations ruled out by the sort of inquiry I was exploring, not because they're unreasonable, but in order to make a strategic point. The explanations I was considering are consistent with some religious truths – God could have made the world and then left evolution to run its course, or he could have worked *through* evolution somehow. But the obvious explanation for belief – that there's lots of evidence for the truth of these religious claims – is not taken seriously by most scholars (even most religious scholars) as an option.[16]

---

[15] I'm grateful to Dominic Johnson for a list of recent exemplary work on these topics. For supernatural agency (apart from Atran, cited above): Barrett (2004), Bering (2006a, pp. 453–462, 2006b, pp. 142–149). For social solidarity: Wilson (2002), Sosis and Alcorta (2003, pp. 264–274). And for inter-group conflict: MacNeill (2004, pp. 43–60), Johnson (2008), Sosis et al. (2007, pp. 234–247), Bushman et al. (2007, pp. 204–207).

[16] Though there are exceptions. In Alston (1991) defends the possibility of religious experience as a source of knowledge. And, as John Hare pointed out to me, John Hick and Karl Rahner each has a view of this form, in which the truths of religion are expressed differently in different socio-cultural contexts.

This simple point – that modern scientific explanations of the causes and effects of religion are at a minimum methodologically irreligious and, at a maximum, committed to the falsehood of religious claims – does not, however, distinguish them much from the explanations that *religious* people offer for the causes or effects of religion. Whatever your religion is, you're not going to be able to explain the causes and effects of all the other religions that are inconsistent with yours by supposing them to be true either. So, granted that there are serious incompatibilities in truth-claims among religions – and that is something most people, religious or not, can agree on – some of the effects of some religions are going to have to be explained in a way that does not presuppose their truth.

It could, in principle, be that the reason my co-religionists and I believe in our religion is boring: God revealed the truth to us. But there will still be interesting questions about why other people believe in all the other (false) religious views (including atheism). Similarly, if it turns out that religions make social groups cohesive, we won't want to explain that by the indwelling of the spirit except in the case of our own religion. So I suggest that we can avoid some unnecessary controversy by agreeing that we will start, on the cognitive side, by trying to explain the extraordinary persistence of those religious claims that are false, and, on the group formation side, by looking at the social effects of religious membership in groups whose religions are mistaken. Since everyone of every faith agrees that there are examples of these things, even if their own religious faith is not among them, there will be scope for shared explorations here. Such an inquiry does not, in short, rule out the possibility that *some* religious claims are correct; it presupposes only the uncontroversial claim that the claims of all the religions cannot all be correct. It is irreligious only if to be religious you have to accept the claims of all religions.

One final point, which addresses on the cognitive side. Suppose you think that most religious metaphysics is false. It is mostly false, one might think, in similar kinds of ways. From that point of view the interesting question might be not about the evolution of religion – which is, from a certain point of view, not so surprising. The interesting question is about the evolution of irreligion. Here we are lacking some very basic data. For example, we don't know, so far as I am aware, what proportion of humanity has been in some sense religious, since, let's say, the evolution of language. Paleo-anthropologists are inclined to take many things in the archeological record as evidence of religion. Whether they are right depends, of course, in part on what you take to be the necessary conditions for religiosity. I have not tried to answer that question. My hope is only that I have made it seem harder than it might first appear.

Acknowledgement I am grateful to John Hare and Dominic Johnson for comments on an earlier version. I have incorporated ideas of theirs and responded to queries they raised, where possible. (Naturally, the responsibility for the results is all mine).

References

Achebe C (1982) Interview with Anthony Appiah, DAN Jones and John Ryle. In: Times Literary Supplement, London, p 209
Alston WP (1991) Perceiving God: the epistemology of religious experience. Cornell University Press, Ithaca
Appiah KA (2005) The ethics of identity. Princeton University Press, Princeton
Appiah KA (2008) Causes of quarrel: what's special about religious disputes? In: Banchoff T (ed) Religious pluralism, globalization, and world politics. Oxford University Press, New York, pp 41–64
Atran S (2002) In Gods we trust: the evolutionary landscape of religion. Oxford University Press, New York
Barrett JL (2004) Why would anyone believe in God? Altamira Press, MD
Bering JM (2006a) The folk psychology of souls. Behav Brain Sci 29(5):453–462
Bering JM (2006b) The cognitive psychology of belief in the supernatural. Am Sci 94:142–149
Bloor D (1991) Knowledge and social imagery, 2nd edn. Routledge, London
Burton R (1880) The Kasidah. Privately Printed, London, p 10
Bushman BJ, Ridge RD, Das E, Key CW, Busath GL (2007) When God sanctions killing: effect of scriptural violence on aggression. Psychol Sci 18(3):204–207
Dawkins R (2008) The God delusion. Mariner Books, New York
Dennett D (1989) The intentional stance. MIT, Cambridge
Evans-Pritchard E (1951) Witchcraft, oracles and magic among the Azande. Oxford University Press, Oxford
Harris S (2005) The end of faith: religion, terror and the future of reason. W. W. Norton, New York
Horton R (1967) African traditional thought and western science. Africa 37:50–71, 155–87
Horton R (1993) Patterns of thought in African and the west: essays on magic, religion and science. Cambridge University Press, Cambridge
Johnson DDP (2008) Gods of war: the adaptive logic of religious conflict. In Bulbulia J, Sosis R, Genet C, Genet R, Harris E, Wyman K (eds) The evolution of religion: studies, theories, and critiques. Collins Foundation Press, Santa Margarita, CA
MacNeill A (2004) The capacity for religious experience is an evolutionary adaptation to warfare. Evol Cogn 10:43–60
Phillips H (2007) What good is God? New Scientist 2619
Sosis R, Alcorta C (2003) Signaling, solidarity, and the sacred: the evolution of religious behavior. Evol Anthropol 12:264–274
Sosis R, Kress H, Boster J (2007) Scars for war: evaluating alternative signaling explanations for cross-cultural variance in ritual costs. Evol Hum Behav 28:234–247
Turnbull C (1968) The forest people. Touchstone, New York
Tylor E (1899) Primitive culture: researches into the development of mythology, philosophy, religion, language, art and custom. Henry Holt and Company, New York, pp 424–426
Whitehouse H (2008) Cognitive evolution and religion: cognition and religious evolution. In: Bulbulia J, Sosis R, Genet C, Genet R, Harris E, Wyman K (eds) The evolution of religion: studies, theories, and critiques. Collins Foundation Press, Santa Margarita, CA
Wilson DS (2002) Darwin's cathedral: evolution, religion, and the nature of society. University of Chicago Press, Chicago
Wiredu K (1980) Philosophy and an African culture. Cambridge University Press, Cambridge, p 2

Part III
Cooperation and Problems of the Commons

Building Trust to Solve Commons Dilemmas: Taking Small Steps to Test an Evolving Theory of Collective Action

Elinor Ostrom

Abstract Extensive field research has found that when users of a resource do gain good feedback about the effect of their actions on a resource and can build norms of reciprocity and trustworthiness, they are frequently able to craft new institutions to solve puzzling dilemmas. We need to ask: How do different kinds of institutions support or undermine norms of reciprocity and trustworthiness? The finding from many field studies throughout the world that monitoring and graduated sanctions are close to universal in all robust common-pool resource (CPR) institutions is important as it tells us that without some external support of such institutions it is unlikely that reciprocity alone will allow individuals to solve CPR problems over time. On the other hand, the sanctions are graduated rather than initially severe. The current theory of crime deterrence – based on strict expected value theory – does not explain the graduated nature of these sanctions. But if people can learn to value trust and reciprocity and use them as fundamental norms for organizing their lives, they can agree on a set of rules that they agree to follow. Then graduated sanctions are a way of informing those, who have made an error or faced some emergency temptation, that others are watching and, if someone else were to break a rule, they would likely be observed. Thus, continuing to follow a positive norm of reciprocity is reasonable and it is then feasible to build trust over the long term.

Problems of the commons exist in a wide variety of settings ranging in size and complexity from the family (e.g., the household budget and the kitchen sink) to the global scale (e.g., loss of biodiversity and global warming). Game theory is a useful theoretical tool for representing a simplified, core social dilemma facing a set of individuals sharing a commons. Game theorists, who assume that individuals base decisions on immediate returns to self, frequently use the Prisoner's Dilemma (PD) game to represent the problem of the commons (Alcock and Mansell 1977; Richards 2001; but see Cole and Grossman forthcoming). The individuals in such a game are assumed to have complete information about the strategy space they

E. Ostrom
Workshop in Political Theory and Policy Analysis, Indiana University, 513 North Park Avenue, Bloomington, IN 47408-3895, USA

S.A. Levin (ed.), *Games, Groups, and the Global Good*, Springer Series in Game Theory, DOI: 10.1007/978-3-540-85436-4_13, face and the outcomes that will be obtained depending on their own and others' actions. On the other hand, the pure theory is about individuals who do not know one another, do not share a common history, and cannot communicate with one another. In this model, game theory predicts that individuals jointly using a commons will overharvest, leading to Hardin's (1968) "Tragedy of the Commons."

An important asset of game theory is that it provides a clear theoretical prediction that can be tested in carefully designed, laboratory experiments (Gardner and Ostrom 1991). When a set of anonymous subjects makes decisions without communication about appropriation from a one-shot or finitely repeated, common-pool resource in a laboratory setting based on Gordon's (1954) bioeconomic theory, they behave broadly as game theory predicts (Ostrom et al. 1992, 1994). They overharvest. In fact, they overharvest on average even more than predicted, even though there is considerable heterogeneity in the decisions of individual subjects. In a new experimental design that represents a more complex resource, where subjects can make many more harvesting decisions than in earlier experiments, participants still overharvest dramatically on average when they cannot communicate (Janssen et al. 2008; Janssen and Ostrom 2008).

Individuals in a wide diversity of field settings also overharvest as conventional game theory predicts. Ocean fisheries provide a clear example of rampant overharvesting (Berkes et al. 2006; Myers and Worm 2003; Pauly et al. 2002; Clark 2006). Deforestation in many countries also reflects massive overharvesting (Kaimowitz and Angelsen 1998; Rudel 2005; Moran and Ostrom 2005; Ostrom 2008). Overharvesting of resources, biodiversity loss, and global warming all tend to reinforce the belief that game-theoretical predictions of the outcomes from commons dilemmas – especially for large groups of anonymous users – are correct.

This is, however, not the end of the story. Making one simple change in the design of a laboratory experiment, allowing participants to engage in face-to-face communication (cheap talk), enables them to reduce overharvesting substantially (Ostrom and Walker 1991). Game theory predicts that individuals who have only self-regarding preferences will not change their behavior as a result of cheap talk alone. When given a chance to communicate, most subjects first try to figure out what is the best joint strategy. Subjects, who are most successful, use communication to help build a group identity and commitment to follow their agreed-upon strategy (Simon and Schwab 2006). Behavior changes dramatically and subjects greatly increase their joint payoffs. Sally (1995) has undertaken a meta-analysis of 35 years of published experiments on PD games and found that discussion among the subjects significantly influences the rate of cooperation in repeated experiments. Communication also enables some subjects to develop ingenious strategies to use attributes of their environment to achieve high returns (Janssen and Ostrom 2008).

Further, when given opportunities for engaging in costly punishment, subjects punish others who continue to overharvest – contrary to game-theoretical predictions based on a model of the individual with only self-regarding motives (Ostrom et al. 1992). Multiple studies have now found that individuals facing dilemmas are willing to sanction those who do not cooperate as well as punish out of presumed retribution (Falk et al. 2005; Henrich et al. 2006). Further, when given an opportunity to devise their own sanctioning rules, those who adopt their own rules tend to follow these rules closely, achieve higher joint returns, and the use of punishment drops to almost zero (Ostrom et al. 1992). Parallel to laboratory findings, field researchers have recorded a large number of empirical settings where those directly involved in a commons have themselves devised, adopted, and monitored rules over time that have led to robust common-pool resource institutions (McCay and Acheson 1987; Ostrom 1990; NRC 1986, 2002; Dietz et al. 2003; Dolšak and Ostrom 2003; Acheson, 2003; Coleman and Steed 2009; Schlager and Ostrom 1992; Ghate et al. 2008; Shivakumar 2005; see also Janssen and Ostrom 2006).

1 A Theoretical Puzzle

Why do individuals conform to the game-theoretical predictions based on self-regarding motives in some social dilemma settings, but not in others? A frequent answer to this question is that the *context* of the situation makes a difference as to whether individuals cooperate to solve social dilemmas or simply pursue short-term material interests. But what does the "context of a situation" mean? What contextual factors change the way humans behave? If this were a minor puzzle, it would not be worth devoting a major effort to its analysis. Given the harm that individuals can bring to small- and medium-sized social and ecological systems as well as to the global commons, the effort to untangle this puzzle is worth substantial effort! And substantial effort will be required to untangle this puzzle at multiple scales ranging from small groups in a laboratory experiment to large populations sharing common-pool resources.

A central problem is that human behavior is heterogeneous even in extremely simplified, experimental laboratory settings. As has been known for some time, not all subjects in a finitely repeated PD game defect (Lave 1965; Rapoport and Chummah 1965). Further, some subjects in a finitely repeated PD game mutually cooperate. Axelrod (1984, 1986) helped to sort out how this could happen. As Axelrod established, if individuals adopt a "tit-for-tat" strategy of cooperating on the first move and then copying what the other player does on future rounds, players may get out of the trap of "always defect" leading to lower payoffs for both. This practical strategy – and its many useful variants – does not, however, guarantee mutually productive outcomes. If one player defects on the first round of play (or thereafter), they may again be trapped in a series of mutual defections. Further, tit-for-tat is not effective in larger groups.

Puzzling over the diverse findings from social dilemma experiments, some researchers focused on an even simpler game than that of a 2-player PD game – the Dictator game – in an effort to try to understand individual behavior in an even simpler setting (Güth and Huck 1997). In Dictator experiments, one subject is allocated a fund that they can share with another unknown second subject, or keep all of it for themselves. Scholars have again been surprised by the variety of actions taken by subjects given this limited structure. Some subjects keep all of the funds initially allocated to them, as predicted by self-regarding game theory. Others share a varying portion with the unknown second player, but usually not more than half of the initial allotment (Bohnet and Frey 1999; Eckel and Grossman 1996; Hoffman et al. 1996; Frohlich and Oppenheimer 1992; Burnham 2003).

Another very simple social dilemma experiment with puzzling outcomes is the "Investment Game," which is also referred to as the "Game of Trust." In the regular form of this game developed first by Berg et al. (1995), one player is assigned the position of "first mover." The first mover is assigned a private endowment that they can either keep or allocate to an anonymous second mover. In the standard game, the first player makes a decision that can create a joint surplus to be shared with the second player because each $1 sent to the second mover by the first mover is tripled by the experimenter. The second mover is not known to the first player (nor, when double-blind procedures are used, to the experimenter). The dilemma is that to create the larger joint pool, the first mover must either trust that the second mover will reciprocate the first mover's generous transfer or, potentially have preferences that positively weight the second mover's payoff (Cox 2004; Cox and Deck 2005).

The self-regarding, game-theoretical prediction for this design is that the first player will send no funds to the second player since they will predict that the second player will return zero funds to them. In the first experiment on the Game of Trust, Berg and colleagues (1995) were surprised to find that a substantial proportion of first movers sent funds to the second movers. Only two of the first movers followed the game-theoretical prediction of sending nothing, and many of the second movers returned funds to the first player. This experiment has been replicated extensively by scholars across the world (see Cox et al. 2009 and extensive cites therein). Substantial heterogeneity of individual behavior was found even within this extremely simplified experimental setting. The findings vary from setting to setting, but rarely provide strong support for the game-theoretical prediction (Faysse 2005).

Thus, a core puzzle that we face in thinking about *Games, Groups, and the Global Good* is that scholars find an immense variety of outcomes in the experimental lab and in the field rather than conformance to theoretical predictions based on a model of the individual who maximizes own short-term interests (see Bowles 2008, who reviews evidence from more than 40 laboratory experiments conducted by scholars from all parts of the world that challenge the presumption that humans seek only material benefits for self). As a researcher who has used multiple empirical and theoretical methods to try to understand how groups of individuals can overcome commons dilemmas, I have learned that no single variable has the same effect in all settings. Sometimes increasing the size of a group (within a range) has a positive effect on levels of cooperation (Agrawal 2000; Agrawal and Goyal 2001; Isaac et al. 1994) and sometimes a negative effect (Richards and Andersson 2001; Poteete and Ostrom 2004; Diekmann 1986). Sometimes assigning assured property rights to a resource leads users to adopt long-term strategies that are consistent with sustaining a resource (Blomquist 1992). In other settings, assigning private property rights leads to accelerated harvesting (Mwangi 2007). Sometimes repeating a situation has a positive effect (when participants know with whom they are interacting), and sometimes a negative effect (contributions in repeated public good games fall off over time in repeated games without communication) (Isaac and Walker 1998). Heterogeneity within groups also generates positive, neutral, or negative impacts on levels of cooperation (Baland and Platteau 1996, 1999).

2 The Challenges Ahead

Thus, we face immense challenges in an effort to move beyond the "tragedy of the commons" theory that has dominated the thinking of both scientists and policymakers since Hardin's (1968) dramatic article in *Science*. We now know that individuals who find themselves in social dilemma situations vary immensely in the behavior they adopt and subsequently in the outcomes they obtain. A major challenge is building a more general theory of individual behavior that helps us understand the vast heterogeneity in behavioral patterns and outcomes obtained by individuals facing other social dilemmas.

In my own effort to start developing an alternative "behavioral theory of collective action" (Ostrom 1998, 2003), I posited that building trust among participants that the other participants are trustworthy and reciprocators is an essential core of future theories (see Fig. 1). Without the development of trust and reciprocity, those who cooperate in dilemma settings may be "suckers" who contribute to the unearned benefits of others. It is not easy, however, to use the conventional game-theoretic model of individual behavior to explain reciprocity and trust. To move ahead in understanding commons dilemmas (as well as other social dilemmas), it is important to recognize that the innards of the human animal are more complex than represented in self-regarding game-theoretical models (Fehr and Gintis 2007).

A second major challenge also relates to broadening our theoretical lenses at a level above that of individual decision making. Many theoretical models of resource systems rely on simple stick-figures to represent a bioeconomic model of a resource system, but this is rarely a rich enough representation to be the foundation for effective solutions (Clark 2006: 15). We must learn how to think about social-ecological systems as complex adaptive systems (Levin 1999; Folke et al. 2005) and develop a coherent way of representing them building a common language across the social and ecological sciences for analyzing the immense number of nested variables that

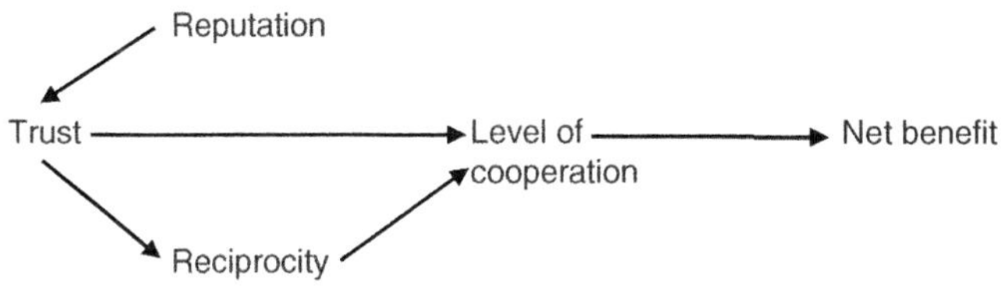

Fig. 1 The core relationships in repeated social dilemmas
Source: Adapted from Ostrom (1998: 13)

affect the structure of a commons or other situations in which individuals find themselves. It is somewhat reassuring to see that a growing number of ecologists and social scientists are now working together in a series of efforts to build more cumulative work on complex and adaptive social-ecological systems (Gunderson and Pritchard 2002; Adger and Jordon 2008).[1] This second challenge is, however, beyond what I can effectively tackle in this chapter, but I have addressed it in recent work (Ostrom 2007). Thus, I will focus my attention on the first challenge of building a better microtheoretical understanding of human decisions in dilemma settings.

3 The First Challenge

The self-regarding model of individual choice has been a powerful engine of theoretical predictions in a variety of contexts. It continues to be a useful model in highly competitive situations. It is a close theoretical cousin to the maximization of fitness that underlies biology (Maynard Smith 1982; Maynard Smith and Szathmáry 1997). The challenge is to develop a theoretical approach for analyzing individual preferences outside of highly competitive environments that does not deny the usefulness of the self-regarding model in competitive settings (Satz and Ferejohn 1994). This challenge is one that a number of researchers are tackling (Cox 2004; Cox et al. 2007; Bolton and Ockenfels 2000; Fehr and Schmidt 1999). Fortunately, we can build on developments in *behavioral* game theory where scholars are broadening the model of the individual used in self-regarding game theory (Camerer 2003; Camerer and Fehr 2006).

3.1 A Core Set of Assumptions about Human Behavior

In light of extensive field and experimental research, I have come to use a basic set of assumptions about individual human behavior in diverse contexts. These include:

1. Humans make decisions within diverse domains of decision-making units that extend from small groups, to communities, to nations, to international organizations.
2. Within all decision-making domains, fallible individuals make decisions that are intended to increase net benefits to themselves and potentially to others, but at times at the expense of others.

[1] See the special feature of *PNAS* of September 2007, where fourteen scholars from multiple fields took a major initiative to move "Beyond Panaceas" and to help develop a framework for diagnosing problems of social-ecological systems and reduce the tendency to think that there are optimal solutions for all kinds of commons problems characterized by huge diversity in their attributes, history, productivity, and scale.

3. Individuals learn from their experiences within a particular domain and from culturally transmitted experiences about the effects of their joint actions on particular resources.
4. Human decisions in all domains are affected by whether they share preferences and goals with others involved, the assets they posses, the information they obtain, the incentives and disincentives they face, the internal learning and choice processes used, and the time horizon invoked.
5. Individuals may adopt norms of behavior, such as "cooperate with others who cooperate with you" or "be trustworthy," in domains where individuals interact repeatedly with others and/or gain information about their past behavior.
6. Decisions at any one domain affect the information, incentives, and time horizon (and, perhaps the cultural values, resources, internal choice processes, and time horizon) of others in that domain, at present and future time periods, and sometimes at other tiers.
7. Human choice is interdependent within tiers and across time and space. Impacts may be horizontal, upward, and downward.
8. Physical and biological processes also affect the information, the incentives, and the time horizon that are used in human choice as well as being affected by human choice.

These assumptions are broader than needed to explain specific behavior in a highly structured setting, but I prefer to use a general set of underlying assumptions that do not need to be changed extensively to fit particular settings even though specific behavior varies from setting to setting. In a highly competitive situation, the above assumptions are consistent with using a theory of individual maximization of expected material values to self, as Adam Smith long ago recognized (Alchian 1950). But as Smith himself also recognized, individuals who interact with one another in other settings, may develop broader sets of preferences than just their own immediate material well being – and they may actually do better as a result of their broader preferences.

3.2 Focusing on the Fifth Assumption: The Possibility of Adopting Norms

Most of the eight assumptions laid out above related to a broader theory of human behavior nested in linked domains of interaction are not controversial. Nor can we use them to make specific predictions of how people will behave in a particular context since they are very general. Assumption 5 – related to the possibility of humans adopting norms – is a crucial assumption for developing an alternative theory of human behavior that can be used to explain outcomes in competitive situations as well as in a variety of social dilemma situations. The fifth assumption is more controversial than the other seven assumptions listed above. Scholars using self-regarding game theory would not make this assumption. Nor would they need it when modeling highly competitive market settings. It would be hard to explain cooperation in social dilemma situations, however, unless humans have the basic capabilities to adopt norms of behavior that may lead them at times to adopt costly strategies that are not in their short-term material advantage (even though such norms are frequently in their long-term net benefit).

In Crawford's and my efforts (1995/2005) to analyze why individuals follow norms of behavior (such as telling the truth, being trustworthy, and using reciprocity), we posited the addition of a delta parameter to the preference function that represented the costs and benefits of following a norm that individuals felt they must or must not do. The function could take on a positive value that would reflect the pride that an individual felt when following a norm or a negative value to represent the shame when breaking a norm. The growing evidence from neuroeconomics that some individuals gain real pleasure from following norms such as trusting and trustworthy behavior is consistent with our effort to include overtly the concept of norms in the preference functions of individuals (Rilling et al. 2002; McCabe and Smith 2001; Fehr et al. 2005) and even from punishing violators of social norms (de Quervain et al. 2004). Further, the work in evolutionary game theory (Axelrod 1986; Gintis 2000; Skyrms 1997; Boyd et al. 2003), cultural evolution (Boyd and Richerson 1985), and the relationship between biological and cultural evolution (Ehrlich and Levin 2005) provides a coherent explanation for how delta parameters may have evolved.

Assuming that some individuals may learn to adopt and use norms of trust and reciprocity substantially alters the way one thinks about social dilemmas. Norms of reciprocity involve returning positive actions with positive responses and negative actions with negative responses. If individuals do not believe that the others with whom they are relating are trustworthy, then the best they can do is to act in a manner consistent with accepted theory of self-regarding preferences. On the other hand, if individuals trust that at least some others will reciprocate cooperation with cooperation, then it may pay – especially in settings where the costs are not too high initially – to explore this possibility by trying cooperative actions and seeing what happens. If others do not reciprocate, one immediately returns to noncooperation and tries to exit and find other situations that are more productive (Axelrod 1997; Axelrod and Cohen 2000). If others do reciprocate, it may be possible to achieve substantial long-term benefits. Once such a pattern is initiated, gaining a reputation for being trustworthy and reciprocating cooperation is an asset that can increase individual-level outcomes (as well as increase joint returns) (Ahn and Ostrom 2008).

Assuming that individuals may invest in a reputation for being trustworthy, can gain trust, and can use reciprocity, does not automatically lead to more optimistic predictions in regard to the provision of public goods and the regulation of common-pool resources. It does, however, refocus the analysis from an assumption that individuals are hopelessly trapped in a situation from which they cannot extract themselves without an external authority deciding what should be done and imposing that decision on participants. Asking what "the" government should do assumes that external actors will always come up with wise decisions and implement them effectively and fairly. The perspective of this chapter leads the analyst to inquire how individuals facing commons problems can gain trust that others are trustworthy and that a cooperator will not be a sucker who contributes while others continue to free ride.

3.3 Unpacking How Context Affects Trust and Reciprocity

In earlier efforts to explain the varying levels of cooperation observed in the field and the laboratory, I developed a relatively simple framework in my Presidential Address to the American Political Science Association and then revised it in a book that James Walker and I edited on *Trust and Reciprocity* (Ostrom 1998, 2003). Figure 1 focuses in on a core set of relationships that characterizes any repeated social dilemma. An individual may cooperate simply from their own norms, but learning the reputation of others increases trust that their own efforts to initiate reciprocity and cooperation will lead to higher cooperation and net benefits.

Figure 2 adds two elements. First, I posited that attributes of biophysical variables (such as those characterizing a resource system), attributes of a community (such as those related to shared norms and sense of common history and future), and institutional variables (related to the rules used to structure relationships) impact on the core set of relationships.[2] Second, I posited that feedback relationships among the variables would strengthen or weaken the levels of cooperation over time. The framework was intended to stimulate further theoretical development beyond narrow assumptions regarding human behavior and to be able to explain behavior in different, simple contexts as well as the more complex contexts in the field where attributes of resources, culture, and institutions have a big impact.

This effort to begin to "unpack" the meaning and components of "context" helps to explain the variability of behavior observed across social dilemmas. In some social dilemma contexts, one should expect low levels of cooperation. In a lab

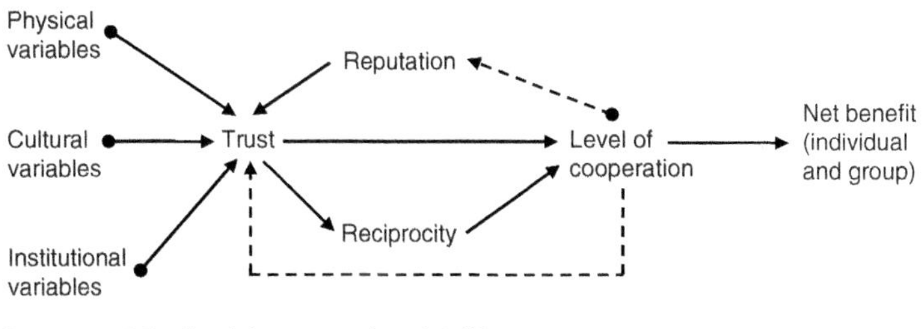

Fig. 2 Context and feedback in repeated social dilemmas
Source: Adapted from Ostrom and Walker (2003: 51)

[2] These three broad clusters of variables are core elements used in the Institutional Analysis and Development (IAD) framework that has evolved from more than three decades of field research conducted by scholars associated with the Workshop in Political Theory and Policy Analysis at Indiana University. For an extended explanation of the IAD framework see Ostrom (2005).

setting where (1) the resource is a finitely repeated common-pool resource experiment, (2) the cultural setting is minimal because the subjects' identities are unknown to each other and no communication is allowed, and (3) the "institution" involved is open access, one should predict, as Hardin did years ago, that most participants will overharvest. As discussed above, this is what is found in the lab under controlled settings. When we made a minor change to the institutional rules of our experiments to allow repeated communication, without changing the other contextual factors, individuals used the opportunity to work out joint strategies to improve their outcomes and to chasten one another if information about aggregate harvesting levels indicated that one or more individuals had broken a verbal agreement (Ostrom et al. 1992).

It is a repeated finding that when institutional rules are changed to allow communication in a social dilemma, while holding other contextual variables constant, subjects use the opportunity to elicit norms, assess each other's trustworthiness, and increase cooperation rates (Sally 1995). Bicchieri (2002) evaluated two possible explanations for the effectiveness of communication in a review of experimental studies of social dilemmas and concluded that communication was more important in eliciting social norms than in simply building group identity. Related to eliciting social norms, several experimental researchers have examined to what extent subjects in dilemma situations are able to assess the trustworthiness of others (Hayashi and Yosano 2005; Cosmides 1989). Humans appear to have a high level of "social intelligence." Those who have participated in some form of communication with one another in a controlled laboratory setting do gain a relatively accurate judgment of the likelihood that others in an experiment will be trustworthy or not (Yamagishi and Kosugi 1999; Kikuchi et al. 1996).

3.4 Some Small but Important Steps to Understand the Impact of Specific Contextual Variables on Cooperation

A recent study tests the core relationships among participants posited in Figs. 1 and 2 for "predicting" results of new experiments. Ebenhöh and Pahl-Wostl (2008) provide further evidence that the level of cooperation in a social dilemma – even in experiments without communication – depends on attributes of the context of a situation that encourage or discourage the development of norms, such as reciprocity, and their sustainability in repeated situations. Ebenhöh and Pahl-Wostl used data from a series of one-shot Dictator games (Hoffman et al. 1996; Güth and Huck 1997) and a set of Games of Trust (Berg et al. 1995; Cox 2004) in an agent-based model (ABM) intended to generate individual agent behavior related to the amount of funds sent to a second subject by a "dictator" in the first type of game and by the "trustee" in the second type of game.

As mentioned above, in Dictator experiments, the first subject is allocated a fund that they may (or may not) share between themselves and a second player who is unknown to them and makes no decisions. Ebenhöh and Pahl-Wostl find that the subjects in settings with the highest anonymity – achieved through double-blind procedures – allocated the least amount to the second subject (Hoffman et al. 1994). In such a setting, only individuals with strong internal norms would give funds. Nothing in the context would motivate them otherwise. Levels of anonymity vary across Dictator experiments conducted by different scholars and affect behavior (Frohlich and Oppenheimer, 2000; Frohlich et al. 2001). While the identity of the specific individual assigned the second position is not revealed to the first player in all these experiments, some scholars do not use double-blind procedures. Without double-blind procedures, the person assigned to hold the Dictator position could potentially worry that a decision to keep all of the funds would adversely affect the evaluation made of them by the researcher who would know their identity. Further, one set of Dictator experiments was conducted in a school where subjects knew that the other participants were from their own school even though they did not know specifically who the other person was (Güth and Huck 1997). Ebenhöh and Pahl-Wostl found that as the assurance of anonymity was lessoned across diverse experiments, from the double-blind experiment to an experiment conducted in a school, the average fund allocated by the subject assigned the Dictator position to the unknown "other player" rose systematically.

Ebenhöh and Pahl-Wostl then examined how much the initial act by a first player in a Game of Trust influences a second player's decisions. They reasoned that the second player is making a decision about allocating funds to the first player, which is similar to the decision made by the first player in a Dictator game. This enabled them to investigate the additional effect of the reputation for cooperation that the first player gains in a Game of Trust as a result of the amount of funds (including no funds) they decide to allocate to the second player. Does the initial "investment" of the first player and their subsequent reputation affect the level of reciprocity extended by the second player? Ebenhöh and Pahl-Wostl answer positively.

As shown in Fig. 3, Ebenhöh and Pahl-Wostl modified the core section of Fig. 2 to represent the relations they posited among the individual-level variables in their ABM. They added two additional norms of being fair and being cooperative to the norm of reciprocity that I had posited. Then, they ran their ABM to assess its power for predicting the results from five sets of Dictator experiments and two sets of Trust experiments. Their model helped to explain the different outcomes across experiments. Agents, who hold initial, but not necessarily strong, commitments to a set of norms, find added strength in basing costly actions on those norms depending on the specific structure of the situation in which interactions occurred. In experiments where anonymity was fully protected and thus no chance to build a reputation, for example, the level of cooperation is lowest. They concluded that "cooperation is highest, if considerations based on fairness, on reciprocity, and on cooperativeness suggest the same behavior" (2008: 246). They describe their approach as "looking at decision processes in very small steps," so as to "abstract more complicated decision situations and simplify them step by step. The simple decisions can be building blocks from which more complex decision mechanisms are made up" (2008: 247).

Boero et al. (2008) have also taken a recent "small" but important step to test how small changes in the contextual structure of a repeated 2-person Game of Trust

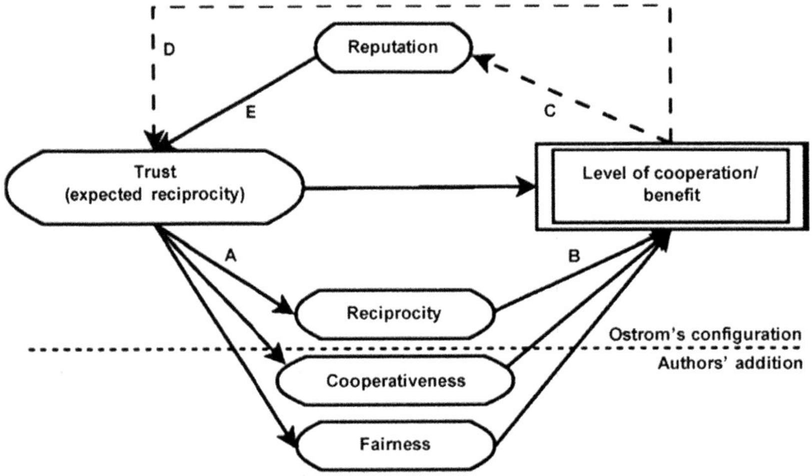

Fig. 3 Adapted and extended from Ostrom (2004: 51). High trust enhances reciprocity (**a**), cooperativeness, and fairness. High reciprocity enhances cooperation in a group and therefore the benefit in a social dilemma (**b**), which in turn creates a good reputation (**c**) and increases trust (**d**). If a good reputation is established, this also increases trust (**e**). *Dashed arrows* indicate the feedback of previous interactions within a group
Source: Ebenhöh and Pahl-Wostl (2008: 231)

affect outcomes. Keser (2003) had earlier shown the importance of reputation in a repeated game where the first mover posted a rating of the second mover after receiving the funds (if any) returned by the second mover. Boero and colleagues designed two experiments to push Keser's findings still further. In the first ten rounds of both experiments, players were anonymously shifted in terms of their position as Trustor or Trustee as well as with whom they were paired. In the first experiment, they divided the groups into three designs: (Design 1) the first mover made a public rating of the second mover after receiving funds, (Design 2) the second mover made a public rating of the first mover after receiving funds, and (Design 3) both players rated each other after receiving funds. In the first experiment, the ratings that had been made about one subject were told to the other subject before that subject made a decision. Boero and colleagues found that significantly larger amounts were sent by the player being rated in the first and second designs and by both players in the third design than in the first ten rounds of this experiment.

In their second experiment, the researchers revealed information about ratings only *after* a subject had made a decision to invest or to return funds (rather than *before* as in the first experiment). The second experiment thus removed the incentive to invest in reputation so as to gain immediate financial returns, since ratings could no longer influence the decision of the other subject. Boero and colleagues found that first and second movers invested in reputation in both experiments by sending significantly more funds when they were rated as contrasted to the rounds when they were not rated. In the first experiment, one could explain the investment in reputation solely by the significantly higher incomes obtained. The second experiment provides evidence that individuals invest in their own reputation when an opportunity to do so is present, even if it does not lead immediately to higher monetary returns.

Colleagues in Colombia have conducted a series of 64 common-pool resource experiments in rural villages to assess the impact of telling subjects that a choice leads to optimal results and backing this with the imposition of either a high or a low fine assessed if a subject chosen randomly did not follow the prescribed rule (Rodriguez-Sickert et al. 2008). The structure of the experiment is similar to those conducted earlier at Indiana University (Ostrom et al. 1994). In a series of eight, five-person, common-pool resource experiments, they found that when the villagers were presented a simple game with no institutional structure – an open access game – they overharvested substantially, as had been found in earlier experiments.

Rodriguez-Sickert and colleagues then explored the effect of three designs. In all of these designs, the experimenter told the subjects the optimal harvesting strategy for the mathematical structure of the resource. In 14 experiments, they imposed a high fine on a subject caught extracting more than the prescribed optimal. In another set of 25 experiments, they imposed a low fine on those discovered not to be following the recommended harvesting level. (Subjects were selected randomly to be monitored and the fines were imposed privately so that whoever had broken the rule was not revealed to others.) The experimenters found that behavior improved substantially when fines were established but that subjects did not fully achieve the recommended optimal. To their surprise, they found no difference between the experimental results from the high-fine design as contrasted to the low fine. The latter finding was quite surprising in that the fine was so small as to not affect final payoff substantially.

In a third design, they allowed subjects to vote on whether they would adopt a fine or not. Those who voted against adopting a prescribed amount of extraction and imposing a fine for breaking the prescription did better on the first few rounds after the decision, but rapidly returned to the level of overharvesting as they had done before the vote. Those who voted for the fine did about the same as the subjects in experiments when the optimal-level prescription was imposed from outside. This goes back to the problem of trust and reciprocity. If people do not trust that others are following a rule or a moral norm, they will tend to stop following them themselves. In an earlier fourth design where subjects were given an opportunity for face-to-face communication, Cárdenas et al. (2000) found that subjects used this opportunity to cajole each other into adopting and following close to an optimal strategy.

Rodriguez-Sickert and colleagues posit that being told the optimal extraction, and either having a fine imposed for not doing the optimal or enabling communication among rounds, was a way of establishing a moral sense that this was what they "should" do. The size of the fine did not affect the moral impact of being told that they should all do X especially when they found that moving toward doing X was a very substantial improvement.

In future work, I plan to return to further development of the framework that led to the above figures. In this chapter, I have focused more on attributes of institutions that affect the structure of an experiment and the observed norm-related behavior. I have not discussed much about attributes of resource systems and communities that I also see as affecting the context, even though these are discussed extensively in Ostrom (1990, 2005). I thought it was essential in a book that is focusing on games and the common good to synthesize some of the core experimental research conducted by other scholars on social dilemmas that strongly support Assumption 5 above. At this point, I hope the reader agrees that it is no longer controversial to assume that individuals may adopt norms of trustworthiness and reciprocity depending on specific attributes of a situation.

4 What are We Learning about Norms and the Context of Social Dilemmas?

In the introduction, I posed a "theoretical puzzle" for those interested in *Games, Groups, and the Global Good*: Why do individuals conform to the self-regarding game-theoretical predictions regarding behavior in social dilemmas in some situations but not in other situations? While a complete answer to this question is not provided in this chapter, we can move ahead of sheer puzzlement.

4.1 What are We Learning about Norms?

One part of a simple answer to this question is the growing evidence that human behavior does not uniformly conform to the underlying assumptions about human behavior utilized in self-regarding game-theoretical analyses. Many game theorists and policy analysts basing recommendations on these models assume that *all* individual behavior is self-centered and focused on maximizing short-term material benefits. In other words, they assume that individuals do *not* use norms in making decisions that affect material outcomes. While it is clear that the traditional game-theoretical assumption is wrong in some contexts, we cannot now assume that all individuals adopt norms of reciprocity, fairness, and cooperation at least initially in all repeated situations. We have to develop the fifth assumption listed above still further and build on the substantial theoretical efforts that have contributed to a broader understanding of human behavior (Fehr and Schmidt 1999; Camerer 2003; Cox et al. 2007; Crawford and Ostrom 1995/2005; Ostrom 1998).

4.2 What are We Learning about the Micro Context?

Results from extensive experimental research are generating knowledge that is now enabling us to move beyond simply claiming that "context" makes a difference. Now it is possible to identify core elements of "micro context" of a particular situation that are likely to enhance or detract from the probability that individuals will use norms of reciprocity, fairness, and cooperativeness. In regard to variables related to the context of simplified experiments that affect the core relationship identified in Figs. 1–3, we are gaining considerable confidence regarding the effect of ten specific variables. In situations characterized by the union of the first three variables listed below, very low levels of cooperation are consistently observed (Frohlich and Oppenheimer, 2000). Even in such settings, some individuals draw upon deeply held, personal norms and cooperate with others, even though no external pressure is expressed by others (or even feared to be expressed):

1. One-shot interactions
2. Full anonymity – current actions taken by an individual cannot be attributed to that individual by anyone else
3. No information is available to one participant about the others involved

Simply moving from one-shot to repeated experiments, while holding anonymity and lack of information constant by switching partners for every decision, does not change the prediction of minimal levels of cooperation (see Frohlich et al. 2001; Macy and Skvoretz 1998; Janssen 2008). Assumption 5 only posits that individuals *may* adopt norms. The likelihood of this occurring is decreased when individuals do not interact with others (or at least learn about the past actions of others in similar situations).

When the structure of a situation includes repeated interactions, the level of cooperation achieved is likely to increase in those contexts in which the following attributes occur:

4. Information about past actions is made available
5. Repeated interactions occur with the same set of participants
6. Participants can signal one another by sending prestructured information
7. Prescriptions are adopted and enforced that when followed do lead to higher outcomes
8. Participants are able to engage in full communication (via writing or "chat room" without knowing the identity of the others involved)
9. Participants are able to engage in full communication with known others (via face-to-face discussions or other mechanisms)
10. In addition to communication, participants can sanction (or reward) each other for the past actions they have taken
11. Participants can design their own rules related to levels of cooperation and sanctions that are to be assigned to those who do not follow agreed-upon rules

The substantial number of carefully designed experiments that have been conducted by researchers in many laboratories (as well as in related field experiments) provides a solid empirical foundation for this initial list of "contextual" variables that have repeatedly been found to affect levels of cooperation. Frohlich and Oppenheimer (2000) sought to unpack the concept of context a few years ago and their unranked list has substantial overlap with the above.

5 What is Next on the Agenda?

Now that we are able to specify 11 micro-situational variables that affect the probability of individuals achieving higher levels of cooperation and outcomes, we need to ask two further questions: How does unpacking the attributes that affect the "context" of a situation improve our more general understanding of human adoption of norms? How can we improve the likelihood of increasing cooperation related to the sustenance of common-pool resources?

Regarding our general understanding of social dilemmas, experimental and field research substantially increases our confidence in the possibility of humans using norms and the development of institutional rules to overcome or reduce the negative outcomes associated with social dilemma settings. This extensive body of research provides strong support for the assumption that individuals have an evolved capacity to adopt norms. This capacity is somewhat similar to the evolved capacity of humans to learn a language. The specific content of norms as well as the specific content of learned norms varies from culture to culture and in light of the effort of a family to instill norms in offspring. Individuals also learn that conformance to particular norms is expected in diverse types of social situations. The norms that they try initially to follow vary by types of situation: when their behavior is totally anonymous they may "cheat" on their own norms; when what they have done or who they are is known to others, they are more likely to draw on their own norms to act cooperatively in the initial rounds of interactions; if they find others to be cooperative, that reinforces their commitment to follow norms; and, if they can communicate with others about strategies and norms, they can build on each other's actions and experiences and use verbal sanctions against those who do not cooperate.

There is obviously much more to be done in the experimental lab and in the field. In the lab, we need to start adding other players who have asymmetric powers to approximate more closely some of the important field conditions present in many dilemma situations. Rarely do all of the participants using a resource have identical investment or harvesting power. Colleagues and I have started several experimental programs where we are exploring various aspects of asymmetrical powers in both common property and private property settings and we hope to encourage others to explore asymmetry more fully in the future. Another important development is using more complex formats in experimental designs so that the decision environments move closer to the kinds of ecological systems that exist in the field. Marco Janssen has designed several computer-based experiments that simulate plants regrowing when other plants are left around them, water flowing from upstream to downstream locations, and trees that grow at a slower rate than other resources. This research program has already produced some exciting findings and is testing some of these in field experiments (Janssen et al. 2008; Cardenas et al. forthcoming).

Now what do these developments have to do with improving the likelihood of sustaining common-pool resources in the field? As long as scholars and policy analysts accepted the narrow game-theoretic assumption that individuals maximize only individual material returns, they presumed that local users of a resource were relatively helpless and would not cooperate to solve collective action problems. The only way to "solve" problems of the commons in this view was to impose either private property or government ownership on the users of a common-pool resource. A substantial portion of academic literature was devoted to presenting arguments in favor of specific property systems that were logically shown to lead to optimal mathematical results to solve particular resource problems.[3]

A change in the model of human behavior enables analysts to ask questions related to which structural variables enhance the likelihood that individuals will cooperate with each other and seek out innovative ways of solving collective action problems fitted to a local setting (Faysse 2005; Xepapadeas 2005). Using a broader theory of human behavior, we should expect that aspects of a resource, the community in which people find themselves, and the rules that are being used have a big impact on whether individuals facing a social dilemma do indeed adopt norms (Ostrom 2005). We should be asking how different institutions support or undermine norms of reciprocity instead of simply presuming that central authority is necessary to enforce rules related to cooperation on participants (Bowles 2008; Frey 1994, 1997).

Understanding the importance of norms does help us to explain a repeated finding related to the kinds of sanctions used in field settings. Findings from multiple studies in many countries is that in long-surviving common-pool resources, the local users monitor the harvesting practices of the others using a resource and use *graduated* sanctions when they discover someone who is breaking a rule (see Hayes 2006; Gibson et al. 2005; Ostrom and Nagendra 2006; Ghate and Nagendra 2005; Ostrom 1990). Without monitoring backed up by graduated sanctions, users cannot be assured that others are following agreements.[4] If some users worry that they are being a sucker while others are taking advantage of them, they are likely to adopt more self-seeking strategies, and the levels of cooperation needed to sustain a resource of time can rapidly come undone.

When humans learn to value trust and reciprocity and use them as fundamental norms for organizing their lives, it is possible for them to agree on a set of rules that they agree to follow. Then graduated sanctions are a way of informing those, who have made an error or faced some emergency temptation, that others are watching and, if someone else were to break a rule, they would likely be observed (Ostrom 1990). With user monitoring and graduated sanctions, continuing to follow a positive norm of reciprocity is reasonable and generates higher outcomes.

---

[3] For an in-depth discussion of the problems of recommending panaceas, see the Special Feature of *PNAS* mentioned in note 1 above.

[4] One of the more successful efforts to create a private property-rights system for inshore fisheries has been developed over time in British Columbia after some initial efforts to create rules to limit fishing rights were not successful (Clark 2006). One of the costly but important attributes of the newer system of rules that has been evolved is that there is a monitor on every boat that is recording where the boat goes, the amount of fish harvested, any by-catch that is thrown over the side of the boat, and the amount of fish sold. The monitoring system is costly, but it does appear that over time fishers have begun to see the logic of the rule system that has been developed, agree to reduce their overharvesting, and know that others are being held to follow the rule.

6 Conclusion

As long as many scholars continue to presume that all humans are self-interested maximizers in *all* contexts, the importance of building trust and reciprocity among users of a commons is not viewed as important. What has been viewed as important for many scholars is devising optimal external rules to impose on resource users so that they will stop overharvesting from a commons. Sufficient research now supports an assumption that humans *may* endogenously adopt norms of trustworthiness and reciprocity in contexts where there is a higher probability that they share a common future, their actions are known or reported to others, and cooperative actions do lead to increased payoffs. This assumption makes a big difference in how one understands the microrelationship among those relying on a commons.

We must also ask serious theoretical and empirical questions about the confluence of ecological and social variables as these affect the structure of settings that produce incentives to overuse or to sustain commons. Further, we need better to understand how diversity in human motivations may have evolved over time rather than a narrowly self-interested set of preferences (Goetze 2008; Masters 2008; Levin, this volume). In addition to further development of an empirically supported microtheory of human behavior, it is also essential that social and ecological scientists begin to develop a common language that can be used to build joint theories and undertake empirical tests of the impact of the size of a resource, the type of resource units produced; and the productivity, predictability, and equilibrium properties of a resource. This will, of course, require major efforts beyond this chapter.

Acknowledgment An earlier version was presented at the symposium on "Games, Groups, God(s), and the Global Good" held at Princeton University, October 4–6, 2007, and for the Tanner Lecture at the University of Michigan, November 2, 2007. The support of the MacArthur Foundation and the National Science Foundation is gratefully acknowledged. I deeply appreciate the wonderfully helpful comments by Giangiacomo Bravo, Marco Janssen, Simon Levin, Tun Myint, Claudia Pahl-Wostl, and James Walker, and the excellent editing of Patty Lezotte.

References

Acheson J (2003) Capturing the commons: devising institutions to manage the Maine lobster industry. University Press of New England, Hanover, NH

Adger N, Jordon A (eds) (2008) Governing sustainability. Cambridge University Press, Cambridge

Agrawal A (2000) Small is beautiful, but is larger better? Forest-management institutions in the Kumaon Himalaya, India. In: Gibson C, McKean M, Ostrom E (eds) People and forests: communities, institutions, and governance. MIT, Cambridge, MA, pp 57–85

Agrawal A, Goyal S (2001) Group size and collective action: third-party monitoring in common-pool resources. Comp Polit Stud 34(1):63–93

Ahn TK, Ostrom E (2008) Social capital and collective action. In Castiglione D, van Deth J, Wolleb G (eds) The handbook of social capital. Oxford University Press, Oxford, pp 70–100

Alchian AA (1950) Uncertainty, evolution, and economic theory. J Polit Econ 58(3):211–21

Alcock JE, Mansell D (1977) Predisposition and behaviour in a collective dilemma. J Conflict Resolut 21(3):443–57

Axelrod R (1984) The evolution of cooperation. Basic Books, New York
Axelrod R (1986) An evolutionary approach to norms. Am Polit Sci Rev 80(4):1095–1111
Axelrod R (1997) The complexity of cooperation: agent-based models of competition and collaboration. Princeton University Press, Princeton, NJ
Axelrod R, Cohen MD (2000) Harnessing complexity. Free Press, New York
Baland J-M, Platteau J-P (1996) Halting degradation of natural resources: is there a role for rural communities? Clarendon, Oxford
Baland J-M, Platteau J-P (1999) The ambiguous impact of inequality on local resource management. World Dev 27:773–788
Berg J, Dickhaut J, McCabe K (1995) Trust, reciprocity, and social history. Games Econ Behav 10(1):122–142
Berkes F, et al (2006) Globalization, roving bandits, and marine resources. Science 311: 1557–1558
Bicchieri C (2002) Covenants without swords: group identity, norms, and communication in social dilemmas. Rationality and Society 14(2):192–228
Blomquist W (1992) Dividing the waters: governing groundwater in Southern California. ICS, San Francisco, CA
Boero R, Bravo G, Castellani M, Squazzoni F (2008) Reputational cues in repeated trust games. Brescia, Italy: Università di Brescia, DSS Papers SOC 01–08
Bohnet I, Frey BS (1999) The sound of silence in prisoner's dilemma and dictator games. J Econ Behav Organ 38(1):43–58
Bolton GE, Ockenfels A (2000) A theory of equity, reciprocity and competition. Am Econ Rev 90:166–193
Bowles S (2008) Policies designed for self-interested citizens may undermine the moral sentiments: evidence from economic experiments. Science 320 (June 20):1605–1609
Boyd R, Gintis H, Bowles S, Richerson PJ (2003) The evolution of altruistic punishment. PNAS 100:3531–3535
Boyd R, Richerson PJ (1985) Culture and the evolutionary process. University of Chicago Press, Chicago
Burnham TC (2003) Engineering altruism: a theoretical and experimental investigation of anonymity and gift giving. J Econ Behav Organ 50(1):133–144
Camerer CF (2003) Behavioral game theory: experiments in strategic interaction. Princeton University Press, Princeton, NJ
Camerer CF, Fehr E (2006) When does economic man dominate social behavior? Science 311 (January 6):47–52
Cárdenas J-C, Janssen M, Bousquet F Forthcoming Dynamics of rules and resources: three new field experiments on water, forests and fisheries. In: List J, Price M (eds) Handbook on experimental economics and the environment. Edward Elgar, New York
Cárdenas J-C, Stranlund JK, Willis CE (2000) Local environmental control and institutional crowding-out. World Dev 28(10):1719–1733
Clark CW (2006) The worldwide crisis in fisheries: economic models and human behavior. Cambridge University Press, Cambridge
Cole DH, Grossman PZ (forthcoming) Institutions matter! why the herder problem is not a prisoner's dilemma. Theory and Decision
Coleman E, Steed B (2009) Monitoring and sanctioning in the commons: an application to forestry. Ecol Econ 68(7):2106–2113
Cosmides L (1989) The logic of social exchange: has natural selection shaped how humans reason? studies with the Watson selection task. Cognition 31:187–276
Cox J (2004) How to identify trust and reciprocity. Games Econ Behav 46:260–281
Cox J, Deck CA (2005) On the nature of reciprocal motives. Econ Inq 43(3):623–636
Cox J, Friedman D, Gjerstad S (2007) A tractable model of reciprocity and fairness. Games Econ Behav 59(1):17–45
Cox J, Ostrom E, Walker J, Castillo J, Coleman E, Holahan R, Schoon M, Steed B (2009) Trust in private and common property experiments. South Econ J 75(4):957–975

Crawford SES, Ostrom E (1995) A grammar of institutions. Am Polit Sci Rev 89(3) (September):582–600. Now revised as chapter 5 in Elinor Ostrom, Understanding Institutional Diversity (Princeton University Press, Princeton, NJ, 2005)
de Quervain DJ-F, Fischbacher U, et al (2004) The neural basis of altruistic punishment. Science 305 (August 27):1254–1258
Diekmann A (1986) Volunteer's dilemma: a social trap without a dominant strategy and some empirical results. In: Diekmann A, Mitter P (eds) Paradoxical effects of social behavior: essays in honor of Anatol Rapoport. Physica, Vienna, pp 188–197
Dietz T, Ostrom E, Stern P (2003) The struggle to govern the commons. Science 302(5652):1907–1912
Dolšak N, Ostrom E (eds) (2003) The commons in the new millennium. MIT, Cambridge, MA
Ebenhöh E, Pahl-Wostl C (2008) Agent behavior between maximization and cooperation. Rationality and Society 20(2):227–252
Eckel CC, Grossman P (1996) Altruism in anonymous dictator games. Games Econ Behav 16(2):181–191
Ehrlich PR, Levin SA (2005) The evolution of norms. PLoS Biol 3(6):e194
Falk A, Fehr E, Fischbacher U (2005) Driving forces behind informal sanctions. Econometrica 73(6):2017–2030
Faysse N (2005) Coping with the tragedy of the commons: game structure and design of rules. J Econ Surv 19(2):239–261
Fehr E, Fischbacher U, Kosfeld M (2005) Neuroeconomic foundations of trust and social preferences: initial evidence. Am Econ Rev 95(2):346–351
Fehr E, Gintis H (2007) Human motivation and social cooperation: experimental and analytical foundations. Annu Rev Sociol 33:43–64
Fehr E, Schmidt K (1999) A theory of fairness, competition, and cooperation. Q J Econ 114: 817–851
Folke C, Hahn T, Olsson P, Norberg J (2005) Adaptive governance of social-ecological systems. Annu Rev Environ Resour 30:441–473
Frey BS (1994) How intrinsic motivation is crowded out and in. Rationality and Society 6(3): 334–352
Frey BS (1997) A constitution for knaves crowds out civic virtue. Econ J 107:1043–1053
Frohlich N, Oppenheimer J (1992) Choosing justice: an experimental approach to ethical theory. University of California Press, Berkeley
Frohlich N, Oppenheimer J (2000) How people reason about ethics and the role of experiments: content and methods of discovery. In: Lupia A, McCubbins MD, Popkin SL (eds) Elements of political reason: cognition, choice, and the bounds of rationality. Cambridge University Press, New York, pp 85–107
Frohlich N, Oppenheimer J, Moore BJ (2001) Some doubts about measuring self-interest using dictator experiments: the costs of anonymity. J Econ Behav Organ 46(3):271–290
Gardner R, Ostrom E (1991) Rules and games. Public Choice 70(2):121–149
Ghate R, Jodha N, Mukhopadhyay P (eds) (2008) Promise, trust, and evolution: managing the commons of South Asia. Oxford University Press, Oxford
Ghate R, Nagendra H (2005) The role of monitoring in institutional performance: forest management in Maharashtra, India. Conserv Soc 3(2):509–532
Gibson C, Williams J, Ostrom E (2005) Local enforcement and better forests. World Dev 33(2):273–284
Gintis H (2000) Game theory evolving: a problem-centered introduction to modeling strategic interaction. Princeton University Press, Princeton, NJ
Goetze D (2008) Public goods, sharing genes, and the formation of large groups. Politics Life Sci 26(2):7–25
Gordon HS (1954) The economic theory of a common property resource: the fishery. J Polit Econ 62 (April):124–142
Gunderson LH, Pritchard L Jr (eds) (2002) Resilience and the behavior of large-scale systems. Island Press, Washington, DC

Güth W, Huck S (1997) From ultimatum bargaining to dictatorship – an experimental study of four games varying in veto power. Metroeconomica 48(3):262–299
Hardin G (1968) The tragedy of the commons. Science 162:1243–1248
Hayashi N, Yosano A (2005) Trust and belief about others: focusing on judgment accuracy of others' trustworthiness. Sociol Theory Methods 37:59–80
Hayes T (2006) Parks, people, and forest protection: an institutional assessment of the effectiveness of protected areas. World Dev 34(12):2064–2075
Henrich J, et al (2006) Costly punishment across human societies. Science. 312 (June 23):1767–1770
Hoffman E, McCabe K, Shachat K, Smith VL (1994) Preferences, property rights, and anonymity in bargaining games. Games Econ Behav 7:346–380
Hoffman E, McCabe K, Smith VL (1996) Social distance and other-regarding behavior in dictator games. Am Econ Rev 86(3):653–660
Isaac RM, Walker JM (1998) Nash as an organizing principle in the voluntary provision of public goods: experimental evidence. Exp Econ 1:191–206
Isaac RM, Walker JM, Williams A (1994) Group size and the voluntary provision of public goods: experimental evidence utilizing large groups. J Public Econ 54(1) (May):1–36
Janssen M (2008) Evolution of cooperation in a one-shot prisoner's dilemma based on recognition of trustworthy and untrustworthy agents. J Econ Behav Organ 65:458–471
Janssen M, Goldstone R, Menczer F, Ostrom E (2008) Effect of rule choice in dynamic interactive spatial commons. Int J Commons 2(2):288–312
Janssen M, Ostrom E (2006) Adoption of a new regulation for the governance of common-pool resources by a heterogeneous population. In: Baland J-M, Bardhan P, Bowles S (eds) Inequality, cooperation, and environmental sustainability. Princeton University Press, Princeton, pp 60–96
Janssen M, Ostrom E (2008) TURFS in the lab: institutional innovation in real-time dynamic spatial commons. Rationality and Society 20(4):371–397
Kaimowitz D, Angelsen A (1998) Economic models of tropical deforestation: a review. Center for International Forestry Research, Bogor, Indonesia
Keser C (2003) Experimental games for the design of reputation management systems. IBM Syst J 42:498–506
Kikuchi M, Watanabe Y, Yamagishi T (1996) Accuracy in the prediction of others' trustworthiness and general trust: an experimental study. Jpn J Exp Soc Psychol 37(1):23–36
Lave LB (1965) Factors affecting cooperation in the prisoner's dilemma. Behav Sci 10:26–35
Levin SA (1999) Fragile dominion: complexity and the commons. Perseus Books, Reading, MA
Macy MW, Skvoretz J (1998) The evolution of trust and cooperation between strangers: a computational model. Am Sociol Rev 63:638–660
Masters RD (2008) Historical change and evolutionary theory. Politics Life Sci 26(1):46–74
Maynard Smith J (1982) Evolution and the theory of games. Cambridge University Press, Cambridge
Maynard Smith J, Szathmáry E (1997) The major transitions in evolution. Oxford University Press, Oxford
McCabe K, Smith V (2001) Goodwill accounting in economic exchange. In: Gigerenzer G, Selten R (eds) Bounded rationality: the adaptive tool box. MIT, Cambridge, MA, 319–342
McCay BJ, Acheson JM (1987) The question of the commons: the culture and ecology of communal resources. University of Arizona Press, Tucson
Moran E, Ostrom E (2005) Seeing the forest and the trees. MIT, Cambridge, MA
Mwangi E (2007) Socioeconomic change and land use in Africa: the transformation of property rights in Maasailand. Palgrave Macmillan, New York
Myers RA, Worm B (2003) Rapid worldwide depletion of predatory fish communities. Nature 423 (May 15):280–283
NRC (National Research Council) (1986) Proceedings of the conference on common property resource management. National Academy Press, Washington, DC

NRC (National Research Council) (2002) The drama of the commons. In: Ostrom E, Dietz T, Dolšak N, Stern P, Stonich S, Weber E (eds) Committee on the human dimensions of global change. National Academy Press, Washington, DC
Ostrom E (1990) Governing the commons: the evolution of institutions for collective action. Cambridge University Press, New York
Ostrom E (1998) A behavioral approach to the rational choice theory of collective action. Am Polit Sci Rev 92(1):1–22
Ostrom E (2003) Toward a behavioral theory linking trust, reciprocity, and reputation. In: Ostrom E, Walker J (eds) Trust and reciprocity: interdisciplinary lessons from experimental research. Russell Sage Foundation, New York, 19–79
Ostrom E (2005) Understanding institutional diversity. Princeton University Press, Princeton, NJ
Ostrom E (2007) A diagnostic approach for going beyond panaceas. PNAS 104(39):15181–15187
Ostrom E (2008) The challenge of common-pool resources. Environment 50(4) (July/August):8–20
Ostrom E, Gardner R, Walker J (1994) Rules, games, and common-pool resources. University of Michigan Press, Ann Arbor
Ostrom E, Nagendra H (2006) Insights on linking forests, trees, and people from the air, on the ground, and in the laboratory. PNAS 103(51):19224–19231
Ostrom E, Walker J (1991) Communication in a commons: cooperation without external enforcement. In: Palfrey TR (ed) Laboratory research in political economy. University of Michigan Press, Ann Arbor, 287–322
Ostrom E, Walker J, Gardner R (1992) Covenants with and without a sword: self-governance is possible. Am Polit Sci Rev 86(2):404–417
Pauly D, et al (2002) Toward sustainability in world fisheries. Nature 418:689–695
Poteete A, Ostrom E (2004) Heterogeneity, group size and collective action: the role of institutions in forest management. Dev Change 35(3):435–461
Rapoport A, Chummah AM (1965) Prisoners' dilemma: a study of conflict and cooperation. University of Michigan Press, Ann Arbor
Richards D (2001) Reciprocity and shared knowledge structures in a prisoner's dilemma game. J Conflict Resol 45:621–635
Richards K, Andersson K (2001) The leaky sink: persistent obstacles to a forest carbon sequestration program based on individual projects. Clim Policy 1:41–54
Rilling JK, Gutman DA, Zeh TR, Pagnoni G, Berns GS, Kilts CD (2002) A neural basis for social cooperation. Neuron 35(2):395–405
Rodriguez-Sickert C, Guzmán RA, Cárdenas JC (2008) Institutions influence preferences: evidence from a common-pool resource experiment. J Econ Behav Organ 67:215–227
Rudel TK (2005) Tropical forests: regional paths of destruction and regeneration in the late twentieth century. Colombia University Press, New York
Sally D (1995) Conversation and cooperation in social dilemmas: a meta-analysis of experiments from 1958 to 1992. Rationality and Society 7:58–92
Satz D, Ferejohn J (1994) Rational choice and social theory. J Philos 91(2):71–87
Schlager E, Ostrom E (1992) Property-rights regimes and natural resources: a conceptual analysis. Land Econ 68(3):249–262
Shivakumar S (2005) The constitution of development: crafting capabilities for self-governance. Palgrave Macmillan, New York
Simon A, Schwab D (2006) Say the magic word: effective communication in social dilemmas. Working Paper. Indiana University, Workshop in Political Theory and Policy Analysis, Bloomington
Skyrms B (1997) Chaos and the explanatory significance of equilibrium: strange attractors in evolutionary game dynamics. In: Bicchieri C, Jeffrey R, Skyrms B (eds) The dynamics of norms. Cambridge University Press, Cambridge, pp 199–222
Xepapadeas A (2005) Regulation and evolution of compliance in common pool resources. Scand J Econ 107(3):583–599
Yamagishi T, Kosugi M (1999) Character detection in social exchange. Cogn Stud 6(2):179–190

How Democracy Resolves Conflict in Difficult Games

Steven J. Brams and D. Marc Kilgour

Abstract Democracy resolves conflicts in difficult games like prisoners' dilemma and chicken by stabilizing their cooperative outcomes. It does so by transforming these games into games in which voters are presented with a choice between a cooperative outcome and a Pareto-inferior noncooperative outcome. In the transformed game, it is always rational for voters to vote for the cooperative outcome, because cooperation is a weakly dominant strategy independent of the decision rule and the number of voters who choose it. Such games are illustrated by 2-person and n-person public-goods games, in which it is optimal to be a free rider, and a biblical story from the book of Exodus.

1 Introduction

A cornerstone of *democracy* is fair and periodic elections. While there is an ongoing debate about how best to conduct elections (Brams 2008), here we assume that voters choose between two alternatives, and the alternative with more votes wins.

We say that voting in a democracy *resolves conflict* if the electorate considers (1) the voting process fair and (2) the outcome chosen acceptable. There may not be a consensus among the voters that the alternative chosen is the best, but as long as some agreed-upon minimum number of voters (e.g., a majority) supports some alternative, this outcome will be implemented.

In this paper, we focus on choices that are costly to implement. For example, if voters in a referendum decide to finance a public project, the cost of this project will be reflected in higher taxes they must pay. We assume that citizens must pay taxes, though later we consider the problem, especially for developing countries, that laws may be difficult to enforce.

Suppose the public project to be financed is renovation of a public park, which can benefit everybody – but more so those who use the park frequently than those

S.J. Brams (✉)
Department of Politics, New York University, New York, NY 10003, USA
e-mail: steven.brams@nyu.edu

S.A. Levin (ed.), *Games, Groups, and the Global Good*, Springer Series in Game Theory,
DOI: 10.1007/978-3-540-85436-4_14,

who don't. In this case, some would argue that those who use the park frequently should pay more for its renovation, such as through the Central Park Conservancy in New York City, which solicits voluntary contributions. But this voluntary approach leads to a *public-goods* or *free-rider problem*, which we model as an n-person prisoners' dilemma (PD).

2 Resolution by Voting in a 2-Person PD

To render the subsequent analysis as transparent as possible, we begin with a 2-person PD, wherein one player is a wealthy individual who can make a big contribution to the renovation of the park. Suppose his or her contribution is expected to equal the contributions made by the rest of the public, whom we treat as a single player. In the ranking of payoffs to the players below, we assume that the wealthy individual and the rest of the public both prefer partial renovation without contributing (4) to full renovation with contributing (3) to no renovation without contributing (2) to partial renovation with contributing (1), as shown in the payoff matrix below:

Rest of public ⇒ ⇓ Wealthy individual	Contribute	Don't contribute
Contribute	Full renovation: (3,3)	Partial renovation: (1,4)
Don't contribute	Partial renovation: (4,1)	No renovation: <u>(2,2)</u>

Key: (x, y) = payoff ranking to (wealthy individual, rest of public), where 4 = best, 3 = next best, 2 = next worst, and 1 = worst
Nash equilibrium underscored

Each player's strategy of don't contribute *strictly dominates* its strategy of contribute, because it is better whichever strategy the other player chooses. Each player, therefore, has an incentive to be a *free rider*, obtaining the benefit of the public good without contributing to it.

But the choice by both players of don't contribute leads to the next-worst outcome of (2,2), which is the unique *Nash equilibrium* – neither player would have an incentive unilaterally to depart from it lest it do worse (by obtaining 1).[1] The dilemma is that (2,2) is worse for both players than the cooperative outcome of (3,3), wherein both players contribute. But the latter outcome is not a Nash equilibrium – each player would have an incentive unilaterally to depart from its strategy associated with it (to obtain 4) – rendering it unstable.

[1] The Nash equilibrium is actually the pair of pure strategies of the players associated with (2,2), not the outcome itself, but for convenience we identify Nash equilibria by the outcomes they produce. We do not consider mixed strategies in this or other 2 × 2 games that we discuss later, because preference information is in ordinal rankings rather than cardinal utilities, precluding expected-utility calculations that underlie the determination of mixed strategies.

To be sure, (3,3) may be stabilized under certain conditions – for example, in tournament play (Axelrod 1984), in strategies that evolve over time (Skyrms 1996; Nowak 2006), or when players are farsighted (Brams 1994).[2] Farsighted thinking, which nonhuman animals seem incapable of, is epitomized by Theodore Sorensen's statement about the deliberations of the Executive Committee (ExCom) during the October 1962 Cuban missile crisis:

> We discussed what the Soviet reaction would be to any possible move by the United States, what our reaction with them would have to be to that Soviet reaction, and so on, trying to follow each of those roads to their ultimate conclusions (Holsti et al. 1964, p. 188).

Because of such farsighted calculations on both sides, this crisis subsided and war was averted, though some argue that the game played resembled chicken (game 8 in Fig. 1) more than PD.[3]

Farsighted thinking aside, what resolution does democracy, and voting in particular, offer in PD, chicken, and the other difficult games we present later? Assume that the players in the preceding prisoners' dilemma can first vote on whether to contribute or not contribute to financing the renovation of the park. If a majority (i.e., both players) must vote to finance the park in order that it be renovated, then their choices and the resulting outcomes are shown in the game below:

Rest of public ⇒ ⇓ Wealthy individual	Vote to finance	Vote not to finance
Vote to finance	Full renovation: <u>(3,3)</u>	No renovation: (2,2)
Vote not to finance	No renovation: <u>(2,2)</u>	No renovation: (2,2)

Key: (x, y) = payoff ranking to (wealthy individual, rest of public), where 4 = best, 3 = next best, 2 = next worst, and 1 = worst
Nash equilibrium underscored

Notice that the option that the park be partially renovated does not appear in the payoff matrix. Instead, the outcomes are starker: The park is either fully renovated or not renovated, which renders the cooperative outcome of full renovation the unique

[2] Farsightedness offers a very different resolution of PD than tournament play or evolution. Pinker (2007, p. 71) distinguishes the former from the latter by arguing that "natural selection [in evolution] is like a design engineer in the sense that parts of animals become engineered to accomplish certain things, but it is not like a design engineer in that it doesn't have long-term foresight." Presumably, only humans possess this foresight and can anticipate that if they move from (3,3), it will not necessarily induce their best outcome of (4,1) or (1,4) but, instead, may trigger a countermove by the player receiving 1 to (2,2). Because this outcome is worse for both players than (3,3), (3,3) is a "nonmyopic equilibrium" in PD if the players start at this outcome and think ahead (Brams and Wittman 1981; Kilgour 1984; Brams 1994).

[3] As in PD, the cooperative outcome in chicken is a nonmyopic, but not a Nash, equilibrium. In fact, the game that best models this crisis, and its resolution, is probably neither PD nor Chicken but a different game (Brams 1994, pp. 130–138).

Class 1 (4 games)

1 (27)

(3,4)	(1,2)
(4,1)	(2,3)

2 (28)

(3,4)	(1,3)
(4,1)	(2,2)

3 (32)

(3,3)	(1,4)
(4,1)	(2,2)

Prisoners' Dilemma

4 (48)

(3,4)	(1,1)
(4,2)	(2,3)

Class 2 (4 games)

5 (22)

(3,3)	(2,4)
(4,1)	(1,2)

6 (35)

(4,3)	(2,4)
(3,1)	(1,2)

7 (50)

(4,3)	(2,4)
(3,2)	(1,1)

8 (57)

(3,3)	(2,4)
(4,2)	(1,1)

Chicken

Class 3 (3 games)

9 (29)

(4,3)	(1,4)
(3,2)	(2,1)

10 (31)

(4,3)	(1,4)
(2,2)	(3,1)

11 (46)

(3,4)	(2,1)
(4,2)	(1,3)

Fig. 1 Eleven difficult games

Key: (x, y) = payoff ranking to (row, column), where 4 = best, 3 = next best, 2 = next worst, and 1 = worst
Cooperative outcomes in boldface; Nash equilibria underscored
Numbers of games in parentheses are those given in Brams (1994, pp. 217–219)

Pareto-optimal Nash equilibrium;[4] moreover, it is supported by *weakly dominant* strategies of the players.[5] This transformation may be viewed as a mapping of two of the four outcomes in the PD (full renovation and no renovation) into the new game, with voting determining which outcomes these two outcomes replace.

Note that this solution does not assume any kind of reciprocity or trust among players, as discussed in Sugden (2009). The "team reasoning" that he prefers to

[4] This equilibrium is *Pareto-optimal*, or efficient, because no other outcome is better for both players than (3,3). Although the lower-right (2,2) outcome is also a Nash equilibrium, and therefore stable, the players would have no reason to choose it over the (3,3) outcome, to which it is Pareto-inferior.

[5] Why "weakly"? Unlike PD, each player's cooperative strategy associated with (3,3) is not strictly better, whichever strategy the other player chooses: If the other player votes not to finance, either voting to finance or voting not to finance leads to the same outcome of (2,2). Because of this "tie," voting to finance is not *always* better than voting not to finance.

invoke to solve a sequential PD, based on the "collective intentions" of players, is also not required, because individual rationality alone is sufficient to induce the cooperative outcome in the transformed voting game. In the parlance of economics, voting internalizes (i.e., renders rational) the positive externality of cooperation (i.e., voting to finance the public good).

3 Resolution by Voting in an n-Person PD

To extend this resolution of a 2-person PD to an n-person public-goods game, assume there are $n \geq 2$ players and two strategies, Cooperate (C) and Defect (D), that each player can choose. If k players cooperate, the payoff to each cooperator is the amount $c(k)$, where $k = 1, 2, \ldots, n$, and the payoff to each defector is the amount $d(k)$, where $k = 0, 1, \ldots, n-1$.[6] An n-person game that satisfies the three properties given below mimics the characteristics of the 2-person PD:

3.1 Properties of n-Person PD

1. *The payoffs $c(k)$ and $d(k)$ are increasing in k.* That is, when more players cooperate, all benefit – whether they chose C or D – because more of the public good is provided.
2. *For each* $k = 1, 2, \ldots, n, c(k) < d(k-1)$. That is, comparing the situations in which there are (1) k cooperators and (2) $k-1$ cooperators after the defection of a cooperator, each of the defectors in the latter situation receives a greater payoff than each of the cooperators in the former situation, given that the strategies of all other players are fixed.
3. $c(n) > d(0)$. That is, when all players choose D, the resulting outcome is *Pareto-inferior*, or worse for all players, than the outcome in which all cooperate.

Property 2 implies that, for each player, C is a strictly dominated strategy. To see this, fix a player and suppose that $k-1$ other players choose C and the remaining $n-k$ choose D. Then the focal player will receive $c(k)$ for choosing C and $d(k-1)$ for choosing D. Because this observation holds for every value of k, D strictly dominates C for every player.

It follows that the unique Nash equilibrium in the n-person PD is for all players to choose D and receive $d(0)$. Because this strategy profile is supported by strictly dominant strategies, the resulting all-D Nash equilibrium is especially stable. But by property 3, the nonequilibrium outcome of all-C, at which all players receive $c(n)$, is strictly preferred by all players to $d(0)$. Thus, this n-person PD has a unique

[6] Because $c(k)$ and $d(k)$ are indexed differently, we can compare $c(k)$ and $d(k-1)$ over all k, as we do in property (2) below.

strictly dominant strategy of D for each player, but when all players choose it, a strictly Pareto-inferior outcome results.

The resulting n-person PD has all the problems of the 2-person PD and more. When there are only two players, they may well stabilize the cooperative outcome by implementing an enforcement mechanism, such as regular inspections in an arms-control agreement, that transforms the PD into a more benevolent game, with the cooperative outcome as the unique Nash equilibrium.

But if there are many players,[7] this becomes far less feasible – short of transforming the game into a voting game, as we will show next. Whereas the voting game we described in Sect. 2 required that only two players agree to contribute to renovation of the park, we now propose that a decision rule be fixed which determines whether a public good is provided. More specifically, we assume that with the introduction of voting by the players, the n-person PD is played according to the rules given below:

3.2 Rules of Transformed n-Person PD

1. A decision rule r, satisfying $0 < r \leq n$, is fixed and announced to all players.
2. The players vote, independently and simultaneously, for either C or D.
3. If the number of players that vote for C, m, satisfies $m < r$, then the all-D outcome is implemented, so all players receive $d(0)$. But if $m \geq r$, then the all-C outcome is implemented, so all players receive $c(n)$.

It is easy to check that a player's choice of C or D only affects its payoff when exactly $r - 1$ other players choose C. In this case, the player receives $c(n)$ for choosing C and $d(0)$ for choosing D; by property 3, the player prefers $c(n)$.

Because voting for C sometimes results in a better outcome and never results in a worse outcome, it is a weakly dominant strategy, as it is in the transformed 2-person PD. Thus, the all-C outcome, supported by the players' weakly dominant strategies of voting for C, is the unique Pareto-optimal Nash equilibrium in the transformed n-person PD.[8]

4 Example of an n-Person PD

Suppose there are $n = 10$ players, and the payoff functions to the cooperators and the defectors are $c(k) = 10k - 50$ and $d(k) = 10k$. It is easy to show that the three properties of an n-person PD are satisfied:

---

[7] In the preceding example, we treated the "rest of the public" as a single player, but if the game is among many similar players, then it is properly modeled as an n-person PD. To ameliorate the problem of defections in such a game, wealthy individuals often commit to match the donations of small contributors, thereby enhancing the incentive of these individuals to contribute by guaranteeing that their donations will be increased by some factor.

[8] Hardin (1971) shows that all-C is a *Condorcet choice* when pitted against any other strategy combination – that is, a majority of voters would prefer it, except in the case of a tie – but he does not provide a procedure that would implement all-C.

1. The payoffs to the players are increasing in k.
2. $c(k) = 10k - 50 < d(k-1) = 10(1-k)$, which simplifies to $-50 < -10$ and so is satisfied.
3. $c(n) = 100 - 50 > d(0) = 0$, which simplifies to $50 > 0$ and so is satisfied.

Let $k = 1, 2, \ldots, 10$. The payoff for being the kth cooperator, $c(k)$ – as opposed to defecting and there being one less cooperator, $d(k-1)$ – are shown for representative values of k in the table below:

No. of cooperators ⇒	$k = 1$	$k = 2$	$k = 5$	$k = 9$	$k = 10$
$c(k)$	–40	–30	0	40	50
$d(k-1)$	0	10	40	80	90

Notice that $k = 5$ cooperators make the value of cooperation, $c(5) = 0$, equal to the value of defection by everybody, $d(0) = 0$, in the n-person PD. Thus, five cooperators is the *breakeven number* at which funding the project has the same value for the cooperators as not funding it.

Whereas all-D at $d(0)$ is the Nash equilibrium in the n-person PD, all-C at $c(10)$, which gives a payoff of 50 to each player, is not an equilibrium. The latter outcome is unstable because if one player defects from all-C, he or she receives a payoff of $d(9) = 90$. In fact, as we know from the previous analysis, every player has a strictly dominant strategy of defecting in the n-person PD, however many cooperators there are.

Now assume simple-majority rule is used in the transformed n-person PD (i.e., $r = 6$), so if there are five or fewer cooperators, no project is funded. But if there are six or more cooperators, everyone, including the defectors, gets a payoff of 50. If we depict the game as a ten-dimensional array in which each of the 10 players can choose between C and D, then C weakly dominates D for each player, whatever the value of r is, but the contingency in which C makes a difference (by raising a player's payoff from 0 to 50) changes when r changes.

Although the value of r does not affect the weak dominance of C, it would be strange indeed if r were not at least a simple majority (six in our example), because less than a majority of cooperators could implement a project, perhaps against the wishes of a majority. (In the extreme case, it would be a single player – who plays the role of a dictator, in effect – that would call the shots.) Accordingly, we propose that r be at least a simple majority in the transformed n-person PD.

In fact, a simple majority may be preferable to a qualified majority, because a simple majority is more robust against defectors. Thus in our example, selecting $r = 6$ means that even if up to four players choose D (for whatever reasons), the majority would still triumph, whereas this would not be the case for a greater r. In particular, if $r = 10$ (unanimity), one defector can undermine the choice of C by the other nine players.

Finally, we introduce a note of caution on the link between voting and democracy. While free and fair elections are a key to democracy, our solution to the free-rider problem is also applicable to oligarchies, wherein only few members of an elite (e.g., the 10 players in our previous example) vote. Insofar as elites are elected to councils or legislatures, however – as occurs in a representative democracy – we think it fair to say that voting by these voting bodies resolves the free-rider problem in a way akin to voting by all the citizens in the electorate.

5 A Biblical Tale

A story from the Hebrew Bible illustrates how a group, aided by a charismatic leader, may resolve an n-person PD when individuals alone cannot not do so.[9] The story begins after Moses descends from Mount Sinai and discovers that the Israelites, who had grown restive during his absence of 40 days and 40 nights, had built, with the complicity of Aaron (Moses's brother), a golden calf that they worshiped.

Observing the revelry of the Israelites at the base of the mountain, Moses is enraged and destroys the Ten Commandments. But he must also deal with another problem – the extreme anger of God, who is infuriated by the idolatry of the Israelites and threatens to destroy them:

> "I see this as a stiffnecked people. Now, let Me be, that My anger may blaze forth against them and I may destroy them, and make of you a great nation." But Moses implored the LORD his God, saying, "Let not Your anger, O Lord, blaze forth against Your people, who You delivered from the land of Egypt with great power and a mighty hand. Let not the Egyptians say, 'It was with evil intent that He delivered them, only to kill them off in the mountains and annihilate them from the face of the earth."' (Exod. 32:9–12)

Moses offers a cogent reason why the Israelites should be spared, asking God to

> turn from Your blazing anger, and renounce the plan to punish Your people. Remember Your servants, Abraham, Isaac, and Jacob, how You swore to them by Your Self and said to them: I will make your offspring as numerous as the stars of heaven, and I will give to your offspring this whole land of which I spoke, to possess forever. And the LORD renounced the punishment He had planned to bring upon His people. (Exod. 32:14–15).

Thus God, realizing the enormous investment he has made in His chosen people, does not brush aside His handiwork out of pique.

Although God relents, Moses must still convince Him that His decision to save His chosen but "stiffnecked" people, who had "acted basely" (Exod. 32:7), is not

[9] This story is adapted from Brams (1980, 2003, pp 94–98), but the interpretation of Moses's resolution of an n-person PD via a kind of referendum is new. Passages from the Bible are drawn from *The Torah: The five books of Moses* (1962). Schelling (1978, ch. 7) gives several contemporary examples of n-person PDs, such as whether a hockey player should wear a helmet, which was not mandated by the National Hockey League (NHL) until the 1990s. Prior to 1990, most players refused to wear helmets because it put them at a strategic disadvantage, limiting their peripheral vision, though they were at a substantially greater risk of serious head injury. The dilemma was resolved not by a secret vote of the players, which arguably would have led to the requirement of helmets in the 1970s, but by a public outcry caused by head injuries, which put pressure on the NHL. Even so, players who entered the league before the helmet requirement were exempted; the last player to refuse to wear a helmet retired in 1997.

a foolish one. After wringing a confession out of Aaron for his part in the idolatrous affair, Moses looks with horror on the Israelites, who are "out of control" (Exod. 32:25).

Moses averts catastrophe by seizing the initiative: "Whoever is for the LORD, come here" (Exod. 32:26). Moses's gamble pays off, at least for one tribe:

> And all the Levites rallied to him. He said to them, "Thus says the LORD, the God of Israel: Each of you put sword on thigh, go back and forth from gate to gate throughout the camp, and slay brother, neighbor, and kin." The Levites did as Moses had bidden, and some three thousand of the people fell that day. And Moses said, "Dedicate yourselves to the LORD this day – for each of you has been against son and brother, that He may bestow blessing upon you today." (Exod. 32:26–29).

I interpret Moses's summons to "come here" as less a command than a desperate plea for a sizeable number – if not a majority – of the Israelites to rally to the side of the LORD and renounce their sinful behavior. In effect, Moses, acting as a political entrepreneur, asks for the Israelites to vote in a referendum on his leadership.

If only a few Israelites had heeded Moses's plea and supported him, their numbers would not have been sufficient to persuade God that the Israelites were willing to turn from their idolatrous ways and worship Him as their rightful God, "who brought you out of the land of Egypt!" (Exod. 22:8). But Moses wants not just a vote of confidence but also seeks the annihilation of all dissidents.

This serves his and God's purpose by wiping out the last vestiges of idolatry among the Israelites. That the faithful are spared reinforces God's message since the time of Adam and Eve – He is stern in punishing sinners – but He is also merciful in protecting those who redeem themselves.

Effectively, Moses's solution to the n-person PD – whereby D is for the Israelites to continue to worship the golden calf and C is for them to return to the God of Israel – is to eliminate the outcome in which some Israelites choose D and some choose C. True, it is nowhere specified that if r Israelites choose C, C will be implemented. To prevent defections from this outcome, Moses deemed it necessary that those who chose D be decimated. This is a gruesome way to achieve consensus, but it is hardly unknown in recent times.

The solution worked, at least for a while (the Israelites become restive again). However, we strongly recommend voting, without the sacrifice, as a more civilized way to resolve n-person PDs.

6 Other Difficult Games

The hypothetical example we discussed in Sect. 4 illustrates a public-goods or common-pool game (Ostrom et al. 1994), in which there is a free-rider problem unless a mechanism like voting is introduced to transform the game into one that encourages cooperation – and the resulting outcome is indeed implemented. In the biblical example in Sect. 5, no Israelite alone has an incentive to support Moses – knowing that his or her faith in God will not appease Moses or save the Israelites from the wrath of God – but if Moses can turn the game into a referendum on his leadership and rally a sufficient number to his side to show their collective commitment, then he can snuff out idolatry, especially if those that refuse to go along are eliminated. Note that this kind of commitment is public, whereas voting about the provision of public goods will generally be private.[10]

PD is only one of the 57 distinct 2×2 strict ordinal *games of conflict,* in which there is no mutually best (4,4) outcome.[11] How many of these games can be transformed into more cooperative games through voting?

Define a *cooperative outcome* in a 2×2 strict ordinal game to be one in which each of the two players obtains either its best (4) or its next-best (3) outcome. Call the players' strategies associated with this outcome *cooperative strategies*. Call the other player strategies *noncooperative strategies*, and the outcome associated with these the *noncooperative outcome.* A 2×2 strict ordinal game is *difficult* if it satisfies the following three conditions:

1. There is only one cooperative outcome.
2. The cooperative outcome is not a Nash equilibrium, so at least one player has an incentive to defect from it.
3. The noncooperative outcome is Pareto-inferior to the cooperative outcome, so both players would prefer the cooperative outcome to it.

Obviously, 2-person PD meets these conditions, but so do the ten other games shown in Fig. 1.[12] The 11 games, which constitute 19% of all the 2×2 conflict games, can be broken down into three classes:

[10] The privacy of a voting booth is important if voters might be under social pressure to vote differently if their votes were known. To be sure, this social pressure might be critical to the passage of certain kinds of legislation, such as that backed by a political party that can punish defectors when there is a roll call vote. Perhaps the support that Moses, who was a Levite, received from his fellow Levites was reinforced by the public nature of those rallying to his side gave him. By contrast, the ringleaders on a ship who pledge in writing to participate in a mutiny are immediately identifiable, and subject to severe punishment, if they are discovered before the mutiny and were the first to sign the pledge. The institutional solution that mutineers devised to prevent the discovery of the ringleaders was to write their names in a circle ("round robin") (Leeson, 2007).

[11] If we include games with (4,4) outcomes, there would be 21 additional games, making for a total of 78 distinct 2×2 strict ordinal games (Rapoport and Guyer 1966,1976); see Robinson and Goforth (2005) for a further elaboration of these games and their properties. We exclude the games with (4,4) outcomes, because these outcomes are the unique Pareto-optimal Nash equilibria in them, rendering (4,4) the likely outcome that players would choose without the need for voting. The one exception is a game variously referred to as stag hunt, assurance, or coordination (Skyrms 2004):

(4,4)	(1,3)
(3,1)	(2,2)

If either player in this game chooses its second strategy, it assures itself of a minimum of 2, whereas choosing its first strategy may lead to 1. Thus, a player's second strategy is, in a sense, less risky; its choice by both players yields (2,2), which is a Nash equilibrium, albeit Pareto-inferior to (4,4).

[12] Schelling (1978, ch. 7) offers a different classification of PD and non-PD games, using lines and curves on a graph. Still other classifications of the 78 2×2 strict ordinal games, which include the

1. The Nash equilibria in four games, including PD, are the Pareto-inferior noncooperative outcomes. Either one or both (in the case of PD) players has a strictly dominant strategy associated with this equilibrium, and neither player has a dominant strategy associated with the cooperative outcome.
2. The Nash equilibria in three games, including chicken, destabilize the cooperative outcome by inducing the player(s) receiving a payoff of 3 at the cooperative outcome to defect from it.
3. Three games have no Nash equilibria, with one player having an incentive to defect from each outcome, including the cooperative outcome.

Note that only PD and chicken are *symmetric games*, in which the payoff ranks along the diagonal are the same and the payoff ranks along the off-diagonal are mirror images of each other.

Clearly, the cooperative outcome in all 11 games has a shaky status because it is not a Nash equilibrium. But when these games are transformed into voting games in the manner we illustrated for PD, the cooperative outcomes take on a new status: Each becomes the unique dominat-strategy Nash equilibrium.

Unlike PD, we will not try to illustrate these games with examples. But it is worth noting that whether all players receive the same payoff of 3 at the cooperative outcome, or one set of players receives 3 and the other set 4 so their benefits differ (think of frequent and infrequent users of a public park), neither set has an incentive to defect from this outcome in the transformed voting game.

If this outcome does not receive at least r votes, its failure cannot be attributed to a public-goods or free-rider problem. Rather, it fails because more voters view the provision of the public good as detrimental – that is, they see the cooperative outcome as Pareto-inferior, not Pareto-superior, to the noncooperative outcome. Put another way, it is a public bad, presumably because of its cost, unworthy of their support.

7 Conclusions

Democracy resolves conflict in difficult games like PD and chicken by stabilizing their cooperative outcomes. It does so by transforming them into games in which voters are presented with a dichotomous choice between a cooperative outcome and a Pareto-inferior noncooperative outcome. In the transformed game, it is always rational for voters to vote for the cooperative outcome, because C is a weakly dominant strategy independent of the decision rule r and the number of voters who choose C.

Why, then, is the cooperative outcome not always selected, given that voters have no incentive to be free riders in the transformed game? The answer is that the public

57 games of conflict and 21 games with a mutually best (4,4) outcome, are given in Rapoport and Guyer (1966,1976) and Brams (1977); a topology of such games, and even a new classification in a "periodic table," are developed in Robinson and Goforth (2005).

good may be viewed by too few voters to be worth the cost. This explanation for the failure of cooperation – that a majority see the public good as, in fact, a public bad – is very different from the claim that free riders undercut the provision of public goods in a democracy. They do so only if enough voters view them as public bads.

What is "enough"? We suggested that simple-majority rule is more robust than qualified majority rule, because it is not so vulnerable to defectors who may, perhaps out of ignorance, fail to recognize what a majority see as a genuine public benefit.

Even charismatic leaders like Moses, whose brilliant defense of the Israelites – despite their serious lapses – persuaded God that they deserved a reprieve, cannot act alone. He succeeded by persuading the Levites, in a kind of referendum, to renounce their idolatry and, less defensibly, slaughter those who did not go along.

In a standard 2-person PD, it would be odd indeed to ask the players to vote on whether to select C and, if both do, implement the cooperative outcome. The difficulty of doing so – say, in an arms race – is that there may be no mechanism to enforce cooperation, even when both sides agree to it. Choosing a strategy, and enforcing the outcome that the players support, may be two entirely different matters.

On the other hand, when a government can credibly commit to providing a public good that a majority support, the solution that democracy provides is compelling. However, in situations in which crime or corruption is rampant, or social capital or trust are lacking, voters will need assurances that procedures have been put in place that ensure that a cooperative outcome that a majority supports will actually be implemented.

Enforcement is particularly a problem in developing countries, though lax enforcement of laws occurs in developed countries as well. Thus, while the appeal of democracy is considerable in difficult games, questions about how, practically, to resolve conflicts and implement cooperative outcomes must also be answered.[13]

Acknowledgment We thank Todd R. Kaplan, Christian Klamler, Maria Montero, Brian Skyrms, and Donald Wittman for valuable comments on an earlier version of this paper.

References

Axelrod R (1984) The evolution of cooperation. Basic Books, New York

Brams SJ (1977) Deception in 2 × 2 games. J Peace Sci 2(Spring):171–203

Brams SJ (1980, 2003) Biblical games: game theory and the Hebrew Bible. MIT, Cambridge, MA

Brams SJ (1994) Theory of moves. Cambridge University Press, New York

Brams SJ (2008) Mathematics and democracy: designing better voting and fair-division procedures. Princeton University Press, Princeton, NJ

---

[13] Beginning in the 1960s, the United States and the Soviet Union were able to reach limited arms-control agreements, because both asides could detect violations of these agreements with a sufficiently high probability (e.g., via satellite reconnaissance) and take appropriate countermeasures if a violation were detected. By and large, this deterred both superpowers from violating these agreements.

Brams, SJ, Wittman, D (1981) Nonmyopic equilibria in 2 × 2 games Conflict Management and Peace Sci 6(1):39–62
Hardin R (1971) Collective action as an agreeable *n*-prisoners' dilemma. Behav Sci 16(5):472–481
Holsti OR, Brody RA, North RC (1964) Measuring affect and action in international reaction models: empirical materials from the 1962 Cuban missile crisis. J Peace Res 1:170–189
Kilgour DM (1984) Equilibria for far-sighted players. Theory Decis 16(2):135–157
Leeson PT (2007) Rational choice, round robin, and rebellion: an institutional solution to the problems of revolution. Preprint, Department of Economics, George Mason University, Virginia
Nowak MA (2006) Evolutionary dynamics: exploring the equations of life. Harvard University Press, Cambridge, MA
Ostrom E, Gardner R, Walker J (1994) Rules, games, and common-pool resources. University of Michigan Press, Ann Arbor, MI
Pinker S (2007) The Discover interview. Discover 71:48–52
Rapoport A, Guyer MJ (1966) A taxonomy of 2 × 2 games. Gen Syst 11:203–214
Rapoport A, Guyer MJ (1976) The 2 × 2 game. University of Michigan Press, Ann Arbor, MI
Robinson D, Goforth D (2005) The topology of 2 × 2 games: a new periodic table. Routledge, New York
Schelling TC (1978) Micromotives and macrobehavior. W. W. Norton:New York
Skyrms B (1996) The evolution of the social contract. Cambridge University Press, Cambridge, UK
Skyrms B (2004) The stag hunt and the evolution of social structure. Cambridge University Press, Cambridge, UK
Sugden R (2009) Neither self-interest nor self-sacrifice: the fraternal morality of market relationships. In: Levin S (ed) Games, groups, and the global good. Springer, New York
The Torah: The five books of Moses (1962) Jewish Publication Society, Philadelphia

Two Strategic Issues in Apologizing

Barry O'Neill

Abstract Social norms are typically embedded in networks of supporting norms that call on other parties to confer punishments or rewards depending on compliance with the original norm. Apologies are "all-purpose" supporting norms since the prospect of having to apologize deters many kinds of transgressions. One puzzle is how a network can avoid an infinite hierarchy of norms. A repeated game model of apologizing shows how a small number can be arranged in loops of mutual support. A second puzzle is why an apology bundles together so many speech acts – it acknowledges that one committed an offense and that it caused risk or harm. It expresses remorse and promises that there will be no repetition. Sometimes the actor is ready to perform only some of these, but recipients typically want full apologies, and there seem to be no single words for the subsets. A possible explanation is that the elements are synergistic. A game model hypothetically reduces an apology to just a promise not to do it again, but those apologizer-types who are less scrupulous about keeping a promise would be more ready to make one, so making a promise is itself grounds for disbelief and in the end none are made. Adding a requirement to show remorse confers credibility and produces an equilibrium that includes promising.

Apologies are delivered between individuals, between groups within a society, and between nations. They vary greatly in their subject, style, and success in reconciliation. Even some international ones seem trifling, for small mistakes of protocol, as when the first President Bush apologized for the US Marine Band carrying Canada's flag upside-down at a Toronto baseball game (Smith 1992). Others have influenced the long-term relations between states. After World War II, Germany entered a period of "amnesia," emphasizing its own suffering more than the suffering it had inflicted, but by the late 1950s it had started a program of penitent actions that included apologies (Herf 1997). Japan's World War II deeds were more easily denied internally since they happened outside its territory, and its leaders have been more reluctant to make a clear apology or adopt the internal policies that would follow, such as rewriting school history texts or avoiding symbolic displays

B. O'Neill
Department of Political Science, University of California, 405 Hilgard Avenue, Los Angeles, CA 90095-1472, USA

S.A. Levin (ed.), *Games, Groups, and the Global Good*, Springer Series in Game Theory,
DOI: 10.1007/978-3-540-85436-4_15, © Springer Physica-Verlag Berlin Heidelberg 2009

that others interpret as glorifying the past (Negash 2006; Lind 2008). Accordingly, Japan's acceptance among its neighbors has lagged behind Germany's. Its recent quest for a permanent seat on the United Nations Security Council was opposed officially by South Korea and by hundreds of thousands of Chinese signers of an internet petition (Gross 2005).

Sometimes an apparently sincere international apology has been refused. In 1998, US planes bombed the Chinese embassy in Belgrade, killing three Chinese citizens. The United States assured China that it was a mistake and apologized, but China rejected the statements as insincere, possibly due to a different cultural expectation of how to say one is sorry. The surrounding elaboration in some American statements delineated just what the US had and had done wrong (United States 1999), and the Chinese audience may have taken this as attempting to excuse the action, instead of focusing on the harm done to them. Very little research on Chinese apologizing has appeared in English, but a difference in such expectations separates American and Japanese styles (Sugimoto 1997). In the United States an apology satisfies a duty to oneself by recognizing the exact misdeed and giving the receiver a correct apprehension of future behavior, but in Japan there is a tendency to express a sincere surrender to the other side's viewpoint, without excuses. In 2001, a US spy plane was forced to land in China and the two states again fell into a dispute over apology language (Zhang 2001).

Orthodox opinion among neorealist scholars of international relations would hold that apologies are ephemeral; instead it is the struggle for power and interest that determines international affairs. However, international decisions are made by human beings influenced by emotions. Even within an unemotional and strategic framework, apologies are important since their symbolism can determine players' mutual expectations and thereby select an equilibrium when several exist (O'Neill 1999).

World events show the need to understand the emotional aspect of apologies and how they can misfire when delivered across cultures. The focus here, however, will be on some basic issues that must be understood first. Why have the institution of apologies at all? The next section will suggest that they function within a larger system of norms, and are compromises between deterring violations and restoring relationships. A norm is often supported by a network of norms, which impose duties on the whole group, either to punish a violation of the original norm or to reward compliance with it. A question is how such a system of norms might work without stretching off to infinity. The model illustrates that a finite number (in this case, only five) can be stable if they are properly arranged them in loops. The duty to apologize includes the recursive feature of normative systems. It supports others and itself as well: someone who fails to apologize acquires a further duty to apologize for the failure.

The section also compares game-theoretical analysis of norms with the leading formal method, deontic logic. The latter has considered "contrary-to-duty obligations," which prescribe what to do next in case the actor has violated a norm. Apologizing is a prototypical example. We argue that the debate has suffered by overlooking the strategic aspect of apologies.

The next section asks why an apology combines several speech acts and why recipients are often not satisfied unless they get them all. An apologizer is admitting that the event happened, that it caused risk or harm, and that he or she was the agent, as well as expressing remorse, promising not to do it again and in some cases to undo the damage, and perhaps asking for forgiveness. Sometimes the party can do only some of these acts with sincerity, so why are apologies all or none? The section suggests a partial answer: that the promise component alone would fail since those who are most willing to make a promise are the ones least likely to keep it. The recipient, knowing this, would disbelieve the promise. A game model suggests that requiring an expression of emotion helps prove the apologizer's sincerity.

1 The Duty to Apologize Within a Normative System

1.1 Systems of Social Norms

A *social norm* can be defined as a rule calling for a certain kind of behavior in a certain kind of circumstance. The grounds for the behavior are typically moral, but following the norm confers practical benefits on the group, generally if not in every case, and the party subject to the rule would be sensible to follow it, again at least in general. The typical motivation comes from other norms that are in place calling on the group to reward compliance and/or punish violations (O'Neill 1999). The latter have been termed *metanorms* (Axelrod 1986) or *supporting norms* (Crawford and Ostrom 1995), and they sometimes call for the group to behave in ways that would normally be wrong. No one should be deprived of their freedom, but the government has an obligation to do that if the person has committed a crime. (This pure stance, that compliance with a social norm is motivated only by the group's response pursuant to other social norms, is exaggerated and adopted here partly for simplicity. Often the supporting norm is an internalized one, so that it is the party's conscience that motivates good behavior.)

If the party's behavior were self-rewarding or self-punishing, the practice would be a *convention*. If I always meet my friend at the Grand Central clock but this time I go to Times Square and miss him, it is my own act that harms me. I am not being punished by others in response, so the practice is better termed a convention. Some rules are both norms and conventions – driving on the wrong side of the road is punished by the police and is self-punishing as well.

Supporting norms are full-fledged norms as much as the ones they support so they too must be supported by other norms, and this implies a network. From the network viewpoint understanding a norm means more than knowing what it calls on the party to do. We must know how other actors in the group and even the party himself should respond to compliance or violation, and know how the supporting norms are supported. For apologizing we can ask: How should others respond when someone has failed to apologize? If the non-apologizer is to be ostracized but one group member refuses to do that, how should the group respond to the latter? If the individual makes a sincere apology, does the recipient have a duty to accept it? What is one committing to by accepting an apology, and is accepting it the same as forgiving? The answers may depend on the context, the offense and the recipient of the apology.

Game theory becomes relevant since a group member will generally follow a norm if he or she believes that the others are motivated to follow the supporting norms. The qualification "generally" is needed because, as defined above, norms are associated with *types* of situations rather than specific games (O'Neill 1999). These situations show a typical pattern of utilities, but there are exceptions. Though normative behavior is usually the equilibrium, sometimes the payoffs motivate the actor to violate – to break a promise, steal or murder. Indeed overlapping norms may exist that justify a violation of some of them when all factors are considered. The examples here are for a typical simple payoff pattern.

1.2 A Simple Normative System for Apologies

The skeleton of a system around apologies can be shown by a repeated game. It has two players who at each stage simultaneously choose one of three moves, with the following payoff consequences to the mover:

Transfer (T)	Transfer 12 units to the other at a cost of 6
Withhold (W)	Transfer 0 units to the other at a cost of 0
Self-punish (S)	Transfer 0 units at a cost of 1

Matrix 1. The stage game

	Transfer	Withhold	Self-punish
Transfer	6, 6	–6, 12	–6, 11
Withhold	12, –6	0, 0	0, –1
Self-punish	11, –6	–1, 0	–1, –1

The stage game, Matrix 1, is a prisoner's dilemma augmented with a third row and column. The added moves seem pointless since they are strongly dominated by the second moves and all their outcomes are Pareto-inferior, but they will influence the equilibria when the game is repeated.

The players move at $t = 1, 2, 3, \ldots$, and know all past moves. Each has the goal of maximizing the present value of its payoff stream and they use the same discount rate $\delta \in (0, 1)$. Thus, if both played T forever each would receive 6, 6, 6, ..., and value that at $6 + 6\delta + 6\delta^2 + \ldots = 6/(1 - \delta)$.

A strategy in the game tells a player what to do at each stage for any possible history of what they both have done so far. Our task will be to assign the players a pair of strategies such that neither player has an alternative yielding a higher present value, given the opponent uses its assigned strategy. This property must hold for any situation they might find themselves in, even one arising from moves that were contrary to the strategies. That is, we will look for a subgame perfect Nash equilibrium.

Rather than consider strategies directly we take an approach due to Abreu (1988), which is both computationally convenient and fits with the concept of a normative network. The equilibrium will be given indirectly by specifying three paths of play. A *path of play* is an infinite sequence of pairs of moves by the players – an example would be TT, WW, TT, WW, ... An equilibrium is then defined by three paths, an initial path and two punishment paths, one for each player. The *initial path* states their joint play if they follow the equilibrium. A player's *punishment path* specifies the pairs of moves that both make if that player deviates from the current path, whether the latter is the initial path or someone's punishment path. At any point in the game a deviation from the current path has the same result: it switches play to the start of the deviator's punishment paths. A failure to appropriately punish the other, for example, sends the game to the start of one's own punishment path. A simultaneous deviation by both players is ignored and the current path continues. Together the three paths are known as a *simple strategy profile* (SSP). From an SSP one can derive corresponding strategies for the players, but the reverse is not true – not every pair of strategies can be represented by three paths. However, in terms of observed behavior in equilibrium the two methods are alike: if a certain sequence of moves arises from a subgame perfect Nash equilibrium it also arises from an SSP (Abreu 1988).

We will introduce further constraints on an equilibrium. The first is that it be in pure strategies, and the second that it yield mutual cooperation, TT forever. As in repeated PD games, mutual "Always Withhold" constitutes an equilibrium, but it is socially undesirable and so cannot be considered a norm. Third, for the sake of simplicity the equilibrium must treat the players identically. It must be stationary over time in the sense that its prescription cannot depend on the stage independent of actions players have taken. Finally, after a deviation their paths must return to mutual cooperation TT reasonably soon. We require that it happen within two moves. This condition is prompted by the earlier idea that norms are represented by game-types rather than games proper, and so will sometimes be violated. They should therefore be "non-grim" – their violation should not lead to permanent harm.

One SSP, called *Apologize-and-Restitute*, constitutes a subgame perfect equilibrium that satisfies the conditions. Also, for the particular payoffs used it will be shown to be the most robust such equilibrium in the sense that it produces cooperation at the lowest discount rate (Appendix). Its definition is as follows:

The initial path is TT, TT, TT, TT, ...
Row's punishment path is SW, TW, TT, TT, ...
Column's punishment path is WS, WT, TT, TT, ...

The initial path gives players 6 forever, but should Row unilaterally go for the immediate payoff of 12 by choosing Withhold, then Row's punishment path has Row choosing self-punish while Column chooses Withhold and at the next stage restituting Column by choosing Transfer while Column chooses Withhold. Then the players resume mutual transfers. Technically they are still on Row's punishment path, but they are making the same moves as if there had been no deviation. The everyday notion of an apology gets translated into punishing oneself, which in the real context would be paying a social cost in face and credibility, then giving restitution, which might mean undoing the damage.

A player is induced to stay cooperative by the fear of having to self-punish and restitute. If that course is called for, the player endures it by the prospect of an imminent return to cooperation. If the player refuses then the punishment path will be restarted, in the sense that the other will choose Withhold at the next two stages. Contrary to Gilbert and Sullivan, we do not make the punishment fit the crime – all deviations are dealt with in the same way, so fewer norms are necessary.

Apologize-and-Restitute is a perfect equilibrium down to $\delta = 0.564$ (as proven in the Appendix). This minimum measures the equilibrium's robustness to the influence of the future, and lower is also better in the sense that players would stay with the equilibrium if payoffs varied somewhat from those assumed. It can be proved that for the payoffs given, Apologize-and-Restitute has a lower minimum than any other equilibrium satisfying the cooperation-in-equilibrium, symmetry and cooperation-again-within-two-moves conditions. Its success follows from the order of the outcomes on the punishment paths – the less costly apology comes first, and the more costly restitution second. A deviator gets $-1, -6, 6, 6, \ldots$ whereas the reverse SSP, "Restitute-and-Apologize", for example, would give $-6, -1, 6, 6, 6, \ldots$ and would not be an equilibrium at such a low discount rate, since a violator would not accept -6 now while waiting two moves for cooperation.

1.3 The Consequence and Support Graphs

The equilibrium was set forth as paths of play, but another way of describing it fits the system-of-norms interpretation somewhat better. Figure 1 shows the equilibrium's *consequence graph* and *support graph* (O'Neill 1999). Each has nodes for the five states that current play might be in. Each state has an associated norm, instructing both players what to do there. The consequence graph indicates where play will

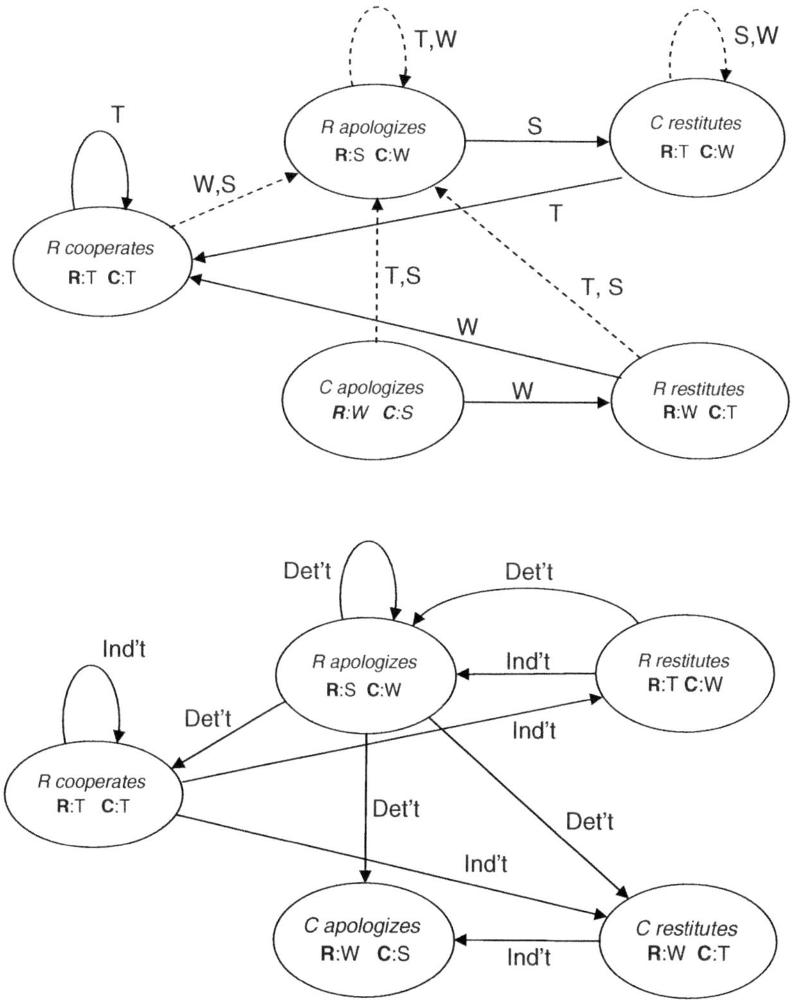

Fig. 1 Row's consequence graph (*top*) and support graph (*bottom*) for apologize-and-restitute. The ovals represent the system's five norms, with arrows showing transitions or deterrence and inducement relations

go next as a consequence of compliance with or violation of the current norm. The support graph (which follows immediately from the consequence graph), shows, for each norm and associated state, which norms at other states support it by giving the player an expectation of a reward or punishment. The graphs in Fig. 1 take Row's viewpoint, and corresponding versions exist for Column. Consider Row's consequence graph (Fig. 1, top.) If play is at the leftmost oval, the state of mutual cooperation, Row is to play Transfer. If Row does so, the game stays in that state, as indicated by the solid arrow. If Row violates the norm by choosing W or S, the game shifts to the oval in the upper middle, as indicated by the dashed arrow, whose norm calls on Row to "apologize." (An assumption behind the rules for Row's moving is that Column follows the appropriate norms).

Turning to the support graph, if Row's discounted present value at "Row Apologizes" is lower than at "Row Cooperates," the prospect of staying at cooperation induces Row to follow the rule at that state, while the (worse) prospect of entering the apology state deters Row from doing an alternative. The norm at the apology state supports the norm at the cooperation state. The incentives at every state generate behavior consistent with the equilibrium. The inducement and deterrent aspects of the norm associated with each state correspond to social rewards and punishments, the defining aspects of norms.

The whole regime comprises five norms – one for what to do in normal play, two for what to do if the other commits a violation of any norm, and two for what to do if you yourself commit a violation of any norm. Note that if Row apologizes (inappropriately) during mutual cooperation (i.e., moves to the bottom left oval), then play goes to Row's punishment path – Row must apologize and restitute for inappropriately apologizing. Also, Row's failing to "accept" Column's restitution by non-normatively playing Transfer at the top right box calls for an apology sequence, as does Row transferring goods to Column when Row should be letting Column apologize. Apologizing for these situations seems odd but the network is set up this way to minimize the number of norms.

Apologizing is a general purpose supporting norm since violations of other norms usually require an apology, and indeed one can argue that failing to apologize itself merits an apology. The apology for not apologizing is usually left understood, to keep the interaction smooth and simple. Still, when President Clinton apologized for the 1930s government-sponsored syphilis experiments in Tuskegee, Alabama, he included both elements (1997), "The American people are sorry – for the loss, for the years of hurt. You did nothing wrong, but you were grievously wronged. I apologize and I am sorry that this apology has been so long in coming."

Apologies are unusual as supporting norms in that they call on the wrongdoer to participate in his own punishment. The stereotypical view of social norms is that the violator is punished by others. One can imagine psychological reasons for this, that the apologizer "owns" the wrong and feels less resentment than if others were inflicting the harm.

1.4 Contrary-to-duty Obligations in Deontic Logic and Game Theory

The strategic approach can be compared with the formal method most commonly used in the philosophy of ethics, deontic logic. The latter is concerned with the logical relationships among statements about obligations of agents or ideal states, and is non-strategic, with no probabilities or utilities.

That literature has seen a debate on *contrary-to-duty obligation*, which arise when a party has violated a duty and thereby incurred a new one (Chisholm 1964).

I am obliged not to commit a murder, one example goes, but if I murder nonetheless, I should do it gently. It seems odd to proclaim an ethical rule for how to do something wrong. Apologizing is such a contrary-to-duty obligation – I should not insult someone, but if I do I should apologize for it. A "paradox" shows a technical problem for a deontic logic analysis. In an example adapted here from Forrester (1984), let the proposition i mean that I insult someone, let a mean that I apologize for it, and let *Oblig a* mean that I am obliged to apologize. We assume as premises:

$$Oblig \sim i \tag{1}$$

$$i \supset Oblig\ a \tag{2}$$

$$a \supset i \tag{3}$$

$$i \tag{4}$$

The first is self-explanatory, (2) is the contrary-to-duty obligation, while (3) states that I cannot apologize for something I did not do. I may say the words, "I'm sorry for insulting you" when I did not, or for that matter I could apologize for having started the Hundred Years War, but these would be vacuous. Premise (4) is that I did indeed insult the person and so am facing my contrary-to-duty obligation. As well as the inference rule of modus ponens, the following is included, a standard one in deontic logic:

$$p \supset q \rightarrow Oblig\, p \supset Oblig\, q \quad \text{(closure)}.$$

For example, if going to work means getting out of bed, it follows that if I am obliged to go to work then I am obliged to get out of bed.

From (2), (4) and modus ponens,

$$Oblig\ a, \tag{5}$$

and from (3) and closure we have,

$$Oblig\, a \supset Oblig\ i. \tag{6}$$

Combining (5) and (6) with modus ponens,

$$Oblig\ i. \tag{7}$$

So (1) and (7) together require me to refrain from insulting and to insult – a contradiction.

The difficulty seems to be that the closure rule is too broad but just how to restrict is debated, with proposals that are either flawed or quite complicated. From a game theory viewpoint the problem is that standard deontic logic has no strategic account of contrary-to-duty obligations. Philosophers sometimes call them "secondary obligations," as if I am bound by (1) but if I happen to violate it I should do (2) as second best. This ignores the strategic issue, that the prospect of facing (2) is a reason to follow (1), so one needs to incorporate goals and beliefs. Contrary-to-duty obligations are like behavior off the equilibrium path, a notion familiar to game theorists, who are comfortable with the idea that hypothetical behavior can determine real behavior. This is not to downplay attempts to solve the problem within deontic logic, only to point out that the strategic approach easily captures this feature.

2 Why are Apologies All-or-Nothing?

Apologies bundle several functions together into one speech act – a set of assertions, an expression of a feeling, perhaps a request, and at least one promise. When someone performs only part of the list they are seen as evading a full apology. One would think it might be useful on some occasions to say, "I honestly don't think I was to blame, but now that I know how you feel, I regret doing it and promise never to repeat it." This might reassure the offended party, yet there is no special word that does it, and often it might be taken as a refusal to "really" apologize.

2.1 Apologizing as Promising

The game described now suggests a reason for the many functions. The element of promising no repetition is important, and the model takes it to the limit by making it the only element of an apology. Performed alone, the promise turns out to be unconvincing. A promise comes more easily from an offender who is less conscientious about keeping it, since it puts less restriction on his future options. The receiver does not know the promiser's personal honesty, so notes the selection effect and puts less credence in any promise. This lowers the belief benefits of promising and thus strengthens the link between dishonesty and willingness to promise. In equilibrium no promise would be believed, so effectively no one, scrupulous or not, makes one. The result is like Akerlof's "market for lemons" (1970) where no matter what you offer for a car, the supposition that the seller accepts the offer tells you that the car is probably not worth that, so no deal is made, even though a price exists that would benefit both parties.

In the game, Player 1 decides whether to apologize, which here means only whether to promise Player 2 that he will not do action X again. If Player 1 makes no promise and does X he will receive payoff $d > 0$, but if 1 does X after promising not to, he will bear a cost of c, for a net payoff $d - c$. The cost c might represent his conscience or his worries about reputation or reprisals from 2. Player 1's motive to make a promise is that he would receive value bp from Player 2's belief p that he will refrain from X. Perhaps 1 wants 2 to take some action that follows from the belief, and b measures his value per unit of 2's probability.

At the time of his decision Player 1 knows the values b and c but not d, while Player 2 is uncertain about all three. Making a promise is, in a sense, betting that the value of d will not be too great. The three are the realizations of random variables B, C, and D, which have mutually independent uniform distributions on [0,1].

The game is played as follows:

Stage 1: Player 1 learns b and c;
Stage 2: Player 1 promises or does not promise to refrain from X;
Stage 3: Player 1 learns d;
Stage 4: Player 1 does or does not do X.

Player 1 receives

$$\delta_A\, b\, \mathrm{P}_2\,[1 \text{ refrains from } \mathrm{X} | 1 \text{ promises}] + \delta_X\,(d - c),$$

where δ_A equals 0 or 1 if 1 does not or does apologize, and δ_X equals 0 or 1 if he refrains from X or does X, respectively. Player 2 has no moves and is assigned no payoffs, but holds a belief P_2 about 1's reliability. Note that the probability in 1's payoff is that held by 2, which Player 1 can infer. It is not conditional on b, c, or d, since 2 is unaware of these. A subgame perfect Bayesian equilibrium will require consistency between 1's moves and payoffs as well as between 1's and 2's beliefs.

Generically the game has exactly one equilibrium: at Stage 2 Player 1 makes no promise, and Player 2, whether he hears a promise or not, believes with probability 1 that Player 1 will do X. At Stage 4, of course, Player 1 simply optimizes, taking the action X only if $d - c > 0$. ("Generically" means that the claim is true except for a set of situations with probability 0. Player 1 types with $c = 0$, for example, will lose nothing from a promise and will be ready to make or not make one.)

The result can be understood by assuming, contrary to fact, that an equilibrium exists where Player 1 promises with non-zero probability. Suppose 2's probability that a promise once given will be kept is T. It is easy to show that making a promise yields 1 an expectation of $bT + (1 - c)^2/2$. (Player 1 gains bT from 2's belief. The second term is 1's prospect from breaking the promise, which will happen with probability $1 - c$ and if it does will yield the player an average gain of $(1 - c)/2$.) Not promising yields $1/2$. If $T = 0$ then promising would be suboptimal for all Player 1-types with $c > 0$, contradicting the assumption. Thus $T > 0$, and Player 1 will promise if $b > [1 - (1 - c)^2]/2T$. As a function of c, this curve starts at $b = 0$ and rises, so the more likely someone is to make a promise the less likely they are to keep it. A calculation of the likelihood of keeping the promise for each hypothesized value of T shows that there is no fixed point in T except $T = 0$.

2.2 *Apologizing as Promising with a Show of Remorse*

The hypothesis is that an apology has further features in part because they mitigate the perverse selection effect of promising. An example is the expression of remorse. One cannot apologize in a deadpan, and like many emotions, remorse is associated with physical displays that are hard to fake. We can model this by postulating that when 1 apologizes he shows a degree of remorse commensurate with his value of c. With a certain probability, here taken as 1/2, Player 2 is able to discern 1's value of c

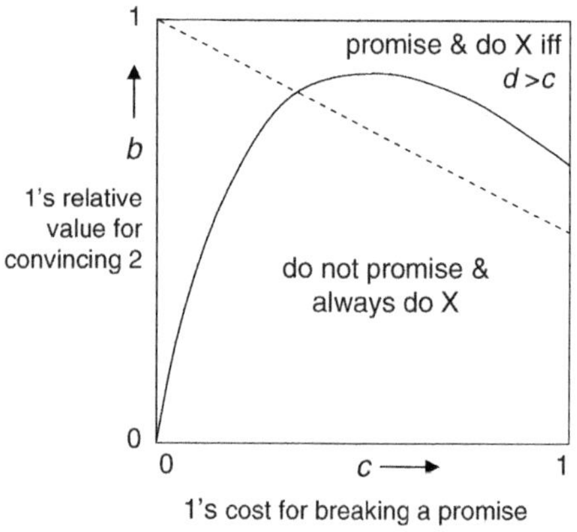

Fig. 2 Equilibria in promising games with partial (*solid line*) and full (*dotted line*) transparency. Types above or below a game's line use the strategies shown

from 1's display. With probability 1/2 Player 2 fails to make a reliability judgment at all, and uses only the fact that 1 has promised as the basis for assessing 1's reliability. This is called the *partial transparency* model since 1's emotional display makes him neither fully transparent nor fully inscrutable.

An equilibrium of this version proceeds as follows. Player 1 promises whenever $b > [1-(1-c)^2]/(c+0.357)$. (The criterion for promising now becomes $bT/2 + bc/2 + (1-c)^2/2 \geq 1/2$, and an hypothesized value of T allows us to calculate Player 2's credence for a promise, which should equal T; the fixed point is $T = 0.357$.) The types of Player 1 who promise are those above the solid curve of Fig. 2. A promise is made 27% of the time and Player 2 holds probability 0.357 that one will be kept. Non-conscientious Player 1 types again tend to promise often, but the rate falls for the middle types. It rises again for the high-c types, who hope that 2 will recognize them as conscientious and grant them belief. Enough of the high 1-types join the mix to justify 2's limited trust and this fact adds to everyone's motivation to promise.

It might be objected that 1 is not really promising since both parties know that 2 is convinced only to degree 0.357. Can one make a promise with common knowledge that it is disbelieved? Do notorious cheaters lose the ability to make (and therefore to break) promises? This position is plausible, but it is really a claim about the meaning of the word, and even if it were accepted it would reinforce the model's point, that the promise component of apologies need bolstering.

2.3 *Efficiency of Promising*

The two models and some related ones can be assessed for their equilibria's efficiency. The results depend entirely on the parameters assumed and cannot be claimed to be widely valid, but help us understand the models' structure.

First consider Player 1's view point and suppose there were no promising because promises were not credible. Player 1 would always do X and would expect $E(D) = 0.500$. Suppose now that all promises were fully credible, i.e., that $c = 1$ was commonly known. Player 1 would promise if and only if $b > E(D)$, and on average would gain $P(B > 1/2)E(B|B > 1/2) + P(B \leq 1/2)E(D) = 1/2 \times 3/4 + 1/2 \times 1/2 = 0.625$

The partially-transparent equilibrium gives Player 1 an expectation calculated as 0.516, which is 12.8% of the way to the ideal of full honesty. In a model with full transparency (where Player 1's promise somehow reveals his type exactly and they commonly know that) then 2 would assign a promiser a credibility of $1 - c$, and 1's prospect of possibly doing X, having promised, confers a benefit of $(1 - c)^2/2$. So 1 would promise if $bc + (1 - c)^2/2 > 1/2$, that is, if $b > 1 - c/2$, (Appendix) shown as a dotted line in Fig. 2. The line's downward slope indicates that the pool of promisers is biased towards the conscientious types. A simple calculation shows that 1's expected benefit is 0.532, which is 25.6% of that from a society of honest people. Both gains are relatively small.

To measure Player 2's benefits, we assume that 2's only goal is that X not happen and 2 has no interest per se in not being deceived. (Assuming this extreme makes more sense than the alternative. A Player 2 who cared only about not being deceived would not want 1 to make a promise at all.) Without promises, 1 will always do X, so 2 is maximally dissatisfied. With partial transparency 1 promises 27% of the time and is faithful with probability 0.357, so the likelihood of X is $1 - 0.270 \times .357 = 0.903$. With full transparency it is calculated to be 0.833, and with complete honesty it is simply the probability of a promise, which is P $(B < 1/2) = 0.500$. Thus partial and full transparency confer 19% and 33% of the benefit 2 would enjoy in a society of honest people.

The emotional display helps mitigate the lemons effect, but other facets of an apology might have a role as well. The apologizer's statement that he did wrong means a loss of face and an on-the-record admission. Unlike a commitment to do something in the future, these are irreversible, so in modeling terms the apologizer is tying his hands or sending a costly of his sincere intentions.

3 Game Theory as a Tool for the Analysis of Norms

In research on group reconciliation, apologizing is usually studied politically or legally and its consequences are treated as matters of parties' emotions (see, for example, the discussion in Kaminski et al. 2006), but the analyses here bring out the strategic issues. Understanding the structure of apologies gives a basis for studying them across cultures, so as to implement them more effectively.

Appendix

Condition for the Apologize-and-Restitute Equilibrium

To determine the critical δ that Apologize-and-Restitute is an equilibrium, we consider the positions that the Row player could be in, comparing the present value of following the SSP vs. choosing the most attractive alternative. By standard arguments Row's present value is calculated under the assumption that Column uses the SSP, that Row deviates at the first stage but returns to the SSP immediately afterwards.

Suppose players are on the initial path. The sequence of play will be TT, TT, TT,... and Row's payoff stream will be $6, 6, 6, \ldots$ for a present value $6/(1-\delta)$. If, for example, Row deviates to W on the first move the play will be WT, SW, TW, TT, TT,... and Row's stream will be $12, -1, -6, 6, 6 \ldots$ for a present value of $12 - 1\delta - 6\delta^2 + 6\delta^3/(1-\delta)$. As long as $\delta > 0.473$, the former value will be greater and Row will not deviate. There are four other positions that Row might face – at the first or second move or Row's own punishment path, or of Column's path, and each of these yields a condition on δ for compliance with the equilibrium. All the conditions must hold for the strategies to be a subgame perfect equilibrium, and the strictest among them is found to be $\delta > 0.564$.

The Robustness of Apologize-and-Restitute

There is a finite number of pure-strategy symmetrical equilibria that produce cooperation and restore it after a violation in two stages. Row has three possible moves for each of the first two stages of Row's punishment path, as does Column, yielding 81 possibilities. By considering cases it can be shown that for these particular payoffs, Apologize-and-Restitute has a lower threshold for maintaining cooperation than any other in the set.

Acknowledgments The author is grateful for support from the Russell Sage Foundation, the Leon Levy Foundation, and the Institute for Advanced Study, Princeton, New Jersey.

References

Abreu D (1988) On the theory of infinitely repeated games with discounting. Econometrica 56:383–96

Akerlof G (1970) The market for lemons: quality uncertainty and the market mechanism. Q J Econ 84:488–500

Axelrod R (1986) An evolutionary approach to norms. Am Polit Sci Rev 80:1096–1111

Chisholm R (1964) Contrary-to-duty imperatives and deontic logic. Analysis 24:33–36

Clinton W (1997) Remarks by the President in apology for study done in Tuskegee. White House, Office of the Press Secretary, 16 May 1997

Crawford S, Ostrom E (1995) A grammar of institutions. Am Polit Sci Rev 89:582–99

Forrester JW (1984) Gentle murder, or the adverbial Samaritan. J Philos 81:193–197

Gross J (2005) Petition against Japanese security council seat. Website, University of Heidelberg. http://www.sino.uni-heidelberg.de/dachs/gross050531.htm. Accessed 30 August 2008

Herf J (1997) Divided memory: the Nazi past in the two Germanys. Harvard University Press, Cambridge

Kaminski M, Nalepa M, O'Neill B (2006) Normative and strategic aspects of transitional justice: Introduction to the special issue on transitional justice. Journal of Conflict Resolution 50:95–302

Lind J (2008) Sorry states: apologies in International politics. Cornell University Press, Ithaca

Negash G (2006) Apologia politica: states and their apologies by proxy. Lexington Books, Lanham, MD

O'Neill B (1999) Honor, symbols and war. University of Michigan Press, Ann Arbor

Smith C (1992) World series: marines rally 'round the maple leaf, easing a flap. New York Times, 21 October 1992

Sugimoto N (1997) A comparison of US – Japanese apology styles. Communic Res 24:349–369

United States (1999) Government printing office. Papers of the presidents, administration of William J. Clinton. Remarks at the White House strategy meeting on children, violence, and responsibility. 10 May 1999

Zhang H (2001) Culture and apology: the Hainan Island incident. World Englishes. 20:383–391

Neither Self-interest Nor Self-sacrifice: The Fraternal Morality of Market Relationships

Robert Sugden

Abstract Economists have traditionally represented the market as a domain in which interactions are characterised by mutual unconcern; the self-interested motivations of individual agents are brought into harmony by the "invisible hand" of the market. Recently, however, economists have started to emphasise the extent to which markets rely on practices of impersonal trust, and to explain trust by hypothesising that economic agents are motivated by social preferences. In this paper, I review a range of social-preference theories and argue that none gives an adequate explanation of trust. These theories represent self-sacrificing motivations of giving and taking, while trust should be understood as cooperation for mutual benefit. Such cooperation is better represented by theories of team reasoning. I argue that the team-reasoning approach can be applied to market relationships in general. It leads to an understanding of market relationships as having moral content without involving self-sacrifice.

Economists have traditionally represented the market as a domain in which human interactions are characterised by *mutual unconcern*: within the bounds imposed by the law of contract, each party to a market transaction consults only his own interests, expecting other parties to do likewise. Such motivation is not seen as blameworthy: in Gauthier's (1986, p. 84) words, a perfectly competitive market is a "morally free zone, a zone in which the constraints of morality ... have no place". Mutual unconcern is understood as an inherent part of the mechanism by which markets tend to promote the common interests of those who participate in them. This conception of market relationships is encapsulated in two of the most famous passages in Smith's *Wealth of Nations* (1776/1976):

> It is not from the benevolence of the butcher, the brewer, or the baker, that we expect our dinner, but from their regard to their own interest. We address ourselves, not to their humanity but to their self-love, and never talk to them of our own necessities but of their advantages. Nobody but a beggar chuses to depend chiefly upon the benevolence of his fellow-citizens. (pp. 26–27)

R. Sugden
School of Economics, University of East Anglia, Norwich NR4 7TJ, UK
e-mail: r.sugden@uea.ac.uk

S.A. Levin (ed.), *Games, Groups, and the Global Good*, Springer Series in Game Theory,
DOI: 10.1007/978-3-540-85436-4_16, © Springer Physica-Verlag Berlin Heidelberg 2009

> [By] directing that industry in such a manner as its produce may be of the greatest value, [the merchant] intends only his own gain, and he is in this, as in many other cases, led by an invisible hand to promote an end which was no part of his intention. Nor is it always the worse for the society that it was no part of it. By pursuing his own interest he frequently promotes that of the society more effectually than when he really intends to promote it. I have never known much good done by those who affected to trade for the public good. (p. 456)

As the first quotation makes clear, Smith does not see it as a matter for regret that market relationships are characterised by mutual unconcern. To the contrary, this feature of markets is crucial in allowing us to satisfy our wants with independence and self-respect – something we would not be able to do if we had to rely on one another's benevolence.

Recently, however, economists have started to emphasise the extent to which markets rely on practices of impersonal trust. One strand in this literature models market relationships as repeated games between individuals who have some continuing knowledge of one another's identities, and so can build reputations for trustworthiness. Smith himself recognised the importance of trust for the workings of markets, and explained trust in commercial societies as a product of merchants' long-term interests in maintaining reputations for honest dealing.[1] On this account, practices of trust are ultimately grounded in self-interest, and pose no fundamental challenge to the idea that the market is a morally free zone.

In this paper, my concern is with a different strand of this literature, which seeks to explain practices of trust in non-repeated or anonymous interactions. Recent work of this kind has usually started from the hypothesis that economic agents are motivated by *social preferences*. I shall use this term to encompass any theory in which individuals' preferences with respect to economic interaction reflect concerns (benevolent or malevolent) about other people's payoffs, concerns about how their own payoffs compare with other people's, concerns about other people's motivations or intentions, desires to punish or reward other people's behaviour, or desires to confirm or disconfirm other people's expectations. Common to all these theories is the idea that an individual is socially oriented by virtue of his willingness to sacrifice his material interests to achieve some "social" objective (equality between his own outcomes and those of others, rewarding others for good behaviour, and so on). In this paper, I shall argue that social-preference theories of trust in economic relationships – or as I shall say for short, *economic trust* – are problematic.

Some theories of social preference are simply inconsistent with patterns of behaviour that are characteristic of real-world economic trust. Such theories may explain other kinds of non-selfish behaviour but, as would-be explanations of trust, they fall at the first fence. Other theories do not fail quite so obviously, but their claims to provide genuine explanations of trust rely on strained and implausible representations of what, at first sight, might seem plausible motivational principles, or else implicitly presuppose a perceived obligation to return trust. I shall argue that

---

[1] For more on this aspect of Smith's economics, see Bruni and Sugden (2000).

these problems are symptoms of a fundamental weakness in the social-preference modelling strategy – the linking of social orientation with self-sacrifice.

A further set of problems concern discontinuities between the hypotheses used to explain economic trust and those by which economics explains behaviour in markets where trust is *not* needed (I shall call such markets *paradigm* markets). Whatever form of social orientation lies behind economic trust, we should expect it to be characteristic of agents in paradigm markets too. If that orientation is some kind of social preference, the received theory of markets needs to be reconstructed to fit the hypothesis that individuals have social preferences with respect to the people they trade with. But, I submit, the received theory is already quite successful in organising our observations of paradigm markets. In broad outline, the patterns of behaviour we observe in these markets are consistent with the hypothesis that individuals act *as if* their attitudes to one another were the mutual unconcern of Smith's shopkeepers and their customers. A social-preference reconstruction of the theory of markets will need to explain why this is so. Because self-sacrifice is intrinsic to the concept of social preference, this is very difficult to do.

I shall argue for an alternative approach to explaining economic trust, based on the concepts of *team reasoning* and *collective intentionality*. The core idea is that the parties to a market transaction understand it as a joint commitment to an activity whose aim is to benefit them all. Mutual trust is a facet of that commitment. This approach allows us to understand trust as the product of individual motivations that are positively oriented towards others without being self-sacrificing. Further, the same motivations can be attributed to agents in paradigm markets without contradicting the main predictions of the received theory of markets.

The team-reasoning approach gives a new perspective on the intentions that lie behind market relationships. It opens up the possibility that, in market transactions *in general*, the parties' attitudes to one another go beyond mutual unconcern. It provides a way of thinking of market relationships, not as morally neutral components of a system with socially beneficial but unintended consequences, but as having moral content in themselves. That moral content is not benevolence, and does not compromise individuals' independence or self-respect. It relies on an understanding of market relationships as *fraternal* in a sense that is expressed in the work of one of Smith's contemporaries, the Italian philosopher-economist Antonio Genovesi.[2]

1 Trust as Gift Exchange

To provide some context for what might otherwise be a rather abstract analysis, I begin with a well-known paper in which Akerlof (1982) proposes the hypothesis that some labour contracts involve "partial gift exchange", and offers this as an explanation of involuntary unemployment. From a macroeconomic perspective, this

[2] For more on Genovesi's account of the fraternal nature of market relationships, see Bruni and Sugden (2008). The original text is Genovesi (1765–1767/2005).

is a variant of "efficiency wage" theory. That is, firms choose to offer wages that are higher than the reservation wages of marginal unemployed workers, even though the workers they employ are, on average, no different from those who would work for less. For the moment, however, I am not concerned with the truth or falsity of that empirical hypothesis. My concern is with Akerlof's concept of partial gift exchange. By looking at some of the difficulties of this concept, I set the scene for an examination of social-preference theories of trust.

Akerlof's central idea is that workers come to feel "sentiment" for the firm that employs them and, as a consequence of this sentiment, "acquire utility for an exchange of "gifts" with the firm" (pp. 543–544). The firm's gift to the worker is a wage in excess of the *reservation wage* – that is, the minimum wage the firm must pay to recruit the labour it needs. The worker's gift to the firm is a level of effort in excess of the minimum that is specified by the employment contract. If firms and workers were entirely self-interested, the reservation wage would be exchanged for the minimum effort; this would be a straightforward market exchange. To the extent that both wages and effort exceed those minimum levels, the labour contract involves an element of gift exchange.

In Akerlof's model, each firm seeks to maximise its profit. Knowing that workers' effort is greater at higher wages, firms pay more than the minimum necessary to fill their jobs, even though this has no effect on the distribution of skills and preferences among the workers they hire. Recognising that they have been paid what they see as a fair wage, workers anthropomorphise the firm and feel positive sentiment for it. Then: "For the same reason that persons (brothers, for example) share gifts as showing sentiment for one another, it is natural that persons have utility for making gifts to institutions for which they have sentiment". In a "donation of goodwill and effort", workers do more than the firm's work rules require of them, validating the expectations that led the firm to pay high wages (p. 550). If workers are motivated in this way and if firms maximise profits, labour-market equilibrium requires that wages are sufficiently high that there is a pool of involuntarily unemployment. (If all firms pay more than the reservation wage, there must be workers who are willing to work for less than firms are paying but who are unable to find jobs.)

One apparent problem with this account is that "firm" and "worker" do not have the obvious equivalence in status that two brothers do. To begin with, the firm is an institution while the worker is a human being. To get over this difficulty, Akerlof has to use the idea of "anthropomorphising". But even if we take the case of an owner-managed firm, substituting the human owner for the institution he manages, there may be large asymmetries in wealth and power between the owner and the worker. Presumably Akerlof thinks that such asymmetries are compatible with gift exchange, and I am inclined to agree. But, as will emerge later, some analyses of trust apply only to cases in which trustor and trustee are approximately equal in wealth.

A second asymmetry between the parties to the gift exchange is, I think, more problematic. In Akerlof's model, the firm can neither impose selective sanctions on individual workers who supply only the minimum level of effort, nor give selective rewards to those who supply more. Thus, from the viewpoint of an individual worker, effort above the minimum level is not a means of increasing income or avoiding dismissal; it is genuinely gratuitous. In the model, this gratuity is represented by including what would now be called a "social preference" term in the worker's utility function; Akerlof calls this term "norms for effort" (pp. 557–558). In contrast, firms simply maximise profit. If a firm pays more than the reservation wage, it does so instrumentally, as a means of activating the workers' norms in a profitable way. Is this really a gift? It seems that the workers are being generous but the firms are not.

Akerlof offers a partial reply to this objection when he argues that "the norm ... for the proper work effort is quite like the norm that determines the standards for gift giving at Christmas. Such gift giving is a trading relationship – in the sense that if one side of the exchange does not live up to expectations, the other side is also likely to curtail its activities" (p. 549). Akerlof is reminding us that gift exchange is a bilateral relationship, not the coincidence of two independent acts of gratuity. That is clearly right, but it does not imply that either side is acting instrumentally, giving only in order to elicit a return gift of greater value. In the case of Christmas presents exchanged between friends, it would surely be more realistic to postulate that, in purely instrumental terms, the exchange involves a net loss to both parties. It seems that such gifts serve an expressive purpose: they *affirm* friendship. It is because friendship is a bilateral relationship that it is affirmed by bilateral gift-giving. In contrast, the firm which pays above the minimum wage is not affirming any sentiment for its workers; it is maximising profit.

Nor is the firm expressing trust in any individual worker, or even in its workforce as a whole. Akerlof's firms do not distinguish between one worker and another; they merely rely on a propensity that is common to all workers – the propensity to feel "sentiment" for firms which provide the appropriate stimuli, and to respond in ways that increase the profits of the firms which do the stimulating. In the equilibrium state of the model, all firms provide the same stimuli and all earn the same profits from doing so. The worker's perception that she is the recipient of a gift reduces to the thought that she would be worse off if she were unemployed.

In the context of Smith's account of market relationships, a further feature of Akerlof's account is worth noticing. In a revealing footnote, Akerlof characterises the gift-exchange relationship between employer and worker as one of "mutual benevolence and dependence", and suggests that such a relationship can often "go together with mutual hostility and militancy" (p. 550). In other words, if market relationships include components of gift exchange, the market may not provide us with the kind of independence that Smith values. This is not a problem for Akerlof, who has set himself the task of explaining how the world is, not how he would like it to be. Still, it allows us to see some of the attractive features of Smith's account of market relationships. If there is more morality in these relationships than Smith thinks, one might hope that the missing element does not turn out to be the combination of benevolence and dependence that Akerlof is analysing.

2 The Trust Game

The essential features of Akerlof's account of the relationship between firms and workers can be represented by the *Trust Game*. This game has a long history in social theory; variants of it are discussed in Hobbes's *Leviathan* (1651/1962, Chaps. 14–15) and in Hume's *Treatise of Human Nature* (1740/1978, pp. 520–521).[3] My discussion will centre on the game shown in Fig. 1.

This game has two players, A and B. A moves first, choosing between the actions *hold* and *send*; *send* is interpreted as an act of trust by A. If A chooses *send*, B chooses between *return* and *keep*; *return* is interpreted as the repayment of A's trust. The payoffs to each combination of actions are shown by the numbers at the end of each path through the game tree, A's payoff being shown first. Payoffs are to be interpreted in what game theorists call the *material* sense. That is, they are expressed in units of some commodity of which, other things being equal, each player can be assumed to want more rather than less; they are not representations of players' all-things-considered preferences. Thus, there is no presupposition that each player seeks to maximise his own expected payoff.

It is important for the interpretation of the game as a model of trust that the payoffs satisfy the inequalities shown in Fig. 1. Specifically: the path (*send*, *return*) gives higher payoffs to both A and B than *hold*, so that the practice of giving and returning trust benefits both parties; (*send*, *keep*) is better for B than (*send*, *return*), so that B has a temptation not to return A's trust; and (*send*, *keep*) is worse for A than *hold*, so that A exposes himself to risk by trusting B. Apart from satisfying these conditions, the particular numbers used in Fig. 1 have no special significance.

The Trust Game can be interpreted as a simplified version of a sequential Prisoner's Dilemma, in which *send* and *return* are the cooperative moves, while *hold* and *keep* are the moves which defect. (The simplification is to remove the second mover's unappealing option of cooperating after the first mover has defected.)

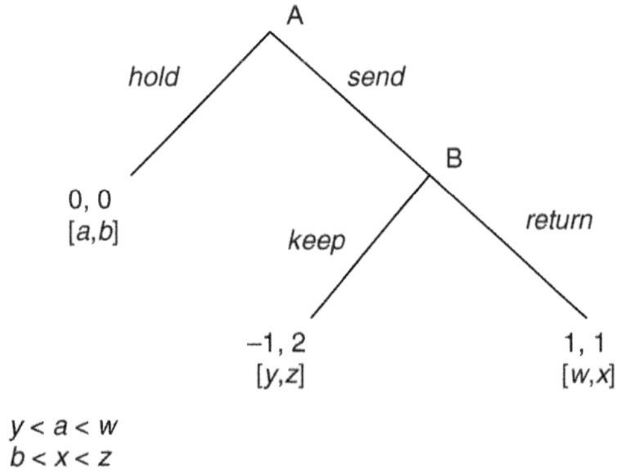

Fig. 1 The trust game

[3] For more on Hobbes's and Hume's trust games, see Sugden (2004, pp. 108–111, 165–169).

Analogously with the sequential Prisoner's Dilemma, if payoffs are measures of subjective utility, *hold* is the unique subgame-perfect equilibrium and is prescribed by the logic of backward induction. (If B seeks to maximise her payoff, she will choose *keep* after A chooses *send*; anticipating this, A maximises his payoff by choosing *hold*.)

Experimental economists have investigated people's behaviour in a variant of the Trust Game which allows both players a wider range of alternative actions. This design was first used by Berg et al. (1995). The experimenters give Player A (the "sender") some amount of money, say $20. A chooses how much of this to keep for himself. The remainder is tripled and transferred to B (the "responder"). B then chooses how much of this sum to keep for herself. The remainder of this sum (without further multiplication) is returned to A. If players maximise payoffs (and if this is common knowledge), B will keep everything she receives; anticipating this, A will send nothing. But this is not what happens in experiments. Senders typically transfer about half of their endowments to the responder. There is a lot of variation in the amounts returned by responders, but the average is typically around one-third of the amount received, so that senders get back about as much as they sent, while responders make substantial gains. This finding is usually interpreted as evidence that people have some motivation to engage in practices of economic trust, even when doing so is contrary to their material self-interest. The challenge for economic theory is to explain what that motivation is, and how it induces the practices of trust that we observe in laboratory experiments and in real markets.

3 Social-preference Explanations of Trust

Over the last decade, a range of alternative theories of social preferences have been proposed, with the aim of explaining observed regularities in economic behaviour that appear to be inconsistent with the previously standard assumption of self-interest. In this section, I investigate how successful these various theories are at explaining trust.

3.1 Inequity Aversion

One approach to the modelling of social preferences is to assume "inequity aversion", as in the very similar theories proposed by Fehr and Schmidt (1999) and Bolton and Ockenfels (2000). The essential idea is that each individual derives subjective utility from his own material payoff and subjective disutility from differences (whether positive or negative) between that payoff and the payoffs of others. In other words, individuals – or at least, some individuals –"are willing to give up some material payoff to move in the direction of more equitable outcomes" (Fehr and

Schmidt, p. 819), "equitable" being understood as equivalent to "equal".[4] Notice that the social orientation of inequity aversion is represented as a self-sacrificing motivation – as the willingness to *give up* material payoffs to generate outcomes in which one's own position relative to others is less unfair. In some cases, this self-sacrifice might be interpreted as gift-giving (to reduce the extent to which he is better-off than B, A incurs material costs to increase B's payoff). In other cases, it seems more like an expression of resentment (to reduce the extent to which she is worse-off than A, B incurs material costs to bring about a larger reduction in A's payoffs).

When applied to two-player games, the two theories have very similar implications. We can ask how player A's utility varies with player B's payoff, A's payoff being held constant. In Bolton and Ockenfels's theory, A's utility depends on the ratio of his payoff to the mean payoff to the two players; it is increasing in this ratio if the ratio is less than unity, and decreasing in the opposite case. In Fehr and Schmidt's theory, A's utility depends on the difference between his payoff and B's; it decreases as the absolute difference increases in either direction, but the rate of decrease is greater for *disadvantageous inequality* (that is, when A has the lower payoff) than for *advantageous inequality* (when A has the higher payoff). In relation to the Trust Game, the relevant implication of both theories is that an individual may be willing to give up material payoff to reduce advantageous inequality.

Using the logic of backward induction, we can start by analysing the choice faced by player B in the Trust Game, given that A has already chosen *send*. Her choice is between the equal payoff distribution (1, 1) and the unequal distribution $(-1, 2)$. Relative to the former, the latter gives B a higher payoff (2 instead of 1), but at the cost of advantageous inequality (a ratio of 4 in Bolton and Ockenfels's theory, a difference of $+3$ in Fehr and Schmidt's). Thus, B will choose *return* if and only if she is sufficiently inequity-averse. If A expects B to choose *return*, his choice is between (0, 0) and (1, 1); the latter is clearly preferable, and so he will choose *send*. The implication is that if B is sufficiently inequity-averse and if A knows this, the path (*send*, *return*) will be followed.

But are such players really following a practice *of trust*? Notice that B's motivation for choosing *return* has no connection with A's choice of *send*. Suppose there had been no first move by A, and that B simply faced a one-person Dictator Game in which she had to choose between (1, 1) and $(-1, 2)$. According to the theory of inequity aversion, B's choice would be the same in both cases, and for the same reason. But in the Dictator Game, A has not trusted B at all. Since B's motivation has been assumed to be the same in both games, we must conclude that in the Trust Game, B does not perceive *return* as the return of trust (or, indeed, as the return of anything). It is simply a gift, an act of self-sacrifice. Correspondingly, A's choice of

[4] I take it that Fehr and Schmidt use the term "inequity" rather than "inequality" because they want to leave open the possibility that, when judging fairness, individuals use principles other than equality of payoffs. But unless some structure is imposed on fairness judgements, the hypothesis that individuals are averse to *unfairness* has very little content. In Fehr and Schmidt's model, the necessary structure comes from the assumption that equality of outcome is the criterion of fairness.

send does not express reliance on B to return trust. Rather, it expresses A's belief that, if an opportunity arises, B will make a gift to him.

Notice also that (*send*, *return*) would not be chosen if the payoffs were changed to make A sufficiently richer than B. Suppose we add 20 to all A's payoffs and 10 to all B's, so that $a = 20$, $b = 10$, $w = 21$, $x = 11$, $y = 19$, $z = 12$. Both games can be interpreted as sequential Prisoner's Dilemmas in which each player's cooperative option requires him or her to give up one unit of payoff in order to benefit the other by two units; the only difference is that in the original game both players start out with equal wealth, while in the variant game A starts out richer. But now if A has chosen *send*, an inequity-averse B clearly prefers the (19, 12) that results from *keep* to the (21, 11) that results from *return*. The implication is that if the would-be trustor is significantly richer than the would-be trustee, trust will not be returned. This, I suggest, is not what we observe in everyday practices of trust. For example, as in Akerlof's account, wealthy employers often rely on the trustworthiness of much poorer employees.

3.2 Rabin Reciprocity

It is often said that the theory of inequity aversion fails to take account of *intentions*. Thus, in the Trust Game, it fails to recognise that B's motivation to choose *return* may be a response to the intentions that are revealed in A's choosing *send*. Rabin's (1993) theory of "fairness" – now generally referred to as a theory of "reciprocity" – provides a formal representation of mutually responsive intentions. The idea is that individuals are motivated by kindness towards people who are being kind to them, and by unkindness to those who are being unkind to them. As in the theory of inequity aversion, these other-oriented motivations are modelled in terms of self-sacrifice: to be kind to another person is to be willing to sacrifice one's material payoff to benefit her, and to be unkind to another person is to be willing to sacrifice one's material payoff to harm her. I shall call this idea *Rabin reciprocity*, to signify that this is not the only way of understanding reciprocity in social life.

Formally, Rabin's theory applies only to normal-form games. Extending it to the general class of games in which players move sequentially is not a trivial task, but the implications of such an extension to the Trust Game are straightforward.[5] Suppose that A has chosen *send*. Then B's choice between *return* and *keep* can be construed as a choice between the payoff vectors (1, 1) and (–1, 2). Relative to *return*, *keep* is better for B and worse for A. On Rabin's definition, this makes *keep* "unkind" and *return* "kind".[6] Now consider B's assessment of the intention that lies behind A's choice of *send*. To make this assessment, she forms a belief about what

---

[5] Dufwenberg and Kirchsteiger (2004) propose a version of Rabin's theory which applies to sequential-move games.

[6] Rabin's definition of kindness is less transparent when the relevant player can choose between three or more efficient payoff vectors, but I do not need to go into these complications.

A believes she will do, and then uses this to reconstruct A's decision problem as a choice between vectors of expected payoffs.

In Rabin's theory, each player derives subjective utility from his own payoffs and from positive correspondence between his intentions and those of the other player (that is, from being kind when the other player is kind, or being unkind when the other player is unkind). Conversely, negative correspondence of intentions (being kind to an opponent who is unkind, or vice versa) is a source of disutility. There is a state of equilibrium if each player is maximising expected utility, given his beliefs, and if the players' beliefs are mutually consistent. So is there an equilibrium in the Trust Game in which the path (*send*, *return*) has probability 1? The answer is "No".

To initiate a proof by contradiction, suppose that A chooses *send* knowing that B will choose *return*. Knowing that *send* will be followed by *return*, A's choice is between (0, 0) and (1, 1). According to Rabin's definitions, choosing (1, 1) is neither kind nor unkind. Formally, this is because, in assessing the options open to a player, Rabin's theory treats dominated payoff vectors as irrelevant. Given the underlying logic of the theory, this seems unavoidable. If all B knows is that A has chosen (1, 1) rather than (0, 0), what inferences can she make about the kindness or unkindness of A's intentions towards her? A's behaviour is entirely consistent with what might seem to be the most natural default hypothesis, namely that he has neither positive nor negative concern about B's payoff. (Admittedly, A has revealed that he is not *so* malevolent that he is willing to sacrifice a unit of payoff to reduce B's payoff by an equal amount; but that is very weak evidence of kindness.) Given that A's choice of *send* is at the zero point of the kindness/ unkindness scale, B receives neither utility nor disutility from being kind or unkind towards A. (In Rabin's model, correspondence or dissonance of the intentions of two individuals enters each individual's utility function as the product of two net kindness terms.) Thus, B will be motivated only by material payoffs, and will choose *keep*. This establishes that there cannot be an equilibrium in which (*send*, *return*) has probability 1. Notice that this conclusion mirrors one of the difficulties in Akerlof's account of gift exchange – that of explaining why, when it is profitable for the firm to pay more than the reservation wage, workers perceive such a wage as including a gift that calls for reciprocation.

Rabin recognises that this implication of his model is unsatisfactory. (Discussing a normal-form version of the Trust Game, with $a = b = 5$, $w = x = 6$, $y = 0$ and $z = 12$, he suggests that, contrary to the predictions of his model, "it seems plausible that cooperation would take place" [p. 1296].) As part of their extension of Rabin's theory to sequential-move games, Dufwenberg and Kirchsteiger (2004) propose an amendment to Rabin's definition of kindness that is explicitly designed to overcome this problem. In the context of the Trust Game, the amendment works as follows. Suppose A knows that, were he to choose *send*, B would choose *return*. Then A's choice is between (0, 0) and (1, 1). Clearly, (1, 1) is better for B; but to interpret A's choice of (1, 1) as kind, we have to treat (0, 0) as, in some sense, eligible from A's point of view. Rabin deems it to be ineligible because it is dominated by (1, 1). Dufwenberg and Kirchsteiger treat it as eligible on the grounds that, *in the set of all conceivable outcomes of the game*, it is not dominated. Specifically, it is not dominated by (–1, 2). Given that (0, 0) is treated as eligible, A's choice of *send* is kind (he has chosen to give B a payoff of 1 when, by choosing a different and eligible action, he might have given her 0). This allows there to be an equilibrium in which A chooses *send* with kind intentions and B reciprocates that kindness by choosing *return*.

It might seem from this example that the Dufwenberg–Kirchsteiger definition of fairness is taking account of the element of trust in A's choice of *send*. The outcome (0, 0) is deemed to be eligible because (*send*, *keep*) gives A a smaller payoff than *hold*; and this inequality expresses the fact that, by trusting B, A is exposing himself to a potential loss. But consider a game that is just like the Trust Game except that the players' payoffs from (*send*, *keep*) are reversed – that is, $y = 2$ and $z = -1$. In this game, too, (0, 0) is non-dominated in the Dufwenberg–Kirchsteiger sense, and so *send* is deemed to be kind. But now *send* guarantees A a higher payoff than *hold*: it is in A's self-interest to choose *send*, irrespective of what B will do. How is that choice kind? Notice also that *send* would cease to be kind if we set $z = 0$. How can an improvement in B's potential payoffs from the action that A chooses make that action less kind? I conclude that the Dufwenberg–Kirchsteiger amendment lacks a coherent motivation.

Rabin shows a better sense of the nature of the problem when he says that the failure of his theory to allow (*send*, *return*) as an equilibrium shows that theorists need to consider modelling "additional emotions" (p. 1296). In other words, Rabin's model gives a coherent representation of the motivations associated with a particular conception of reciprocity, but those motivations fail to provide a satisfactory explanation of trust. Nevertheless, I suggest, the idea that intentions matter – that one person's motivations can respond to another person's intentions – seems to be capturing something of what is involved in returning trust. The problem is that we need a way of representing the intentions of a trustor as non-self-sacrificing and yet worthy of reciprocation.

3.3 *Self-fulfilling Expectations*

Another branch of the theory of social preferences postulates that individuals are motivated to meet other people's expectations about them. More precisely, if person B expects person A to act in a particular way, B derives disutility from actions which reduce A's payoff below the level he was expecting. Again, a "social" orientation is represented in terms of self-sacrifice: B is willing to sacrifice material payoffs to avoid disconfirming A's expectations.

I have proposed one version of this hypothesis in a model of *normative expectations* (Sugden 1998, 2004, pp. 214–218). This is an evolutionary game-theoretic model in which games are played recurrently in a population, and equilibrium emerges from a process of experiential learning. The underlying hypothesis is that, when some stable pattern of behaviour has become established in a population, principles of morality which support that pattern tend to evolve. According to these principles, it is wrong for any individual to deviate from the established pattern of behaviour in ways which (given that others conform to the pattern) can be expected to reduce others' payoffs. It is assumed that each individual derives subjective utility from his own material payoffs, and disutility from any reductions in other people's payoffs for which he is responsible.

A related hypothesis, applicable to non-repeated games, has been proposed in slightly different forms by Bacharach et al. (2001), Pelligra (2005) and Battigalli and Dufwenberg (2007). Bacharach et al. and Pelligra describe the motivation they are modelling as *trust responsiveness*; Battigalli and Dufwenberg use the term *guilt aversion*.

Battigalli and Dufwenberg present this hypothesis in a very general form. In their model, each individual derives disutility from reductions in other individuals' payoffs (relative to their prior expectations) for which he is responsible. It is easy to see that, in the Trust Game, this hypothesis is consistent with an equilibrium in which (*send*, *return*) is chosen with probability 1. Suppose it is common knowledge that both players expect this equilibrium to be played. Looking at the game from B's viewpoint, she knows that A expects his choice of *send* to be followed by her choice of *return*, with a resulting payoff of 1 to him. If instead B chooses *keep*, she gains one units of material payoff, but causes a loss of two units to A. If she is sufficiently guilt-averse, she will choose *return*. Now consider the game from A's viewpoint. He knows that B is expecting (*send*, *return*), which will give her a payoff of 1. If instead he chooses *hold*, he loses one unit of payoff and causes a loss of one unit to B. Clearly, he will choose *send*.

The problem with this theory is to delimit its domain. It is surely implausible to claim that a typical individual is motivated to meet *any* expectation that another person may have about him, however morally unreasonable and however empirically ungrounded. For example, consider the Confidence Game shown in Fig. 2. This has the same structure as the Trust Game, but with the payoffs $a = b = 0$, $w = 1$, $x = -1$, $y = -10$, $z = 0$. Suppose you are B, playing this as a one-shot game. Suppose you know that A's motivation is to maximise his expected material payoff; and suppose he chooses *send*. You can infer that he expects you to choose *return* with a probability of at least 10/11, making his expected payoff no less than the zero he could have had by choosing *hold*. So if you choose *keep*, you reduce his expected payoff by at least ten units. But should you feel any guilt about doing this? A has knowingly incurred the risk of a payoff of –10, taking an action that might benefit

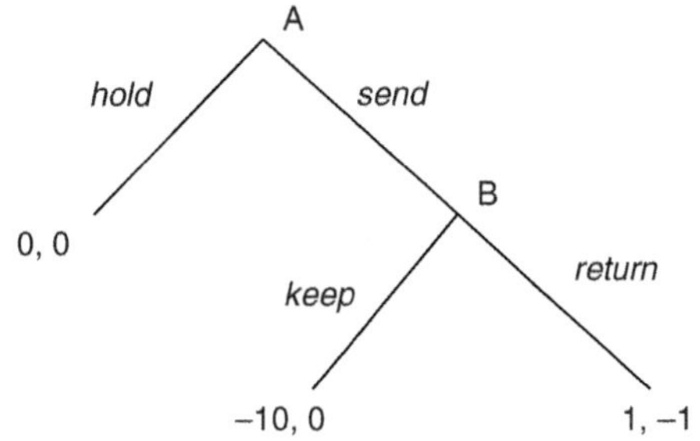

Fig. 2 The confidence game

him at your expense but which cannot possibly benefit you. He is gambling on the expectation that you will gratuitously transfer one unit of payoff to him. If you choose not to do so, why should you feel responsibility for his loss? You might well interpret his choice of *send* as a confidence trick, revealing his belief that you are one of those mugs who are born every minute. If so, you might feel more motivated to disconfirm his beliefs than to confirm them.

Battigalli and Dufwenberg implicitly recognise that their theory has a limited domain, saying that they hope it "will prove useful for a variety of applications concerning economic situations where it seems plausible that decision-makers are affected by guilt" (p. 175). The implication seems to be that, before applying the theory to any particular real-world situation, we have to judge whether this is a case in which individuals feel guilt about disconfirming one another's expectations. This may be defensible as a pragmatic modelling strategy, but it introduces a circularity into any claim that the theory *explains* trust. The problem is that we need to know why players who choose *keep* in the Trust Game feel guilt, while their counterparts in the Confidence Game don't. Intuitively, the answer is that *send* is an act of trust in the first game but not in the second, and that people feel a moral obligation to repay acts of trust. But unless we have a criterion for identifying acts of trust, we are reasoning in a loop.

In their discussions of trust responsiveness, Bacharach et al. and Pelligra recognise this problem but do not provide any formal resolution. Bacharach et al. suggest that trust responsiveness tends to be stronger, the more "kind" or "needful" the trustor is perceived to be, and the more "sympathy" or "respect" the trustee feels for the trustor (pp. 9–10). Pelligra focuses on the trustor's judgement of the character traits of the trustee, as perceived by the trustee. Trust responsiveness is induced when the trustor's action expresses confidence in praiseworthy attributes of the trustee (for example, kindness, fairness or trustworthiness) but not when it expresses confidence in blameworthy attributes (for example, gullibility).

I conclude that the expectations-based approach fails to explain trust. Nevertheless, it captures a significant feature of trust – that the trustee responds not only to the trustor's intentions but also to his expectations.

3.4 Falk–Fishbacher Reciprocity

It might seem that what is needed is a composite model of social preferences which takes account of inequity aversion *and* Rabin reciprocity *and* trust responsiveness. Such a theory has been proposed by Falk and Fischbacher 2006), who describe it as "a theory of reciprocity". To avoid confusion, I shall speak of *Falk–Fishbacher reciprocity*.

This theory is a complex construction. I shall not attempt to summarise it, but merely illustrate its application to the Trust Game. I shall show that, for the parameter values presented in Fig. 1, (*send, return*) is not an equilibrium. The proof is by contradiction, so I begin with the supposition that B expects A to choose *send* and A expects B to choose *return*. Thus, both players expect the outcome (1, 1).

According to Falk and Fischbacher, the fact that A's expected payoff is equal to B's implies that A is being neither kind nor unkind to B; similarly, B is being neither kind nor unkind to A. (Notice that kindness is being interpreted in terms of *relative* payoffs: this is where Falk and Fischbacher are using components from the theory of inequity aversion.) Given B's expectations, A has the opportunity to impose an unexpected loss on her by choosing *hold* rather than *send*, reducing her payoff by one unit. This is where reciprocity comes in. Had B been unkind, A would have wanted to punish her, and so would have derived positive utility from B's loss. Conversely, had she been kind, A would have wanted to reward her, and so would have derived disutility from her loss. But since in fact B has been neither kind nor unkind, A would derive neither positive nor negative utility from imposing a loss on her, while choosing *hold* would lead to a loss of material payoff. So *send* is subjectively utility-maximising for A. This is consistent with the supposition that *send* is chosen.

But now consider the situation from B's viewpoint. She has the opportunity to choose *keep*. This would impose an unexpected loss on A. For the same reasons as applied in A's case, B derives neither positive nor negative utility from this loss. But choosing *keep* would give B an increase in material payoff relative to her expectation. So, contrary to the supposition from which this analysis began, *return* is not subjectively utility-maximising for B. That is, (*send*, *return*) is not an equilibrium: the Falk–Fischbacher theory does not explain trust in this game.[7]

4 Trustworthiness as a Character Virtue

Gintis (2009) proposes an explanation of trust which, although framed in terms of individuals' preferences, is not quite a *social* preference theory by my definition. The core concept in Gintis's approach is that of a *character virtue*, defined as an "ethically desirable behavioural regularit[y] that individuals value for [its] own sake", exemplified by "honesty. loyalty, trustworthiness, promise-keeping, and fairness".[8] For Gintis, virtue-based theories differ from social preference theories in that character virtues "operate without concern for the individual with whom one interacts". An individual who is motivated by virtue is trying to achieve a "desired

[7] In deriving this negative result, it is crucial that $w = x$, i.e. the two players' payoffs from (*send*, *return*) are equal. If instead $w < x$, i.e. (*send*, *return*) gives advantageous inequality to B, then this path can be an equilibrium. Since, according to the Falk–Fischbacher definition, A is being kind to B, B would derive disutility from imposing an unexpected loss on A; this may be sufficient to make *return* optimal for B.

[8] Unless stated otherwise, quotations from Gintis are from Sect 3.12 of the pre-publication text of his book.

state of being" *for himself*, namely the state of being virtuous. Gintis proposes to model states of virtue as arguments in an individual's utility function.

As evidence in favour of this approach, Gintis cites an experiment on lying conducted by Gneezy (2005). This experiment was set up so that some subjects had the opportunity to gain at the expense of others by lying to them in a way that the latter could never detect. The majority of subjects chose not to lie; the frequency of lying was greater, the greater the benefit to the person telling the lie and the less the cost to the person lied to. Gneezy himself speculates that this result might be explained by some variant of the "guilt aversion" approach discussed in Sect. 3.3. Gintis proposes an apparently different explanation – that honesty is a character virtue that individuals are willing to sacrifice material benefits to maintain. Since "trustworthiness" appears alongside "honesty" in his list of virtues, I take it that Gintis would favour a similar analysis of trust. In other words: individuals are motivated to repay trust because they have preferences for the state of "being trustworthy".

But what does it mean to say that a person is motivated by the desire to be virtuous? Recall that, for Gintis, a character virtue is a *behavioural regularity* – for example, the tendency for an individual to act honestly, or to repay trust. To say that a person has a preference for being trustworthy is to say that he has a preference for being a person whose behaviour shows a particular regularity, namely that of trustworthiness. That is an empty statement until we know what trustworthy behaviour is. If a virtue-based explanation of trust is to have content, it must include a specification of what a fully trustworthy person would do in different circumstances. (For example, would such a person choose *return* in the Trust Game at all permissible values of the parameters a, b, w, x, y and z, or only at some? Would she do the same in the Confidence Game?) And this takes us back to the problem that social preference theories are attempting to solve. To make the same point the other way round, just about any theory of social preferences can be re-described in terms of a corresponding character virtue. For example, theories of inequity aversion and reciprocity are different attempts to explain behavioural regularities which, intuitively, seem to be motivated by a desire for fairness. One might equally well say that these theories are different attempts to characterise a virtue of fairness. In this sense, the virtue-based approach is not an alternative to other modes of theorising; it is just another way of describing them.

An advocate of the virtue-based approach might reply that it introduces an additional degree of freedom by allowing trade-offs between virtuous and non-virtuous motivations. I agree that any realistic theory of how behaviour is influenced by social or moral norms – and, in particular, any realistic theory of trust – will need to take account of trade-offs of this kind. But before we can think about trade-offs, we need to know what is being traded-off against what. That requires not only a theory of self-interested behaviour (presumably the maximisation of material rewards), but also a theory of the behaviour of fully trustworthy people. Actual behaviour can then be modelled as some mix of those two theories.

It may be significant that Gintis is committed to the ambitious claim that behavioural game theory is "a general lexicon that applies to all life forms", and sees the assumption that individuals "maximiz[e] consistent preferences" as a fundamental property of that approach (2009, Chap 7, introductory section). Clearly, this claim is threatened whenever a regularity in observed behaviour seems to be best explained by a theory that does *not* assume individual utility maximisation. One way of deflecting such a threat is to re-describe the troublesome behavioural regularity as a virtue for which individuals have a consistent preference. But the cost of this manoeuvre is that utility maximisation ceases to be a genuine explanatory principle.

5 Team Reasoning and Collective Intentions

It seems that the social-preference approach has serious difficulties in explaining trust. We need to consider whether these recurring difficulties stem from some fundamental limitation of that approach.

The essential problem is to explain how, for players of the Trust Game, it could be common knowledge that (*send*, *return*) will be chosen. The intuition is that this pair of actions is a case of trust: A trusts B by choosing *send*, and B repays that trust by choosing *return*. We have to find a way of modelling a psychologically credible motivation which, for individuals on whom it acts with sufficient force, induces *send* and *return*.

As a starting point, it is useful to consider the combination of actions (*send*, *return*) as a composite *practice* of trust. If both individuals conform to this practice, the outcome is (1, 1). If neither conforms to it, the outcome is (0, 0). Looking at the Trust Game in this light, it is an immediately salient feature that the practice of trust is mutually beneficial: by their both following the practice, both players benefit. Further (at least in the version of the game presented in Fig. 1), the benefits of this practice are divided equally between the players. It seems that, for A and B together, the practice is an unambiguously good thing. Their problem is that this practice requires each of them to do something that he would not do if he were motivated only by self-interest and if he expected the other to be motivated similarly. B is required to resist the temptation to choose *keep* rather than *return*, following A's choice of *send*. A is required to incur the risk that B will give way to this temptation. In general terms, what is required is that *each individual plays his part in a mutually beneficial practice, expecting the other to play hers*.

So, if we are trying to explain how practices of trust are sustained, we might want to model individuals as motivated to play their parts in mutually beneficial practices when they can expect others to play theirs. But no such motivation appears in any of the social-preference theories reviewed in Sect. 3. These theories make no use of the concept of mutual benefit.

Recall the motivations that *are* represented in these theories. Inequity aversion is an attitude to the *distribution* of payoffs; it makes no reference to whether, through their interaction, the players are collectively gaining or losing. Each individual has a self-interested motivation to increase his own payoff, but the "social" component of his utility function does not attach any value to *mutual* benefit. This gives us a theory which can represent *giving* (A gives to B to reduce A's advantageous inequality) and *taking* (B takes from A to reduce B's disadvantageous inequality), but which cannot represent *cooperation*.

Rabin reciprocity is an attitude to the kindness of others' intentions. Crucially, kindness is modelled as a form of giving: A is kind to B to the extent that (given his beliefs about what B will do) he sacrifices some of his own payoff to increase B's. Conversely, unkindness is a form of taking, or at least of giving less than one might have done. Thus, positive reciprocity is conceptualised as an exchange of gifts, each of which involves sacrifice on the part of the giver. In some cases, gift exchange is *in fact* mutually beneficial, but even then, mutual benefit plays no part in the definition of kindness. There is no room for the idea that A, by virtue of choosing an action that he believes to be mutually beneficial, is revealing an intention that B will be motivated to reciprocate.

Guilt aversion (and the corresponding emotion in the theory of normative expectations) is an attitude to the validation of other people's expectations. If B is guilt-averse, she is willing to sacrifice payoff to avoid disconfirming A's expectations about what she will do. There is no requirement that A's expectation is that B conforms to some mutually beneficial practice, to which A himself is conforming. In the Falk–Fischbacher model, B's motivation to validate A's expectations is conditional on the kindness of A's action, which takes us back to the idea of gift exchange: A intends that B should be the beneficiary of a gift from him, and B rewards that intention by making a gift in return.

If my diagnosis is correct, we need a theory in which an individual's social orientation is represented as a positive attitude towards mutually beneficial practices. I suggest that the best place to start is with the theory of *team reasoning*.

The concept of team reasoning first appeared in the literature of moral philosophy, in analyses of the distinction between act- and rule-utilitarianism (Hodgson 1967; Regan 1980); it is implicit in Harsanyi's (1977) decision-theoretic analysis of rule utilitarianism. Rule utilitarianism can be thought of as a special case of team reasoning, in which "the team" contains all moral agents who are receptive to rule utilitarian arguments, and their common objective is the overall good of the world. However, the idea of team reasoning can also be applied to smaller groups, in particular to the set of players of a specific game. As far as I know, I introduced this group-based conception of team reasoning to economics (Sugden 1991, 1993); it has been developed furthest by Bacharach (2006). The core idea is that, in a game which is perceived as providing opportunities for cooperation, the players reason as if they were engaged in a problem of collective choice, jointly choosing the *profile* of strategies that has the best consequences for them collectively. In some versions of the theory, "best for them" is interpreted as the maximisation of an objective function, analogous with a social welfare function, which gives positive weight to each player's payoffs. In other versions, it is interpreted as requiring that each player benefits (or at least, does not lose) relative to some non-cooperative benchmark; ideas of distributional fairness are applied only to the division of the cooperative surplus. In my analysis of the Trust Game, I use this second interpretation. In contrast to conventional game theory, in which each player asks himself "What should I do, given what I can expect the others to do?", the players ask themselves the unconditional question "What should we do?"

It is important to recognise that "What should we do?" does *not* reduce to "What should I do, given that my preferences coincide with our collective interests?" To ask the latter question is to treat social orientation as a property of one's preferences, and then to reason in a conventionally game-theoretic way. To ask the former question is to use an altogether different mode of reasoning. This distinction is best seen by thinking about "Hi–Lo" games in which the players' interests are completely aligned but there are two Nash equilibria, one of which dominates the other. As Hodgson (1967) first pointed out, act utilitarianism (and standard game-theoretic reasoning) does not tell the players to choose the dominating equilibrium, while rule utilitarianism (and team reasoning) does. This implication of team reasoning cannot be replicated in conventional game theory by any credible transformation of material payoffs into subjective utilities – or, as game theorists might say, by "getting the payoffs right". If a game has the Hi–Lo structure in material payoffs, it will retain this structure when described in terms of subjective utilities, whether players are self-interested or altruistic.

Having identified the profile of strategies that is best for the players collectively, each player chooses his component of that profile, expecting the other player to choose hers.[9] The precise content and grounding of this expectation are matters of controversy among theorists of team reasoning. However, the general idea is that team reasoning is activated when some property of the relevant interaction leads the players to form the common belief that team reasoning is appropriate; each player is then disposed both to use team reasoning himself and to expect the other player to do the same. If players use team reasoning to determine their choice of strategies, they can be thought of as forming *collective intentions*. That is, each player understands his own intended action as a component of a joint action that is intended by the players together.

Collective intentions are controversial in philosophy. Many different analyses of this concept have been proposed, for example by Tuomela and Kaarlo (1988), Searle (1990), Bratman (1993) and Bardsley (2007). The intuitive idea is that (say) two individuals can act together in ways that cannot adequately be described in terms of those individuals' separate desires and beliefs. In the literature of collective intentions, this idea is expressed by trying to find a sense in which individuals jointly intend the consequences of their acting together. However, it is surprisingly difficult to draw a conceptual distinction between "genuine" collective intentions (for example, the intention of two people to sing a duet together) and the individual intentions that conventional game theory implicitly attributes to players in any Nash

[9] Here I assume that one profile of strategies is uniquely best for the players collectively. If two or more strategy profiles are jointly optimal, the theory does not give a determinate solution. This problem is analogous with other cases of non-uniqueness in conventional game theory, for example in pure coordination games. In such cases, I suggest, the rationality-based principles of game theory need to be supplemented by common conceptions of "salience" or "focality", as first explained by Schelling (1960).

equilibrium (for example, the separate intentions of two players of the Prisoner's Dilemma to defect, when each believes that the other will defect too).

Gold and I have argued that this problem can be resolved by interpreting collective intentions as intentions that are supported by team reasoning (Gold and Sugden 2007). When two individuals i and j engage in team reasoning, each endorses a common objective for their combined actions. Each then chooses to play his part in the combination of actions that best promotes that objective. If such reasoning leads i to perform action a_i and j to perform action a_j, it seems natural to say that i and j are performing (a_i, a_j) together, with the collective intention of achieving their common objective. Because team reasoning does not reduce to individually instrumental reasoning, such collective intentions are conceptually distinct from the individual intentions that game theory conventionally attributes to players in a Nash equilibrium.

Now consider the implications of team reasoning in the Trust Game. Suppose that both players perceive their interaction as one for which team reasoning is appropriate. So each asks "What should we do?" If "best for us" is interpreted in terms of the division of a cooperative surplus, it is natural to treat *hold* as the non-cooperative benchmark, in which case the answer to the question is "Choose (*send, return*)". Having reasoned to this conclusion, and believing that B is reasoning in a similar way, A plays his part in that joint action: he chooses *send*. He does so with the expectation that B will play her part in the same action by choosing *return*, and with the intention that the two players together will bring about (1, 1). Independently reasoning in the same way, B reaches the same conclusion that (*send*, *return*) is "best for us". She expects A to play his part in this joint action, and then she observes him doing so. This observation confirms her prior belief that A is acting with the intention that he and she together will bring about (1, 1). She plays her part in (*send*, *return*) by choosing *return*.

It might be objected that, in this case, there is no observable difference between the behaviour induced by team reasoning and the behaviour that would result if each player were independently motivated by altruistic preferences. For example, suppose that A and B both prefer (1, 1) – the outcome that I have treated as "best for A and B together" – to both (0, 0) and (–1, 2). Then, reasoning individually, B would choose *return* if A chose *send*; and if A expected this response by B, he would choose *send*.

But this argument runs into the same problem as the attempt to explain trust in terms of inequity aversion. Recall the Dictator Game discussed in Sect. 3.1, in which B simply chooses between one action resulting in (1, 1) and another resulting in (–1, 2). Let us call these actions *share* and *take* respectively. If B acts on individual reasoning, and if she prefers (1, 1) to (–1, 2), she will choose *share* in the Dictator Game, just as she chooses *return* in the Trust Game. But *return* is a repayment of trust, while *share* is not.

I submit that an explanation of trust should be compatible with B's choosing *return* in the Trust Game but *take* in the Dictator Game. From the perspective of the theory of team reasoning, the two games are indeed different. In the Trust Game, (*send*, *return*) is a combination of actions by A and B which benefits both of them.

By choosing *send*, A plays his part in this joint action; by choosing *return*, B reciprocates. In the Dictator Game, in contrast, B makes a unilateral choice between two alternative outcomes, one favouring her, the other favouring A.

Notice that, in this account, trust involves the reciprocation of "good" intentions. In this sense, there is a parallel with Rabin reciprocity. However, the intentions that are reciprocated are collective, not individual, and they are directed towards mutual benefit, not gift-giving. A's choice of *send* can reveal an intention for mutual benefit even if (as in the case in which A is confident that B will choose *return*) there is no self-sacrifice on his part. B's choice of *return* involves self-sacrifice in the sense that she could do better for herself by choosing *keep*, but the intention revealed by *return* is still that of mutual benefit. B is not making a gift to A to reward him for a kindness; she is reciprocating his intention of mutual benefit.

Putting this more generally, Rabin reciprocity is an attitude of individuals who are concerned about how each is treating the other; its core ideas are kindness and unkindness, giving and taking. The reciprocity of team reasoning is more outward-looking. Team reasoning is an attitude of individuals who are trying to achieve some common goal together. Each individual is willing to do his part in a collective action directed at this goal, provided he can rely on the others to do their parts. The reciprocation is of individual efforts directed at a common goal.[10]

There is also a parallel between team reasoning and theories of self-fulfilling expectations. In the Trust Game, A's choice of *send* not only confirms B's belief about A's intention of mutual benefit, it also confirms her belief that A expects her to act with the same intention. In choosing *return*, B not only reciprocates A's intention, she also validates A's expectation about her. However, B's motivation to validate that expectation depends on her interpreting it as an expression of A's adherence to a joint intention for mutual benefit. This is responsiveness *to trust*, not to other people's expectations in general.

To illustrate how the concept of team reasoning might be applied in economics, I return to Akerlof's discussion of gift exchange in labour markets. Recall that Akerlof is trying to explain how, in some labour markets, employers pay more than the reservation wage, workers supply more effort than their contracts require of them, and equilibrium is maintained through the existence of involuntary unemployment. I suggest that this combination of phenomena might be explained by team reasoning rather than gift exchange. Suppose that, as Akerlof assumes, it is difficult for an employer to monitor workers' effort levels. Thus, both parties would gain if the employer placed trust in his workers by paying them more than the reservation wage, and if the workers returned that trust by working harder than it was in their immediate self-interest to do. If employer and workers engage in team reasoning, they will see such a practice of trust as a mutually beneficial joint action. If each party acts on the conclusions of this reasoning, each will play his part in that joint

[10] My early theory of reciprocity in voluntary contributions to public goods is reciprocal in this sense, and implicitly assumes a form of team reasoning. In this theory, each individual is motivated to match other people's contributions, but only up to the rate of contribution that she would choose for everyone (Sugden, 1984).

action: the employer will trust the workers and the workers will return his trust. Notice that, in order for this kind of reasoning to be activated, each party must be able to see the employment contract as serving his interests. Thus, it is essential that some significant share of the benefits of the practice of trust accrue to workers. This requires that the wage premium paid by the employer must more than compensate workers for the extra effort they supply. If workers are identical, equilibrium will be possible only with involuntary unemployment. All this is just as in Akerlof's model, except that neither party makes a gift to the other. The relationship between employer and worker is one of cooperation for mutual benefit.

6 The Fraternal Morality of Market Relationships

In the introduction to this paper, I pointed to the discontinuity between social-preference explanations of economic trust and the theories by which economics has traditionally explained behaviour in "paradigm" markets – that is, markets in which mutually beneficial exchange does not require trust. The nature of this discontinuity will now be more obvious. Social-preference theories are structured around the ideas of giving and taking. In the first instance, positive social orientations are represented in terms of motivations *to give* – whether with the aim of reducing inequality, reciprocating other people's gifts, or validating other people's expectations. (In a second-order sense, motivations to take can also play a constructive role, if the taking is a punishment meted out to those who fail to display first-order positive orientations.) In this framework, trust is most naturally understood as a form of gift exchange, as in Akerlof's model of labour contracts. In contrast, the normal understanding of paradigm markets is that each individual is motivated only by his own interests.

I have argued that the practice of trust should be understood, not as gift exchange, but as a joint action carried out with the collective intention of mutual benefit. If this argument is accepted, there is still a discontinuity between the explanation of trust and the normal understanding of exchange relationships in paradigm markets: a collective intention for mutual benefit is not the same thing as two individual intentions directed at self-interest, even if in fact those individual intentions interact to create mutual benefit. But, having understood the possibility of collective intentions, we might have second thoughts about that normal understanding of market exchange. We might ask whether ordinary market relationships can also be understood in terms of collective intentions for mutual benefit.

Take Smith's example of the baker and her customer. On Smith's account, each of these two people is motivated by self-love; their separate motivations lead them to engage in mutually beneficial trade, but mutual benefit is an unintended consequence. The same is true of Smith's merchant, who intends only his own gain but is led by an invisible hand to promote an end which was no part of his intention. Notice how, in each case, Smith contrasts self-love with benevolence. In the first case, the contrast is between the customer who addresses himself to the baker's self-love and the one who appeals to her benevolence. In the second, the contrast is between the merchant who intends only his own gain and the one who affects to trade for the public good. The point of these examples is that the motivations that lie behind market relationships are better characterised as self-love *than as benevolence.* But Smith does not consider the possibility that the baker and her customer, or the merchant and those with whom he trades, might perceive their acts of exchange as joint actions carried out with the intention of benefiting both parties together. While no one would expect a baker to think it her business to give bread away, it does not seem so implausible to imagine a baker who intends that her customers should benefit from the bread she sells them, just as she benefits from the money they pay her. And while it might be true that only a beggar would depend for his dinner on the benevolence of a baker, one might think that neither the baker nor her customer would lose independence by intending that their relationship should be structured by a collective intention directed at mutual benefit.

If market relationships are to be understood in terms of such intentions, and if this understanding is to be broadly compatible with received theories about the workings of markets, it is essential that collective agency comes into existence only with the making of a contract (explicit or implicit). Collective agency is to be seen, not as the motivation for *making* the contract, but as what a contract creates. Thus, before going out to buy bread, the customer does not have an intention to benefit bakers: that would be benevolence, not team reasoning. But when he contemplates a contract with a particular baker, he perceives that contract as committing both of them to a joint intention to seek mutual benefit within the confines of their particular economic relationship. Having contracted with the baker, the customer understands his payment for the bread as something more than the fulfilment of a legal obligation: he intends that the transaction as a whole benefits the baker, and the payment is a necessary part of that.

The idea that market relationships are based on intentions for mutual benefit can be found in embryonic form in the writings of the eighteenth-century philosopher-economist Genovesi. Genovesi's most important work, *Delle Lezioni di Commercio o sia di Economia Civile* (Lectures on Commerce, or on Civil Economy; 1765–67/2005) was written shortly before Smith's *Wealth of Nations*, and develops many of the same themes as its more famous successor. But where Smith (1776/1976, p. 26) starts by assuming a human propensity "to truck, barter and exchange one thing for another", Genovesi grounds his analysis of markets on an assumed human inclination towards mutual assistance (*scambievoli soccorsi*). The concept of "assistance" implies a conscious intention that another person should benefit from one's actions. To understand market exchange as mutual assistance is to interpret its mutually beneficial nature not as an unintended consequence of self-love, but as integral to each person's conception of the market relationships in which he participates.

In the final paragraph of the *Lezioni*, Genovesi summarises the central idea of the whole work. He tells his students that they should study civil economy "to go along with the law of the moderator of the world, which commands us to do our best to be useful to one another" (p. 890).[11] The law of the moderator of the world, as interpreted by Genovesi, is not that each of us should pursue self-interest in the confidence that the interests of society will be achieved as an unintended consequence. But nor is it that each of us should unilaterally pursue the aim of being useful to others. It is that *we* should pursue the aim of being useful *to one another*. The command "Be useful to one another" is addressed to us collectively. It supplies a premise for team reasoning and for the construction of collective intentions.

What difference does it make if we see the market through Genovesi's eyes, as a network of relationships in which individuals act with the joint intention of being useful to one another? One important feature of this conception is that trust is an integral part of market relationships. Or, more accurately, trust is an expression of the same set of motivations as underlie market relationships in their ideal type. Trust is not a theoretically anomalous phenomenon lying outside the core domain of economics, which can be explained only by postulating "social" motivations beyond those needed to explain ordinary market behaviour.

By understanding market relationships in terms of the pursuit of mutual benefit rather than self-interest, we can recognise them as having moral content. That makes the metaphor of the invisible hand less mysterious. Every economist knows how surprising most people find – and how disinclined they are to accept – the claim that when economic agents are motivated only by individual self-interest, markets tend to promote the public good. This scepticism is, I think, linked with the extreme difficulty that most people have in grasping the reality of the idea of gains from trade: they find it more natural to think of economic life as a zero-sum game. When they think about self-interest in the market, the cases that immediately come to mind are ones in which one agent makes opportunistic gains at the expense of another (as when one agent trusts another, but that trust is not returned). But compare the mutual-benefit analogue of the invisible hand claim – that when economic agents are motivated to seek out and realise opportunities for mutual benefit, an economic regime which facilitates that process tends to promote general benefit. That does not seem so surprising, because it makes a closer connection between what economic agents intend and the effects that their actions produce.

Genovesi told his students that the lesson they should learn from economics is to be useful to one another. I take him to be saying that market economies work by realising gains from trade and that, to take advantage of those opportunities, people must be motivated to act together for mutual benefit. Genovesi repeatedly describes relationships based on such a motivation as "friendship", using that word in a sense that is now perhaps better captured by "fraternity". The implication is that market relationships have a particular moral content, and that successful commercial societies – societies in which the potential benefits of exchange are realised – are possible only where that morality is upheld. I hope I have persuaded the reader that, even after the passage of almost 250 years, Genovesi's lesson is worth attending to.

---

[11] The translation from the Italian is by Bruni and me.

Acknowledgements This paper draws on ideas I have developed in joint work with Luigino Bruni. I thank Herbert Gintis and Brian Skyrms for comments on a previous draft. My work was supported by the Economic and Social Research Council (award no. RES 051 27 0146).

References

Akerlof G (1982) Labor contracts as partial gift exchange. Q J Econ 97:543–569

Bacharach M (2006) Beyond individual choice. In: Gold N, Sugden R (eds) Princeton University Press, Princeton

Bacharach M, Guerra G, Zizzo D (2001) Is trust self-fulfilling? An experimental study. Department of Economics Discussion Paper 76, University of Oxford

Bardsley N (2007) On collective intentions: collective action in economics and philosophy. Synthese 157:141–159

Battigalli P, Dufwenberg M (2007) Guilt in games. Am Econ Rev Pap Proc 97:171–176

Berg J, Dickhaut J, McCabe K (1995) Trust, reciprocity, and social history. Games Econ Behav 10:122–142

Bolton G, Ockenfels A (2000) ERC: a theory of equity, reciprocity and competition. Am Econ Rev 90:166–193

Bratman M (1993) Shared intention. Ethics 104:97–113

Bruni L, Sugden R (2000) Moral canals: trust and social capital in the works of Hume, Smith and Genovesi. Econ Philos 16:21–45

Bruni L, Sugden R (2008) Fraternity: why the market need not be a morally free zone. Econ Philos 24:(2008):35–64

Dufwenberg M, Kirchsteiger G (2004) A theory of sequential reciprocity. Games Econ Behav 47:268–298

Falk A, Fischbacher U (2006) A theory of reciprocity. Games Econ Behav 54:293–315

Fehr E, Schmidt K (1999) A theory of fairness, competition and cooperation. Q J Econ 114: 817–868

Gauthier D (1986) Morals by agreement. Clarendon Press, Oxford

Genovesi A (1765–1767/2005) Delle Lezioni di Commercio o sia di Economia Civile. Instituto Italiano per gli Studi Filofící, Napoli

Gintis H (2009) The bounds of reason: game theory and the unification of the behavioral sciences. Princeton University Press, Princeton, New Jersey

Gneezy U (2005) Deception: the role of consequences. Am Econ Rev 95:384–394

Harsanyi J (1977) Rule utilitarianism and decision theory. Erkenntnis 11:25–53

Hobbes T (1651/1962) Leviathan. Macmillan, London

Hodgson D (1967) Consequences of utilitarianism. Oxford

Hume D (1740/1978) A treatise of human nature. Clarendon Press, Oxford

Pelligra V (2005) Under trusting eyes: the responsive nature of trust. In: Gui B, Sugden R (eds) Economics and social interaction. Cambridge University Press, Cambridge, pp 195–124

Rabin M (1993) Incorporating fairness into game theory and economics. Am Econ Rev 83: 1281–1302

Regan D (1980) Utilitarianism and cooperation. Clarendon Press, Oxford

Schelling T (1960) The strategy of conflict. Harvard University Press, Cambridge, MA

Searle J (1990) Collective intentions and actions. In: Cohen P, Morgan J, Pollack M (eds) Intentions in communication. MIT, Cambridge, Massachusetts, pp 401–415

Smith A (1776/1976). An inquiry into the nature and causes of the wealth of nations. Clarendon Press, Oxford

Sugden R (1984) Reciprocity: the supply of public goods through voluntary contributions. Econ J 94:772–787

Sugden R (1991) Rational choice: a survey of contributions from economics and philosophy. Econ J 101:751–785
Sugden R (1993) Thinking as a team: toward an explanation of nonselfish behavior. Soc Philos Policy 10:69–89
Sugden R (1998) Normative expectations: the simultaneous evolution of institutions and norms. In: Ben-Ner A, Putterman L (eds) Economics, values, and organization. Cambridge University Press, Cambridge, pp 73–100
Sugden R (2004) The economics of rights, cooperation and welfare, 1st edn. 1986 Palgrave Macmillan, Houndmills
Tuomela R, Miller K (1988) We-intentions. Philos Stud 53:367–389

GPSR Compliance

The European Union's (EU) General Product Safety Regulation (GPSR) is a set of rules that requires consumer products to be safe and our obligations to ensure this.

If you have any concerns about our products, you can contact us on ProductSafety@springernature.com

In case Publisher is established outside the EU, the EU authorized representative is:

Springer Nature Customer Service Center GmbH
Europaplatz 3
69115 Heidelberg, Germany

Batch number: 10755292

Printed by Printforce, the Netherlands